Gerard Assayag
Hans Georg Feichtinger
Jose Francisco Rodrigues (Editors)

Mathematics and *Music*

A Diderot Mathematical Forum

Springer

Editors:

Gerard Assayag
IRCAM – CNRS
UMR 9912
1, place Igor-Stravinsky
75004 Paris, France

Hans Georg Feichtinger
University of Vienna
Dept. of Mathematics
Strudlhofgasse 4
1090 Vienna, Austria

José Francisco Rodrigues
University of Lisboa
CMAF
Av. Prof. Gama Pinto 2
1649-003 Lisboa, Portugal

Library of Congress Cataloging-in-Publication Data

Mathematics and music : Diderot Forum, Lisbon-Paris-Vienna / Jose Francisco Rodrigues, Hans Georg Feichtinger, Gerard Assayag (editors).
p. cm.
Includes bibliographical references.
ISBN 3540437274 (acid-free paper)
1. Music--Mathematics--Congresses. I. Rodrigues, José-Francisco. II. Feichtinger, Hans G., 1951- III. Assayag, Gerard.

ML3800 .M246 2002
780'.051--dc21

2002070479

ISBN 3-540-43727-4 Springer-Verlag Berlin Heidelberg New York

Mathematics Subject Classification (2000): 00-XX, 01-XX, 03-XX, 11-XX, 42-XX, 68-XX

Springer-Verlag Berlin Heidelberg New York
a member of BertelsmannSpringer Science+Business Media GmbH

http://www.springer.de

Cover design: *design & production,* Heidelberg
Typesetting: Le-TeX Jelonek, Schmidt & Vöckler GbR, Leipzig
Printed on acid-free paper SPIN 10868890 40/3142ck-5 4 3 2 1 0

Preface

Under the auspices of the European Mathematical Society the Fourth Diderot Mathematical Forum took place simultaneously in Lisbon, Paris and Vienna, in 3–4 December 1999. Relationships between Mathematics and Music were presented at this conference in three complementary directions: "Historical Aspects" being addressed in Lisbon at the Fundação Calouste Gulbenkian, the "Mathematical Logic and Music Logic in the XX century" in Paris at the IRCAM and the "Mathematical and computational methods in Music" at the University of Vienna. A main feature of that Forum was its exchange of information through a teleconference among the three cities, which took place under the theme "The relations between Mathematics and Music are natural or cultural relations?" J. Bourguignon representing EMS was the main moderator of this event.

In Western Civilization Mathematics and Music have a long and interesting history in common, with several interactions, traditionally associated with the name of Pythagoras but also with a significant number of other mathematicians, like Leibniz, for instance, to whom music was a "hidden exercise of arithmetics". During the age of Enlightenment, Diderot wrote in the French Encyclopédie (initiated with D'Alembert, who also wrote on music) about and in the Pythagorian tradition: "C'est par les nombres et non par les sens qu'il faut estimer la sublimité de la musique".

The two days conference in Lisbon covered some of the significant aspects of the interactions between Mathematics and Music along History. The participants had the opportunity to attend a concert by the Gulbenkian Orchestra at the end of the first day. This conference was held in co-ordination with the XII National Seminar on the History of Mathematics of the Sociedade Portuguesa de Matemática, that took place in the afternoon of the second day at the University of Lisbon.

The conference in Paris dealt with logic and music in the XX century. At the beginning of this century, two major processes began to grow at the same time in modern thought. Mathematical Logic, after Boole's work in the previous century, developed as a formal calculus with its own conditions of coherence – and, as shown later by Gödel, of incoherence – in an apparent independence from any underlying ontology. Music, freed from the straitjacket of tonality and from its Pythagorian relation to the physical world, began to set its own rules of composition in a self-contained way. It started with the theoretical and aesthetical work of Schoenberg and the Vienna School, continued with the efflorescence of formalization in the Darmstadt period, and got into a state of atomization in the contemporary era, where an original formal operation is sometimes built for a single music work and then forgotten. In the second half of that century, the evolution of Music composition has led to a complex configuration of formal reasoning and search for internal consistency, sometimes calling explicitly the power of Mathematics, as in the case of Babbit, Xenakis and many others.

Thus the question raised in the Parisian part of the Diderot Forum: what do musician talk about when they evoke "musical logic" or "musical reasoning"? Does it have something to do with mathematically expressed logic and its own evolution? If, obviously, the broad use of computation in music involves some relation with logic, is there a residue in music rationality that is unreachable by mathematical or logical means? Being an actor of XX century music, IRCAM (the Institute for Research and Coordination of Acoustic and Music founded by Pierre Boulez) hosted the discussion of these issues.

The Vienna component of the Diderot Forum was actually organized in conjunction with a workshop with a small number of invited speakers and a poster session. Altogether 34 participants of that event provided a written contribution for the Proceedings of that workshop, which have been published separately by the OECG (the Austrian Computer Society), in December 1999, and edited by H.G.Feichtinger and M.Dörfler. Expanded versions of eight of those articles constitute a special issue on "Music and Mathematics" of the Journal of New Music Research (Vol.30/1, 2001). The event ended with a participants concert given in the rooms of the old Bösendorfer piano factory.

The main topics of that workshop where problems of the synthesis of musical sound (such as physical modeling), analysis of musical sounds (such as the transcription problem, time-frequency methods, or the quantitative analysis of instruments and musical interpretations), restoration and denoising of old recordings, instrument optimization and mathematical models for musical sound and rhythm. At almost all levels of musical activities, especially if instruments are involved, some mathematical model can be seen in the background. Such a model may either explaining the kind of sound which is produced by the instrument (as in the famous question: "can you hear the shape of a drum"), or is used as the basis for sound production (successfully implemented in modern key-boards which are based on the principle of physical modeling).

Of course time-frequency analysis (sometimes appearing under the name of Gabor analysis) is an important tool to analyze musical sound, by displaying the "harmonic contents" of a musical signal as a function of time, just like the musical score tells the performer at which instance in time she should produce which kind of harmony. Nowadays we have CDs and DVDs as stable sources for the reproduction of sound on a digital basis, without being aware which kind of mathematical transformations or coding techniques are running in the background. For instance, only recently some new recordings of Caruso have been produced, with the orchestra part being "replaced" by a modern recording, while preserving (resp. restoring to its best) the original voice of that great singer. Clearly, such things have become possible only due the an improved understanding of the mathematical nature of acoustic signals, and how different parts if it can be treated "independently".

The present book collects the sixteen written contributions to this Forum, illustrating with its large variety of articles the rich and deep interactions that exists between Mathematics and Music in a broad and contemporary sense. From the historical perspectives of the first chapters to the modeling and computation of musical sounds, from examples of musical patterns to the cultural aspects, and from the mathematical formalization to the musical logic, this rare collection of paper presents a comprehensive list of topics relating fundamental mathematics, applications of mathematics and the relation of both to society.

The first article by the musicologist Manuel Pedro Ferreira, introduces the historical perspective of the role of proportions in Ancient and Medieval Music. One of the four divisions of the Quadrivium, together with Arithmetic, Geometry and Astronomy, Music was considered a Mathematical Science. From the Greek heritage and the Latin world, up to the late-medieval France, this contribution traces a particular mode of musical thought, based on proportional relationships, that influenced the aesthetics of the Ancients and had impact on the musical composition in the Middle Ages.

Eberhard Knobloch in his contribution on "The Sounding Algebra" shows the role of Combinatorics in the baroque conception of music. He also illustrates how Music was based on a rational foundation and how musicologists and composers of that period believed that beauty and harmony consist in order and their variety stems from composition, combination and arrangement of their parts. Referring to Lullism and its combinatorial art, it is interesting to see how Mersenne, Kircher, Leibniz and, in the 18th century, Euler considerably have contributed to the progress of Combinatorics by studying such mechanical ways of composing.

The paper by the mathematician Benedetto Scimemi shows the use of mechanical devices and numerical algorithms in the 18th century for the equal temperament of the musical scale. In ancient times, before the logarithms and the irrational numbers were theoretically established, music theorists and instruments makers used a number of mechanical devices, geometrical constructions and algebraic algorithms to produce acceptable approximations for the sequence of frequencies for the musical scales. This is exemplified in the Renaissance treatise by Zarlino and, in the "setecento", the craftsmanship of the J.S.Bach's contemporary Strahle and the theoretical works of Schroter and Tartini.

In the article by Jean Dhombres on Lagrange, we discover how a "working mathematician" has contributed to our theoretical understanding of wind instruments and music. One of his first papers of 1759, "Recherches sur la nature et la propagation du son", was written with an objective for the use of music, which was for Lagrange a technique to be explained and therefore a subject for scientific research.

Another aspect of the relationship between Mathematics and Music was shown in the Robin Wilson talk in Lisbon, illustrated with several musical examples in a wide range of styles and musical scales. As we can read in the article "Musical Patterns", many composers have used mathematical devices in their music, namely symmetries and mathematical transformations, such as canon, expansion, retrograde motion and inversion.

François Nicolas, a contemporary music composer with a strong background in science, opens the contributions from the Paris conference by recalling three aspects of logic: a grammar, a tautology factory, a theory of consistency and identifies their partial resonances in the musical field: the syntax of musical language(s), the coherence of large musical forms. Nicolas shows through a series of historical compositional strategies that logic in music is mostly a dialectic one, and concludes that the strategy of each work must be thought within a specific inferential framework rather than a deviation from the broader formal system it inherits.

The text by Marie-José Durand-Richard, an epistemologist and historian of science, recalls the movement of mathematization of logic occurring with the work of George Boole (1815-64). She traces the discussion, still alive, of the place and nature of meaning between the defenders of blind calculation and those of a subjacent ontology. This debate has some resonance with its musical counterpart: is music a formal game or is it based on powerful perceptive cognitive schemes?

More specific, Laurent Fichet, a musicologist, studies music analysis techniques in the 20th century. Several methods have been widely inspired by mathematical processes. Using those that seem more likely to give interesting results, he puts forward different analysis of the 2nd Sonata by Pierre Boulez, which seems to lend itself to a logical approach. The comparison between what these mathematical analyses lead to and what a more intuitive analysis might bring gives a balanced view of the links between musical logic and mathematical logic.

In their contribution Gérard Assayag and Shlomo Dubnov start by observing that many aspects of musical structure are hard to define formally. Nevertheless, music and sound exhibit a great amount of structure and redundancy. The authors use information theory for discovering these hidden structures, specifically by considering statistical relations and dependencies that exist among musical parameters in existing musical and sound material. This "automatic learning" approach raises inferential and inductive relationships between parts of the analyzed music, but this "knowledge" cannot be totally made explicit.

Marc Leman, a defender of a recent discipline called Cognitive Musicology, aims at understanding the nature of musical information processing and imagination in composing, listening and performing. In his text, he sets a bridge between the approach of cognitivists in the seventies, using formal predicate logic as a basic representation system, and the recent physical-physiological theory of human information processing. He also proposes a formalism of "musical images" where logic can be used as a meta-level description system for a clarification of the underlying processing of such images.

In its contribution, Guerino Mazzola, a mathematician who works on a treatise of contemporary Mathematical Music Theory, states that the logic of musical composition, representation, analysis, and performance share important basic structures which can be described by Grothendieck's functorial algebraic geometry and Lawvere's topos theory of logic. He gives an account of these theoretical connections and illustrates their formalization and implementation on music software.

Marc Chemillier, a computer scientist and an ethnomusicologist, asks the question of musical logic in oral tradition societies where complex artistic expressions have emerged. Should one consider that these sophisticated productions are variations of the same universal scheme, that is the playful prolongation of the elementary rationality with which human beings are provided for their survival? And logic in itself, with its highly abstract developments is another prolongation of the same mental resource?

Among the invited talks in Vienna, Gregory Wakefield, who could not contribute to this volume, presented the results of interdisciplinary research on musical sound and visual representation conducted by engineers, musicologists and musicians at the University of Michigan in Ann Arbor. This enterprise is yet another striking example showing that the combination of new mathematical models and electronic media gives artists a chance to enlarge their repertoire of expression and also of training for even classical concerts. It also indicated high promises for a future cooperation between musicians, musicologists, engineers and applied mathematicians.

As a kind of compensation Jean-Claude Risset was able to share his thoughts on "Computing Musical Sound" by a contribution to this book, although he was not able to join the conference in 1999. His contribution gives a summary of the many interactions between music and mathematics through the centuries. In particular, it focuses on the pervasive influence of mathematics on the development of new tools and techniques in music since the advent of digital computers. The phenomena described in Risset's contribution include not only the invention of entirely new classes of sounds with the aid of mathematics, but also the composition of entire pieces of modern music on a basis linked to mathematical foundations.

The article by Erich Neuwirth, although presented in Lisbon with a historical perspective, provides a computational tool-kit based on the mathematical theory of tuning musical instruments. Besides being an illustration of how Mathematics may help to simulate new classes of sounds, it also offers a Mathematica code that allows the listening to the constructed scales and chords.

The contribution by Xavier Serra describes the Musical Communication Chain, by which he understands the processing of musical sound from the conception of a musical idea, through composition and performance to the listener and vice versa. He describes, which types of models have been developed in the past for the different steps in the complex process of production and perception of music. Furthermore he comments on the different degrees of understanding which have been reached in the investigation of the various elements in this process. Concrete topics discussed in some detail the role of the composer, different representations of music, the influence of the player and the instrument, or the perception of listeners.

The final paper by Giovanni De Poli and Davide Rocchesso is entitled "Computational Models for Music Sound Sources". It explains that algorithms for sound generation and transformation are nowadays ubiquitous in multimedia systems, as a result of the progress in information technologies. Despite this fact their performance is rarely satisfactory for the composer, performer or well-trained listener. The article reviews various advanced computational models, based on the physics of actual or virtual objects allowing the user to rely on high-level description of the sounding entities.

The editors wish to take the opportunity to acknowledge the support provided for the conference by the Centro de Matemática e Aplicações Fundamentais, at the University of Lisbon, the Fundação para a Ciência e a Tecnologia, the Fundação Calouste Gulbenkian, the Sociedade Portuguesa de Matemática, the IRCAM-Centre Georges Pompidou, Paris, the CNET-France Telecom, Lannion, and the University of Vienna, as well as the support of the European Mathematical Society and Springer Verlag in the promotion and the publication of this book.

Finally we wish to thank all the authors and participants at the Diderot Mathematical Forum for their contributions and we hope the topics and articles here included will contribute to enhance the understanding and the study of the relationships between Mathematics and Music.

Gérard Assayag, Paris
Hans G. Feichtinger, Vienna
José Francisco Rodrigues, Lisboa

Contents

1 Proportions in Ancient and Medieval Music
M.P. Ferreira 1

The Greek Heritage 2
The Latin World 11
Late-Medieval France 18
The Decline of Proportional Thinking 25

2 The Sounding Algebra: Relations Between Combinatorics and Music from Mersenne to Euler
E. Knobloch 27

Introduction 27
2.1 Mersenne (1635/36) 28
2.1.1 Permutations 29
2.1.2 Arrangements as a Generalization of Permutations 32
2.1.3 Combinations 33
2.2 Kircher (1650) 36
2.3 Leibniz (1666) 43
2.4 The Later Developments in the 18th Century (Euler, Mozart) 44
Epilogue 47
References 47

3 The Use of Mechanical Devices and Numerical Algorithms in the 18th Century for the Equal Temperament of the Musical Scale
B. Scimemi 49

Gioseffo Zarlino 51
Giuseppe Tartini 54
Daniel Strähle 57
Christoph Gottlieb Schröter 59
Conclusion 62
References 62

4 Lagrange, "Working Mathematician" on Music Considered as a Source for Science
J. Dhombres 65

5 Musical Patterns
W. Hodges and R.J. Wilson 79

Canon 79
Modified canon 81
Retrograde motion 83
Inversion 85
The twelve-tone system 86

6 Questions of Logic: Writing, Dialectics and Musical Strategies
F. Nicolas 89

Introduction 89
Variations on the Logical in Music 93
Interlude: Mathematics, Music and Philosophy 102
Musical Proceedings of Logic 103
Conclusion 111

7 The Formalization of Logic and the Issue of Meaning
M.-J. Durand-Richard 113

Introduction 113
7.1 Some Indispensable and Significant Steps Forward in the Mathematicization of Logic 113
7.2 What could be a More Historical Investigation of the Mathematicization of Logic? 116
7.3 The Connection Between Mathematics and Logic. The First Phase: Great-Britain (1812–1854): How is the Permanence of this New World to be expressed? 119
7.4 Peacock's Symbolical Algebra and its Underlying Epistemology 121
7.5 How Boole uses this Symbolical Conception of Mathematics to Algebrize Logic 123
7.6 Philosophical Consequences of this Ontological Conception of Operations 127
7.7 How Frege Claims the Existence of an Objective and Significant Logic 128
7.8 Where is the Issue of Meaning Located in Contemporary Approaches? 132
Conclusion 134
References 135

8 Musical Analysis Using Mathematical Proceedings in the XXth Century
L. Fichet ... 139

References ... 145

9 Universal Prediction Applied to Stylistic Music Generation
S. Dubnov and G. Assayag ... 147

9.1 Introduction ... 147
9.2 Dictionary-Based Prediction ... 152
9.3 The Incremental Parsing (IP) Algorithm ... 154
9.4 Resolving the Polyphonic Problem ... 156
9.5 Experiments ... 158
References ... 158

10 Ethnomusicology, Ethnomathematics. The Logic Underlying Orally Transmitted Artistic Practices
M. Chemillier ... 161

10.1 Sand Drawings from the Vanuatu ... 163
10.1.1 The Guardian of the Land of Dead ... 163
10.1.2 The Logic of the Long Line ... 166
10.2 The Harp of the Former Nzakara Courts ... 170
10.2.1 The Art of Poet-Harpists ... 170
10.2.2 The Plant-of-the-Twins ... 173
10.3 African Asymmetric Rhythms (after the works of Simha Arom) ... 174
10.3.1 Asymmetric Rhythm of the Aka Pygmies ... 174
10.3.2 The Rhythmic Oddity Property ... 176
10.4 Conclusion ... 180
References ... 181

11 Expressing Coherence of Musical Perception in Formal Logic
M. Leman ... 185

11.1 Introduction ... 185
11.2 Formal Logical Accounts of Musical Coherence ... 186
11.3 Reasoning about Musical Coherence ... 187
11.4 Characterizing Coherence ... 188
11.5 Representations of Musical Content ... 188
11.6 Implementation ... 190
11.7 Image-transformations ... 191
11.7.1 The "Primary" Auditory Images ... 192
11.7.2 Pitch Processing ... 193
11.8 Perceptually Constrained Logical Reasoning ... 194

11.9 Discussion ... 197
11.10 Conclusion ... 197
References ... 198

12 The Topos Geometry of Musical Logic
G. Mazzola ... 199

Introduction ... 199
12.1 Galois Theory of Concepts ... 200
12.2 Categories of Local and Global Compositions ... 205
12.3 "Grand Unification" of Harmony and Counterpoint ... 207
12.4 Truth and Beauty ... 209
References ... 211

13 Computing Musical Sound
J.-C. Risset ... 215

Introduction ... 215
Mathematics and Musical Theory ... 215
Digital Sound ... 218
Synthesis Programs ... 220
Additive, Substractive and Non-linear Synthesis Models ... 221
Imitation of Instruments ... 223
Composition of Textures ... 224
Illusions, Paradoxes ... 225
Intimate Transformations and Analysis-Synthesis ... 225
Real-Time Piano-Computer Interaction ... 226
Conclusion ... 227
References ... 228

14 The Mathematics of Tuning Musical Instruments – a Simple Toolkit for Experiments
E. Neuwirth ... 233

References ... 240
Appendix: Mathematica code ... 241

15 The Musical Communication Chain and its Modeling
X. Serra ... 243

15.1 Introduction ... 243
15.2 The Communication Chain ... 243
15.3 Composer ... 245
15.4 Symbolic Representation ... 246
15.5 Performer ... 247
15.6 Temporal Controls ... 248
15.7 Instrument ... 248
15.8 Source Sound ... 250

15.9 Room 250
15.10 Sound Field 251
15.11 Listener 251
15.12 Perception and Cognition 252
15.13 Conclusions 253
References 254

16 Computational Models for Musical Sound Sources
G. De Poli and D. Rocchesso 257

16.1 Introduction 257
16.2 Computational Models as Musical Instruments 258
16.3 Sound Modeling 259
16.4 Classic Signal Models 261
16.4.1 Spectral Models 261
16.4.2 Time Domain Models 264
16.4.3 Hybrid Models 265
16.4.4 Abstract Models: Frequency Modulation 266
16.5 Physics-based Models 267
16.5.1 Functional Blocks 267
16.5.2 Cellular Models 269
16.5.3 Finite-difference Models 271
16.5.4 Wave Models 273
16.6 Non Linear Musical Oscillators 276
16.7 Models for Sound and Space 280
16.7.1 Sound Spatialization 280
16.7.2 Room Modeling and Reverberation 282
16.8 Conclusions 283
References 284

List of Contributors

G. Assayag
IRCAM - Centre Georges Pompidou,
Paris, France
assayag@ircam.fr
http://www.ircam.fr

M. Chemillier
Université de Caen, Caen, France
marc@info.unicaen.fr
http://users.info.unicaen.fr/~marc

G. De Poli
CSC, Dipartimento di
Elettronica e Informatica
Via Gradenigo 6a
35131 Padova, Italy
depoli@dei.unipd.it
http://dei.unipd.it/~depoli

J. Dhombres
EHESS/CNRS
9, rue des Boulangers
75005 Paris
Centre A.Koyré
27, rue Damesme
75013 Paris
dhombres@damesme.cnrs.fr

S. Dubnov
Ben Gurion University,
Beer Sheva, Israel
dubnov@bgumail.bgu.ac.il

M.-J. Durand-Richard
University of Paris 8
Paris, France
mjdurand@paris7.jussieu.fr

M.P. Ferreira
Departamento de Ciências Musicais
Faculdade de Ciências
Sociais e Humanas
Av. de Berna, 26c
1069-091 Lisboa, Portugal

L. Fichet
IUFM, Toulouse, France
Fichet@aol.com

W. Hodges
Queen Mary and
Westfield College
Londond, GB

E. Knobloch
TU Berlin
Ernst-Reuter-Platz 7
10587 Berlin

M. Leman
IPEM – Dept. of Musicology/
Ghent University
Blandijnberg 2
9000 Ghent, Belgium
Marc.Leman@rug.ac.be

G. Mazzola
ETH Zürich/Departement GESS
and
Universität Zürich/Institut
für Informatik
www.encyclospace.org

E. Neuwirth
University of Vienna
Institute of Statistics and
Decision Support Systems
Universitätsstraße 5
1010 Wien, Austria
erich.neuwirth@univie.ac.at
http://mailbox.univie.ac.at/
erich.neuwirth/

F. Nicolas
Francois.Nicolas@ircam.fr
http://www.entretemps.asso.fr/

J.-C. Risset
CNRS/Laboratoire de
Mécanique et d'Acoustique
31, chemin Joseph Aiguier
13402 Marseille, France
jcrisset@lma.cnrs-mrs.fr

D. Rocchesso
DST, University of Verona
Dipartimento di Informatica
Strada le Grazie 15
37134 Verona VR, Italy
rocchesso@sci.univr.it
www.sci.univr.it/~rocchess

B. Scimemi
University of Padua
Padua, Italy

X. Serra
Pompeu Fabra University/
Audiovisual Institute
Barcelona, Spain
xserra@iua.upf.es
http://www.iua.upf.es

R.J. Wilson
The Open University
Milton Keynes

1 Proportions in Ancient and Medieval Music

Manuel Pedro Ferreira

This paper deals with a particular mode of thought embedded in both Ancient and Medieval music theory: a mathematical bent stemming from the idea that Music, based on proportional relationships which embody Number, is an audible symbol of a God-given ontological order. This mode of thought influenced the aesthetics of the Ancients and had a crucial impact on the rationalization of musical composition in the late Middle Ages[1].

Reporting the ideas of ancient Greek philosophers, an anonymous second-century writer known as the Pseudo-Plutarch wrote: "Everything, they say, was constructed by God on the basis of musical harmony"[2]. A similar train of thought led St. Jerome, about two hundred years later, to conclude that "[he who examines] the harmony of the world and the order and concord of all creatures, sings a spiritual song"[3].

These two quotes illustrate the continuity between pagan and Christian modes of thought in the late ancient world concerning the harmonious constitution of the universe[4]. But these quotes also suggest that the concept of Music was then far more comprehensive, and more widely applied, than it is today.

In the Ancient and Medieval world, Music was more than just intentional, organized sound, as we now tend to think; it was, above all, the theoretical knowledge of the principles illustrated by organized sound. These are proportional, mathematical principles, coextensive with those which were thought

---

[1] Written on request to be read at the IV Diderot Mathematical Forum for a non-musicological audience, this paper does not attempt to exhaust or to redefine the subject matter; for different, complementary approaches, see Willi Apel, "Mathematics and Music in the Middle Ages", *Musica e Arte Figurativa nei secoli X–XII*, Todi: Accademia Tudertina, 1973, pp. 135–65, and Christian Meyer, "Mathématique et musique au Moyen Age", *Quadrivium. Musiques et Sciences*, Paris: Éditions ipmc, 1992, pp. 107–21.

[2] Andrew Barker, *Greek Musical Writings: I. The Musician and his Art*, Cambridge: Cambridge University Press, 1984, pp. 248–49.

[3] Oliver Strunk, *Source Reading in Music History. Antiquity and the Middle Ages*, New York: W.W. Norton, 1965, p. 72.

[4] Russell A. Peck, "Number as Cosmic Language", *By Things Seen: Reference and Recognition in Medieval Thought*, ed. David Jeffrey, Ottawa: The University of Ottawa Press, 1979, pp. 47–80 (this essay appears also in *Essays in the Numerical Analysis of Medieval Literature*, ed. Caroline Eckhardt, Lewisburg: Bucknell University Press. 1979).

to rule the created world; Music was therefore regarded as fit to lift the soul from sensorial experience to the contemplation of eternal, cosmic truth.

According to Cassiodorus, writing in the sixth century, Mathematics, "that science which considers abstract quantity", has four divisions: Arithmetic, Music, Geometry and Astronomy. Music "is the discipline which treats of numbers in their relation to those things which are found in sounds, such as duple, triple, quadruple, and others called relative that are similar to these". And he continues: "The parts of music are three: Harmonics, Rhythmics, Metrics. Harmonics is the musical science which distinguishes the high and low in sounds. Rhythmics is that which inquires whether words in combination sound well or badly together. Metrics is that which by valid reasoning knows the measures of the various metres"[5].

Music, as a scientific discipline, was thus understood to be a mathematical science which encompassed the various aspects of ordered sound, including the sound of formal speech and poetry. The instrumental piece, the modulating voice, the poetic song, all were recognized as transient, yet organized phenomena. Sound, being, in St. Augustine's words, "an impression upon the sense", which "flows by into the past and is imprinted upon the memory"[6], could only be apprehended by the intellect as abstract organization, which was identified with number, i.e. proportional conformity. This is why music could be defined as "the science of discrete, non-permanent quantity", as Roger Bacon later put it in his *Communia mathematica*[7].

The Greek Heritage

This view of Music as a theoretical discipline had already, by this time, had a long intellectual history. It had started, at least in the Western world, with Pythagoras, a philosopher who lived in the sixth century B.C. but from whom we do not have a single written line. We are nevertheless told, in no uncertain terms, by later writers, that he speculated about the correspondence between the consonant quality of some musical intervals and the simplicity and manifold mutual relations of the first four whole numbers[8].

---

5 O. Strunk, op. cit., pp. 88–89.

6 O. Strunk, op. cit., p. 93, n. 2. The idea was later taken over by St. Isidore of Seville.

7 Roger Bacon, *Communia Mathematica*, ed. Robert Steele, London, 1940, p. 51: *Sciencia vero de quantitate discreta non-permanente est Musica.* Bacon speaks often of Music in his mathematical writings, although with little originality. See also Robert Belle Burke (trans.), *The Opus Majus of Roger Bacon*, I, Philadelphia, 1928, and Don Michael Randel, "Al-Farabi and the Role of Arabic Music Theory in the Latin Middle Ages", *Journal of the American Musicological Society*, **29** (1976), pp. 173–88 [183–85, 187].

8 Cf. G.S. Kirk & J. Raven, *The Pre-Socratic Philosophers*, Cambridge: Cambridge University Press, 1966 (Portuguese translation, Lisbon: Gulbenkian, 1979,

To illustrate, let us have a stretched string attached to a horizontal ruled scale, with a maximum vibrating lenghth of twelve units, whose sound, when pulled, we will take as reference. A vibrating portion of six units, i. e. half the maximum length, produces another sound an octave above the first. The interval of an octave can thus be identified with the proportion 1:2. A vibrating portion of four units, which corresponds to the proportion 1:3, sounds an octave and a fifth, or twelfth, above. A vibrating portion of three units, corresponding to the proportion 1:4, sounds two octaves above. We have thus three consonances related to three ratios, 1:2, 1:3, and 1:4, which are said to be multiple ratios, for the larger term is a multiple of the lesser one, which may be represented as $n{:}xn$.

If we now take the numbers $1, 2, 3$ and 4 and try to find other possible proportional, unequal combinations among them, we are left with two new ratios, 2:3 and 3:4. In both of them the larger term exceeds the smaller by one unit, illustrating the form $n{:}n + 1$. These ratios are called "epimore" or "superparticular". We will be able to hear the interval corresponding to these ratios, selecting vibrating portions of eight against twelve units for the "hemiolic" 2:3 ratio ("hemiolic" meaning that the whole is exceeded by its half); and, again, nine against twelve units for the "epitrite" 3:4 ratio ("epitrite" meaning that the whole is exceeded by its third). We will then find, respectively, the intervals of a fifth and a fourth; both were considered consonant by the Greeks, not only on account of the interaction between their physical properties and human perception, but also because their relational use within the prevailing stylistic conditions allowed, in the absence of conventional assumptions to the contrary, the recognition of their "harmonious", blending quality.

Thus all the ratios comprised by the numbers $1, 2, 3$ and 4 imply, according to Pythagoras, consonant intervals. Two of them, the twelfth (1:3) and the double octave (1:4), can be decomposed into intervals of octave and fifth; the octave, the fifth and the fourth correspond to 1:2, 2:3 and 3:4, where 1:2 is not only multiple, but also superparticular as are the remaining ratios. Moreover, the successive ratios 1:2, 2:3 and 3:4, when combined using as reference the larger vibrating lenghth (6:8:9:12), form an octave (6:12) with a fourth and fifth inside, both when reckoned from top to bottom (*e. g.* 6:8:12) and from bottom to top (12:9:6).

When these same ratios are combined in sucession (3:4:6:12), they form a double octave (3:12) with an octave down below (6:12), a fifth above it

---

pp. 219–34); Andrew Barker, *Greek Musical Writings, II: Harmonic and Acoustic Theory*, Cambridge: Cambridge University Press, 1989, pp. 28–45; Richard L. Crocker, "Pythagorean Mathematics and Music", in id., *Studies in Medieval Music Theory and the Early Sequence*, Aldershot: Variorum, 1997 (chapter II).

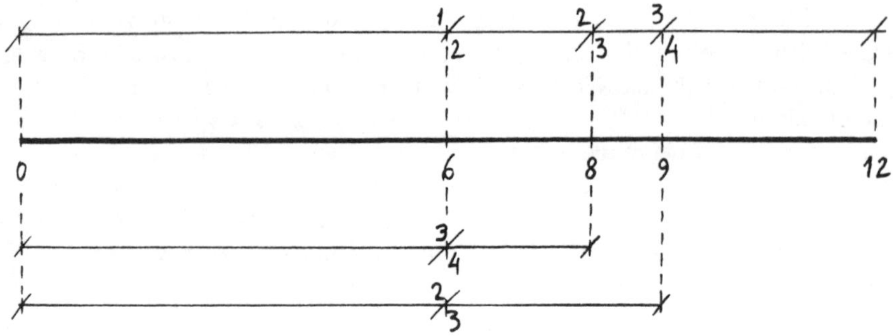

Fig. 1.1.

(4:6) – which also provides a twelfth with the lowest sound (4:12) –, and a fourth at the top (3:4).

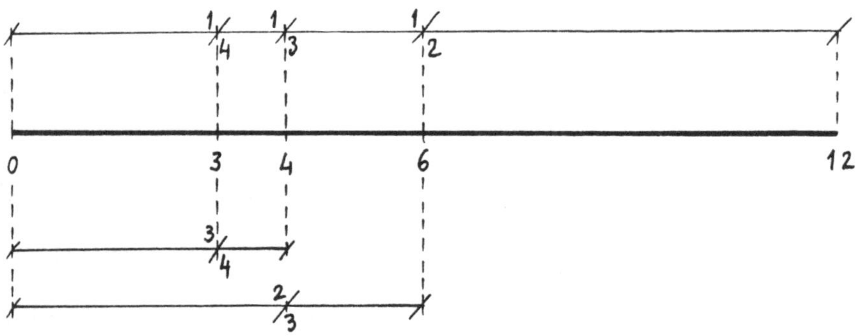

Fig. 1.2.

The correspondence between aural judgment, which recognizes the consonant quality of intervals, and the proportional properties of the first four integers was only part of the story. Pythagoras added two observations: the fact that $1 + 2 + 3 + 4$ equals 10 (ten being the basis of counting operations in all Indo-European peoples)[9]; and the fact that 10 units can be represented on a surface by 10 equidistant dots forming an equilateral triangle with four units on each side (the Pythagorean *tetraktys*):

Although these facts may seem trivial to us, the elegance of the mathematical construct, and its explanatory aesthetic potential, help to understand why it was invested with special mystical significance. Impressed by the

[9] Georges Ifrah, *Histoire Universelle des Chiffres*, Paris: Robert Laffont, 1994, tome I, pp. 73–95. The Greeks were well aware of this fact: cf. G. Kirk & J. Raven, op. cit. (Port. trans., p. 233).

Fig. 1.3.

underlying mathematical logic of consonance, Pythagoras attributed to the Universe an underlying substance which he identified with Number.

In spite of all the above harmonic reasoning, the early Pythagoreans do not seem to have been concerned with the practical aspects of music. But by the time of Philolaus, who lived in the fifth century B.C., musical scales had already been derived from the consonances implied in the *tetraktys.* The difference between the fourth and the fifth, a whole tone (of 204 cents) represented by the proportion 8:9, was projected into the fourth in order to determine its diatonic division[10].

In modern acoustics, the equal-tempered semitone has 100 cents, the tone 200 cents, the ditone or major third 400 cents, the perfect fourth 500 cents, and so on, but these intervals correspond only approximately to the "pure" intervals defined by the ratios between the vibrational frequencies of two pitches; these ratios are the same as the Pythagorean arithmetical proportions[11].

Let me open a parenthesis here to remind you that the basic module of Ancient Greek music, called tetrachord, is comprised in a fourth; its lower and upper limits being fixed, the two movable notes inside divided it into smaller intervals, according to the harmonic genus, which could be enharmonic, chromatic or diatonic (see Fig. 1.4).

In the diatonic genus, the fourth could encompass two whole tones, but there was, of course, a remainder; this was identified as a semitone (of 90 cents) represented by the proportion 243:256, as can be inferred from the table of equivalences below (the ditone implies the ratio 64:81 or 192:243, the fourth, 3:4 or 192:256, therefore the difference between ditone and fourth turns out to be 243:256).

Philolaus calculated intervals other than the tone and semitone by measuring the difference between the sums of basic consonances. The ditone or major third, for instance, which can be construed as the addition of two tones, can also be defined as the difference between the sum of four fifths

[10] A. Barker, *Greek Musical Writings, II*, pp. 36–39; Martin L. West, *Ancient Greek Music*, Oxford: Clarendon Press, 1992, pp. 167–68, 219–20, 235–36.

[11] Cf. Mark Lindley, "Interval", *The New Grove Dictionary of Music and Musicians*, ed. Stanley Sadie, London: Macmillan, 1980, vol. 9, pp. 277–79.

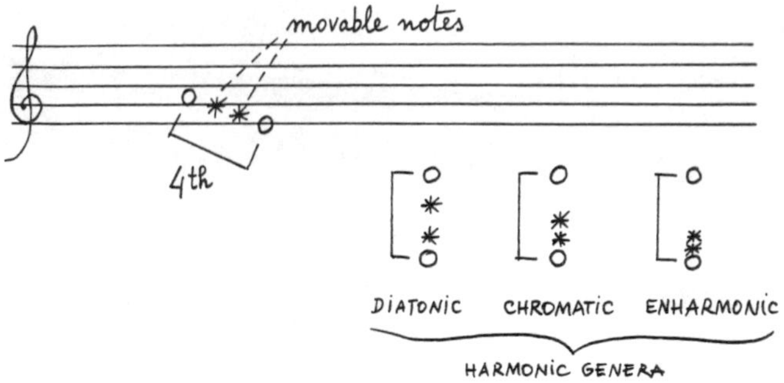

Fig. 1.4.

Table 1.1.

	8			64	(×3 =)	192
tone						
	9	8	72			216
tone						
		9	81			243

	3	(×64 =)	192
fourth			
	4		256

192		
	tone	
216		
		fourth
	tone	
243		
256		

and the sum of two octaves. This can be seen in the following table, in which each column replicates the ratio 2:3, thus representing a continuous series of fifths[12]:

The column headed "16" represents four fifths, reached at "81" below; the double octave relative to "16" is "64"; thus, a major third is given by the proportion 64:81. Similar calculations, probably based on instrumental tuning practices by fourths and fifths, allowed Philolaus to calculate the intervals not only of the diatonic, but also of the enharmonic and chromatic tetrachords[13].

Shortly after Philolaus, the Pythagorean school reached a summit in the work of Archytas, a contemporary of Plato particularly well acquainted with

[12] R. Crocker, op. cit., pp. 195–96.

[13] M. West, op. cit., pp. 167–68.

Table 1.2.

2	4	8	**16**	32	**64**
3	6	12	24	48	96
	9	18	36	72	144
		27	54	108	216
			81	162	324
				243	486
					729

musical practice[14]. The superparticular proportions which, in the *tetraktys*, had stopped at the number four, were allowed to play a larger role in the determination of practical scale divisions, a step no doubt encouraged by the superparticular character of the multiple proportion 1:2 and the relevance of the tone 8:9 as a tetrachordal constituent.

Thus, the incomposite major third in the enharmonic genus, which we had encountered in Philolaus as a 64:81 ratio, was newly described by Archytas as a 4:5 ratio (or 64:80) equivalent to 387 cents. The proportions 5:6 (= 315 cents) and 6:7 (= 267 cents) were acknowledged as existing in the context of a pentachord; later theorists associated these minor thirds with the tetrachordal division of the chromatic genus. Finally, the ratio 7:8 (= 231 cents) came to represent the lower tone of the diatonic genus. In this way, all the superparticular proportions up to number nine were included in Archytas' harmonic scheme[15]:

	multiple							
superparticular	(1:2	2:3	3:4)	4:5	5:6	6:7	7:8	(8:9)
	(1:3)							
	(1:4)							

The ratio 27:28, slightly less than a third of a tone, was also proposed by Archytas as the lower interval of any tetrachord, and 35:36, roughly a quarter-tone, as its complement in the enharmonic genus. Other superparticular proportions, like 9:10 or 15:16, were later associated with tetrachordal divisons by several writers in the Pythagorean tradition such as Erastosthenes, Didymus and the great Ptolemy[16] (Ptolemy's mathematical idealization of tunings known to him consists exclusively of superparticular ratios, and he even proposes a new diatonic division of the tetrachord [9:10, 10:11, 11:12] on account

[14] Martin Vogel, *Die Enharmonik der Griechen, 1. Teil: Tonsystem und Notation*, Düsseldorf, 1963, pp. 50–57; R. Crocker, op. cit., pp. 331–33; M. West, op. cit., pp. 168, 236–38; A. Barker, *Greek Musical Writings, II*, pp. 39–52.

[15] Cf. A. Barker, *Greek Musical Writings, II*, pp. 46–47.

[16] M. West, op. cit., pp. 169–71, 237–40.

of the mathematical elegance of near-equal ratios in equal, superparticular excesses. He finds, however, no compelling reason to deny a consonant quality to the interval of octave plus fourth, represented by the ratio 3:8, which is neither multiple nor superparticular)[17].

Archytas was also the first to classify and explore the musical potential of proportional means. According to him, there are three means in music. One is arithmetic, the second geometric, the third harmonic. And I quote:

"There is an arithmetic mean when there are three terms, proportional in that they exceed one another in the following way: the second exceeds the third by the same amount as that by which the first exceeds the second [*e. g.* 12:9:6 twice subtracts 3]. In this proportion it turns out that the interval between the greater terms is less, and that between the lesser terms is greater [in the given example, 12:9 gives a fourth, while 9:6 gives a fifth]."

"There is a geometric mean when they are such that as the first is to the second, so is the second to the third [*e. g.* 12:6:3 replicates the proportion 2:1]. With these, the interval made by the greater terms is equal to that made by the lesser [in the given example, both 12:6 and 6:3 give an octave]".

Finally, there is a harmonic mean "when they are such that, by whatever fraction of itself the first term exceeds the second, the second exceeds the third by the same fraction of this latter [*e. g.* 12:8:6, where $12 - 8 = 12/3$, and $8 - 6 = 6/3$]. In this proportion the interval between the greater terms is greater, and that between the lesser terms is less" [that is, 12:8 gives a fifth, and 8:6 gives a fourth][18].

Archytas, in addition, penned mathematical demonstrations which reveal his concern with harmonics. A single theorem of his survives; it states that a superparticular ratio cannot be divided by a whole number into equal parts. It reached us in two slightly different versions. The version transmitted by Boethius, possibly closer to the original, mirrors the Greek distinction between the unity (*monas*) and number as originated by the unity (*arithmos*); it runs like this:

"A superparticular ratio cannot be split exactly in half by a number proportionally interposed [...]."

"Let A:B be a superparticular ratio [...]. I take the smallest integers in that same ratio, C:$D[+]E$. Since C:$D[+]E$ is the same ratio and the ratio is superparticular, the number $D[+]E$ exceeds the number C by one of its – that is, $D[+]E$'s – own parts. Let this part be D. I say then that D will not be a number [i. e., a plurality of units], but unity. For if D is a number and is part of $D[+]E$, the number D measures the number $D[+]E$, and thus it will also measure the number E. Whence it follows that it should also measure C. Thus the number D would measure both the numbers C and $D[+]E$, which is

17 A. Barker, *Greek Musical Writings, II*, pp. 306–12; André Barbera, "The Consonant Eleventh and the Expansion of the Musical Tetractys: A Study in Ancient Pythagoreanism", *Journal of Music Theory*, 28 (1984), pp. 191–224.

18 A. Barker, *Greek Musical Writings, II*, p. 42 (adapted).

impossible. For these are the smallest integers in the same ratio as some other numbers, the first numbers so related, and they maintain the difference of unit alone. Therefore D is unity. So the number $D[+]E$ exceeds the number C by unity. For that reason no mean number comes between them that splits the ratio equally. It follows that between those greater numbers that maintain the same ratio as these, a mean number cannot be interposed that splits the same ratio equally."[19]

A: B	C: D+E
9: 12	3: 1+3

Fig. 1.5.

The significance of this theorem, which probably belonged to a now lost series of theorems, is that, the basic consonances being represented by superparticular ratios, they cannot be divided equally by rational numbers – or receive a geometric mean, which is the same. It demonstrates the need for unequal divisions of all the consonances comprised in an octave.

Once this was acknowledged, Archytas showed how arithmetic and harmonic means could be used to generate proportionally integrated scales. According to Andrew Barker, "notes an octave apart are represented by terms in a geometrical progression by doubles (*e. g.*, $6, 12$). If the harmonic and arithmetic means are inserted between those terms (*e. g.*, $6, 8, 9, 12$), the new terms are the inner boundaries of the tetrachords, separated by a tone [...]. When harmonic and arithmetic means are placed between terms in the ratio 3:2, the ratios between means and extremes are 5:4 and 6:5. When they are placed between terms in the ratio 4:3, the resulting ratios are 7: 6 and 8:7 [...] Hence all the ratios underlying Archytas' divisions [...] can be constructed proportionally, by the location of means first in the octave, then in the concords generated by the first construction"[20].

Surprisingly, Archytas' divisions were not generally followed by later theorists; his methodology was, however, adopted by Plato in the *Timaeus*, where the philosopher, besides referring to the proportions of the world's body and to the numbers of time, explains at length the harmonic constitution of the soul of the universe through geometric progressions ($1, 2, 4, 8$ and $1, 3, 9, 27$) and arithmetic and harmonic means. The Platonic harmony, which replicates, albeit in a novel way, the traditional Pythagorean divisions of Philolaus, has

[19] Anicius M.L. Boethius, *Fundamentals of Music*, trans. Calvin M. Bower, New Haven: Yale University Press, 1989, pp. 103–5; square-bracketed additions mine.
[20] A. Barker, *Greek Musical Writings, II*, pp. 48–49

received a great deal of attention[21]; since it does not intend to describe music as practiced, we will leave it aside on this occasion.

So far, we have exclusively dealt with philosophical approaches to harmonics. But there were also music teachers, practitioners with theoretical leanings, and minor musicographers who based their research on empirical data and experiment, rather than abstract philosophical constructs. These were the ones who attempted to find a lesser common denominator for musical intervals, irrespective of rational proportions; who conceived instruments meant for acoustical research, including the monochord adopted by the followers of Pythagoras; and who ended up by creating and developing Greek musical notation. They were also, in a sense, the forerunners of the second most important musical philosopher in Ancient Greece: Aristoxenus, a brilliant pupil of Aristotle[22].

Aristoxenus rejected the Pythagorean-Platonic tradition, including the representation of intervals by numerical ratios; he eschewed any attempt to justify the phenomenon of consonance and chose instead to start from continuous sound as perceived by the ear. He did not make relative pitch depend on string lengths, using instead the twin concepts of string tension and relaxation. This allowed him to deal with all musical intervals, even when these did not correspond to any rational proportion.

Aristoxenus' work cannot be adequately summarized here; nor can we trace its considerable impact on Greek musical theory. What is of particular interest for our present purposes is that Aristoxenus' choices were closely linked to a geometric concept of the tonal space. Intervallic quantity being treated as continuous, it was allowed to be divided freely according to perceptual or practical considerations.

Philosophers, however, tended to despise practicioners; and these latter retaliated by simply ignoring philosophers. The extent to which theoretical awareness influenced Ancient Greek, Roman and Hellenistic musical practice must, in fact, remain largely moot, due to the scarcity of historical records. Variable tetrachordal divisions were no doubt in use; we can suspect that the Pythagorean-Platonic diatonic was not a very common one. Perfect fourths and fifths were important structural features in the melodies,

---

21 See, for instance, J. Dupuis, "Le nombre géométrique de Platon", in id., *Œuvres de Théon de Smyrne*, Paris: Hachette, 1982, pp. 365–400; J.F. Mountford, "The musical scales of Plato's *Republic*", *Classical Quarterly*, vol. 17 (1923), pp. 125–36; Jacques Handschin, "The *Timaeus* Scale", *Musica disciplina*, vol. IV (1950), pp. 3–42; Ernest McClain, *The Pythagorean Plato*, New York: Nicolas Hays, 1978. Selection of texts and commentary in A. Barker, *Greek Musical Writings*, I, pp. 124–69; II, pp. 53–65. For a general presentation of Plato's comments on music, see Warren Anderson, "Plato", *The New Grove Dictionary of Music and Musicians*, ed. Stanley Sadie, London: Macmillan, 1980, vol. 14, pp. 853–57.

22 M. West, op. cit., pp. 4–5, 167–69, 228–33; Annie Bélis, *Aristoxène de Tarente et Aristote. Le Traité d'Harmonique*, Paris: Klincksieck, 1986; A. Barker, *Greek Musical Writings, II*, pp. 119–89.

although their contour seems to have been dominated by adjacent degrees, thirds and unisons. Even the heterophonic instrumental accompaniment of song, while making use of vertical consonances, did not prefer them to other intervals, like sevenths and tritones[23].

The rhythmic proportions of 1:1, 1:2, 1:3, 2:3 and 3:4 were, on the other hand, frequently encountered; sung poetry, based on the alternation of short and long syllables measured according to "feet", implied the use of rhythmic markers, the "arsis" and the "thesis", which, implying bodily movement, literally embodied these ratios. It would be impossible not to notice that the basic rhythmic proportions correspond, melodically, to the unison, the octave, the twelfth, the fifth and the fourth; rhythmics, metrics and harmonics were completely congruent from a mathematical point of view.

The Latin World

It was through proportional rhythm applied to words that Ancient Greek and Roman musical theory made its entrance into Western Christian literary thought. It was at once a remarkably bold and a remarkably ambiguous *debut*, the one responsible being St. Augustine. This bishop, a towering figure of the late fourth-century Latin Church, wrote a voluminous treatise on Music, of which he had planned to write a second volume. What he wrote is concerned only with Rhythm.

St. Augustine deals almost exclusively with literary rhythms; verbal rhythm is seen, however, as a musical phenomenon independent from Grammar: "The science of music", he says, "to which belongs the reasoned measurement of words in themselves and their rhythm, is only concerned to see that the syllable in this or that place be shortened or lengthened according to the pattern of the proper measure. For if you put the word *cano* where there ought to be two long syllables and pronounce the first syllable long although it is really short, it is not a musical offence; for the lengths of the sounds reach the ear as the rhythm demands that they should. But the grammarian insists on a correction being made and directs you to substitute a word whose first syllable is long according to the authority of the ancients, whose traditions he guards"[24].

This observation is full of consequences, for three hundred years later, St. Bede would comment that some hymns of the Ambrosian type have a regular rhythmic flow, in spite of the fact that, grammatically speaking, they escape metric organization. The door was open to the imposition of freely chosen rhythms on sung texts, a possibility that the Ancient world, attached to quantitative Greek and Latin, had denied the poet-composer.

[23] Cf. M. West, op. cit., chapters VI, VII, X.

[24] Gregory Murray, "Gregorian Rhythm in the Gregorian Centuries. The Literary Evidence", *The Downside Review*, 75 (1957), pp. 234–58; published independently by the Downside Abbey, Bath, n.d. [quote on p. 6].

In his musical tract, St. Augustine reveals himself as heir to the Pythagorean concept of music as sounding number. According to Robert O'Connell, for St. Augustine "the sound-embodied numbers, which delight us in measures of verse as they strike our ear, proceed in downward cascade from the eternal numbers, which themselves proceed from God [. . .] The sensible and intelligible orders are [therefore] on speaking terms [. . .] The upward way is in the strictest sense of the term a return to the contemplative delight from which the soul has fallen"[25].

St. Augustine, moreover, expands the concept of musical proportion as found in rhythm to embrace all manifestations of artistic beauty. In his words, "that which is beautiful pleases on account of number, in which, as demonstrated already, we seek equality. This can be found not only in what reaches the ears and in the movement of the bodies, but also in visible forms, of which we usually speak of beauty"[26].

Although the step taken by St. Augustine would later prove influential, he does not display in his musical treatise a direct concern with Christian, ecclesiastic practice. Indeed, only once does he refer in passing to a contemporary musical composition, and this is a hymn written by his friend, St. Ambrose of Milan. The reference is relevant, however, for he analyses the verse *Deus creator omnium* as being composed of four iambic feet with a total of twelve beats; in the *Confessiones* the analysis is more detailed: "this line", he says, "is composed of eight syllables, short and long alternately; [...] each long syllable has double time of each short syllable"[27].

In spite of all his strengths, the impact of St. Augustine's thought on Christian music seems, prior to ninth-century, to have been negligible[28]. The same applies to Pythagorean-Platonic or Aristoxenean harmonic theory, transmitted by a few medieval authors. Between the fifth and the eighth century, the time which saw the emergence of specialized liturgical repertoires of chant, Philosophy was not part of the daily ration of the Latin Church[29].

---

[25] Robert J. O'Connell, *Art and the Christian Intelligence in St. Augustine*, Oxford: Basil Blackwell, 1978, pp. 67–68, 71.

[26] *De musica*, VI, 12, cit. in Wladyslaw Tatarkiewicz, *Historia de la Estetica, II: La estetica medieval*, Madrid: Akal, 1989, p. 65 [my translation]. See also *De vera religione*, XXX, 35: "Since in every art what pleases is conformity, which by itself saves and embellishes everything, this conformity, in fact, requires equality and unity, be it in the semblance of the equivalent parts, be it in the proportion of the unequal ones" (ibid., p. 63).

[27] G. Murray, op. cit., p. 4.

[28] Intellectuals could not, however, ignore St. Augustine. Cassiodorus, in the middle of sixth century, refers to St. Augustine's teaching that "the human voice naturally has rhythmical sounds and proportioned melody in long and short syllables" (*Institutiones*, V, quoted in G. Murray, op. cit., p. 3).

[29] Joseph Dyer, "The Monastic Origins of Western Music Theory", *Cantus Planus. Papers Read at the Third Meeting (Tihany, 1988)*, Budapest: Hungarian Academy of Sciences, 1990, pp. 199–225.

Liturgical singing was developed in clerical contexts where philosophical endeavour was generally not welcome; instrumental practice, which could provide a link with musical theory, was no more fortunate in this respect. The Psalms of the Bible were omnipresent as the basis for worship, either modestly chanted, syllabically sung, or melismatically ornamented. Music was conceived of as an emphatic proclamation of the divine Word; this public proclamation came to be more and more elaborate and ended up by requiring an extremely long and specialized training, but was never regarded as an autonomous, abstract art.

Some musical concepts of Antiquity were assimilated, filtered through the writings of late Roman grammarians and the contributions of Cassiodorus, St. Aldhelm of Malmesbury, St. Isidore of Seville and St. Bede. The grammatical model associated with writing permitted the identification – by analogy with letters, words and periods – of melodic units, basic motives and sections; the grammatical approach to the inflexions of the voice and the corresponding vocabulary allowed for a basic analytical consciousness of relative time values and for a much more ambiguous consciousness of relative height[30].

As in Ancient times, a connection between the theory developed by the Greeks and liturgical practice can be discerned only in very special instances. We have seen that rhythmic alternation between short and long sounds in the proportion 1:2 was used in some hymns. The probable parallel singing in fifths or fourths of the Alleluia in the Old Roman tradition, or the fifth used as a structural interval and, sometimes, as a melodic leap in chant, are other possible instances. Our musical sources being almost entirely lacking for the earlier centuries of the Middle Ages, it is already risky to say this much. These are indeed obscure centuries, the real Dark Ages, and yet, it was then that a new Latin European identity was forged and a new musical art was born.

Historically, the decades around 800 represent a turning point. The Carolingian educational and ecclesiastical policy changed the face of Western music forever. Since the time of Pepin the Short, the Franks had procured the best teachers, had multiplied the copying of books and had raised decisively the reading and writing skills of the churchmen. The adoption in the Frankish Empire of Roman liturgy and its chant necessitated a vast literary and musical effort, which eventually resulted in novel schemes of melodic classification and the invention of neumatic notation; this new notation recorded relative pitch contour and, at first, also the brevity or prolongation of single sounds.

In Fig. 1.6a, we can see an example of early Paleofrank, "graphic" notation, where single sounds are represented by dots (isolated when long,

30 Marie-Elisabeth Duchez, "Description grammaticale et description arithmetique des phenomènes musicaux: le tournant du IXe siècle", *Miscellanea Mediaevalia, Band 13/2: Sprache und Erkenntnis im Mittelalter*, Berlin, New York: Walter de Gruyter, 1981, pp. 561–79.

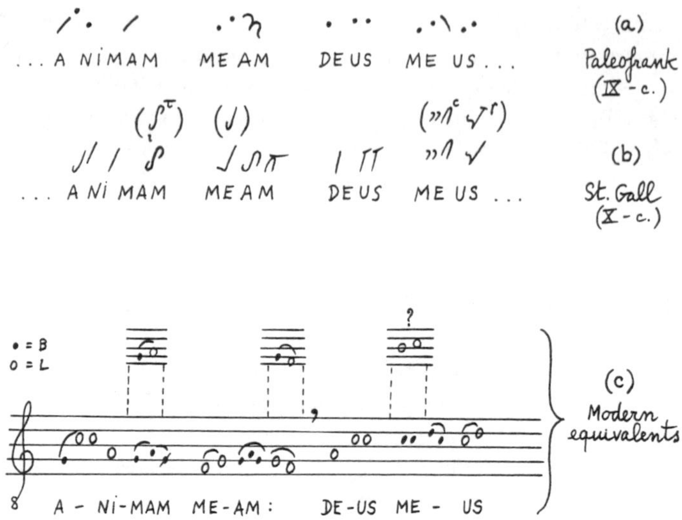

Fig. 1.6.

connected when quick): over the first syllable of *animam*, for instance, the notation implies quick ascending movement involving two sounds and then a third, long upper note. Figure 1.6b presents, for the same passage, an example of tenth-century St. Gall notation, representative of a "gestural" type of writing where the shapes incorporate the additional movement (extra loops and lines without melodic significance) by the hand of the choral director; note the use of horizontal strokes and letters *c* and *t* to indicate the relative speed of notes. Figure 1.6c presents, for comparison, a modern transcription of the same extract, taken from the introit *Ad te levavi* (Paleofrank variants above); black and void note-heads stand for short and long notes, respectively[31].

The extent to which these rhythmic oppositions were conceived as proportional ratios or qualitative (approximative) distinctions has been much debated, and probably a conclusive answer will never be reached[32]. Alcuin,

---

[31] This paragraph is based on the momentous re-examination by Kenneth Levy of the relationship between the different families of neumes ("On the Origin of Neumes", first published in 1987), now available in his book *Gregorian Chant and the Carolingians*, Princeton: Princeton University Press, 1998, pp. 109–40. In the example, the Paleofrank neumes are taken from the Düsseldorf Sacramentary D. 1, while the St. Gall ones derive from MSS 339 and 376 (see also *Le Graduel Romain. Édition critique*, IV/II, Solesmes, 1962, pp. 69–70: over *[ani]mam* most sources have only two notes, both long in MSS Chartres 47 and Laon 239). The Paleofrank variant over *me[us]* finds a parallel in MS Chartres 47 (cf. facsimile in *Paléographie musicale*, vol. XI, Solesmes, 1912).

[32] A summary of the debate up to the early 1960s can be found in John Rayburn, *Gregorian Chant. A History of the Controversy Concerning Its Rhythm*, New

however, at the beginning of the ninth century, refers to the chant practice at Charlemagne's palace school as follows: "[The cantor] prepared the boys for the sacred chant in order that they might sing the sweet melodies clearly, and learn that music consists of prosodic feet and proportions."[33] A later reference by Lupus of Ferrières, to the singing of short and long sounds in psalmody, and a passage in the ninth-century tract *Scolica enchiriadis* (the explanation of the expression *numerose canere*, illustrated by the prosodic notation of a simple antiphon) substantiate, albeit with some ambiguity, Alcuin's report[34].

Early in the tenth-century, the remarks by the anonymous author of the *Commemoratio brevis* – partly echoing the *Scolica enchiriadis* – are particularly striking: "Breves must not be slower than is fitting for Breves; nor may Longs be distorted in erratic haste and be faster than is appropriate for Longs [...] All notes which are long must correspond rhythmically with those which are not long through their proper inherent durations [...] for the longer values consist of the shorter, and the shorter subsist in the longer, and in such a fashion that one has always twice the duration of the other, neither more nor less [...] for without question all music should be strictly measured in the manner of prosody. Teachers must impress this zealously upon their pupils, imparting to the children from the beginning this habit of evenness and strict measure"[35]. The sense of a parallel between prosodic

York, 1964, repr. Westport, Conn.: Greenwood Press, 1981; and Bruno Stäblein, "Thèses equalistes et mensuralistes", *Encyclopédie des musiques sacrées*, ed. J. Porte, vol. 2, Paris: Labergerie, 1969, pp. 80–98. The following years, however, saw the emergence of the semiological school of Eugène Cardine, which completely changed the arena. The old opposition between defenders of metrical chant and equalist chant did not altogether disappear, but a mid-way gradually became accepted, at least in practice. A bridge between the mensuralist distinction of longs and breves and the semiological acknowlegment of different rhythmical values is outlined in Manuel Pedro Ferreira, "Bases for Transcription: Gregorian Chant and the Notation of the *Cantigas de Santa Maria*", in José López-Calo (coord.), *Los instrumentos del Pórtico de la Gloria: Su reconstrucción y la música de su tiempo*, La Coruña: Fundación Pedro Barrié de la Maza, 1993, Vol. 2, pp. 573–621.

33 Nancy Phillips, "Classical and Late Latin Sources for Ninth-Century Treatises on Music", *Music Theory and Its Sources. Antiquity and the Middle Ages*, ed. André Barbera, Notre Dame (Indiana): University of Notre Dame Press, 1990, pp. 100–35 [124].

34 N. Phillips, op. cit., p. 125 (remark by Lupus of Ferrières); *Musica enchiriadis and Scolica enchiriadis*, trans. Raymond Erickson, New Haven: Yale University Press, 1995, pp. 50–53, 69; Nancy Phillips & Michel Huglo, "Le *De musica* de saint Augustin et l'organisation de la durée musicale du IXe au XIIe siècles", *Recherches Augustiniennes*, XX (1985), pp. 117–31 (on the antiphon *Ego sum via*).

35 *Commemoratio brevis de tonis et psalmis modulandis*, ed. & trans. Terence Bailey, Ottawa: The University of Ottawa Press, 1979, pp. 103, 107.

feet and ecclesiastical musical practice was still alive in the early eleventh century, as testified by Guido of Arezzo in his treatise *Micrologus*[36].

After a couple of generations, the Carolingian clerics were also able to start a revival of Classical learning. Some monasteries, and a few Cathedral schools, were able to rise to fame as centers of culture, without losing sight of their liturgical obligations. This allowed the emergence of a new kind of intellectual: one philosophically well read, yet attentive to his clerical, contemplative mission, which was embodied in the daily liturgy. Musical theorists from around the fifth century A.D. like Martianus Capella and Boethius, who transmitted Ancient Greek harmonics, and the roughly contemporary commentaries by Macrobius and Calcidius on Plato's *Timaeus* were again read with devotion[37]. A new kind of music theorist was about to be born.

The ninth-century music theorist devoted his best efforts to assimilating the harmonic science of the Greeks. But he was also concerned with its practical application to liturgical chant. Competing scalar systems were proposed to that end; the traditional Pythagorean-Platonic diatonic division of the octave, which in the Ancient world had never been widely used, suddenly occupied the center of the stage, and with minimal modification was adopted by many as their chief theoretical reference. The authority of Boethius, the Christian thinker in whose writings the Pythagorean octave divisions had been extensively explained, and the growing analytical awareness of melodic intervals, which resulted in diastematic notation, slowly led to the rejection of certain features of chant which were deemed incompatible with the theoretically defined scale, like modulating chromatic notes other than the B flat and the remains of soft diatonic and enharmonic microtones. The traditionalists tried to keep non-diatonic or non-Pythagorean intervals, but eventually lost[38].

---

[36] Cf. *Hucbald, Guido and John on Music – Three Medieval Treatises*, trans. Warren Babb, New Haven: Yale University Press, 1978, pp. 70–73. A few commentaries: Jan W.A. Vollaerts, *Rhythmic Proportions in Early Medieval Ecclesiastical Chant*, 2nd edition, Leiden: E.J. Brill, 1960, pp. 168–94; Richard L. Crocker, "*Musica rythmica* and *Musica Metrica* in Antique and Medieval Theory", in id., *Studies in Medieval Music Theory and the Early Sequence*, Aldershot: Variorum, 1997 (chapter IV); Nino Pirrotta, *Musica tra Medioevo e Rinascimento*, Torino: Einaudi, 1984, pp. 3–19; Calvin M. Bower, "The Grammatical Model of Musical Understanding in the Middle Ages", *Hermeneutics and Medieval Culture*, ed. P. Gallacher & H. Damico, New York: State University of New York Press, 1989, pp. 133–45 [140 ff.].

[37] Michel Huglo, "The Study of Ancient Sources of Music Theory in the Medieval Universities", *Music Theory and Its Sources – Antiquity and the Middle Ages*, ed. André Barbera, Notre Dame, Indiana: University of Notres Dame Press, 1990, pp. 150–172.

[38] Manuel Pedro Ferreira, "Music at Cluny: The Tradition of Gregorian Chant for the Proper of the Mass. Melodic Variants and Microtonal Nuances" (PhD diss., Princeton University, 1997) ProQuest-UMI 9809172 (1998).

At the same time, extempore polyphony (simultaneous melodic strands) associated with liturgical practice was made an object of harmonic analysis. Those practices which did not depend on the consonances acknowledged by the Greeks, like the simultaneous seconds conspicuously used in Lombardy, or the simultaneous thirds possibly current in the British Isles[39], were left outside musical theory, which recognized only the fourth, the fifth and the octave as legitimate harmonic foundations for two-part singing. Passing intervals, however, were allowed, and the major second, used at the end of phrases, was sometimes defended on aesthetic grounds. Two schools of thought, in fact, emerged: one defended the rigid application of theoretical principles; another let perceptual judgment play a major role. The end-result was mixed: the major second survived as appogiatura, the major third was allowed as a passing and preparatory sonority, while the fourth lost ground, leaving the fifth and the octave to dominate the polyphonic texture[40].

In the course of the eleventh-century, the pedagogic role given to the monochord, the rise of diastematic notation, the invention of the staff and of solmization completely changed the way liturgical musical would be learnt and preserved in the centuries to follow. Trained in *pugna numerorum*, an intellectual game which required the participants to calculate geometric, arithmetic and harmonic means[41], the younger, educated clerics could easily grasp Pythagorean-Platonic musical concepts and would no more tolerate current inconsistencies in the modal behaviour of the melodies. The transition between grammaticaly oriented musical conceptions and a systematic theoretical learning based on Greek harmonics had been completed.

It is no coincidence that an Aquitanian trope to the Sanctus dating from the second half of the eleventh-century, *Clangat hodie vox nostra*, displays in its text an uncommon concentration of technical musical terms, as in the verses (in translation) "High-sounding at the octave, ascending in tetrachords through discrete pitches to the high summit of its contours". But it is also significant that the poem reached us with two melodies, of which the later one, still dating from the eleventh-century, strives to associate with the word

---

[39] F. Alberto Gallo, "Esempli dell'*Organum* dei *Lumbardi* nel XII secolo", *Quadrivium*, VIII (1967), 23–26; Paul J. Nixon, "Giraldus Cambrensis on Music: How Reliable Are His Historiographers?", *Proceedings of the First British-Swedish Conference on Musicology, Medieval Studies, 11–15 May 1988*, ed. Ann Buckley, Stockholm: Royal Swedish Academy of Music, 1992, pp. 264–89.

[40] For a summary of the evidence, with bibliography, see Manuel Pedro Ferreira, "Early Cistercian Polyphony: A Newly-Discovered Source", in Lusitania Sacra, v. 13–14 (2001–2002), forthcoming.

[41] Wolfgang Breidert, "Arithmomachia", *Quadrivium. Musiques et Sciences*, Paris: Éditions ipmc, 1992, pp. 169–78. The game was invented in the first half of the XIth century and survived until the XVIth. The mathematician Roger Bacon mentions it, which he calls *Rithmimachia*, in *De communibus mathematice*: cf. David E. Smith, "The Place of Roger Bacon in the History of Mathematics", *Roger Bacon Essays*, ed. Andrew Little, Oxford, 1914, pp. 153–83 [177].

for octave, *Diapason*, a melodic octave; with *altissona*, for "high-sounding", a leap of a fifth; and with *tetracordis*, a melodic motif comprehended in a tetrachord[42]. Intellectual mastery of harmonics came first, but then the tentation proved irresistible to make practice conform to its conceptual frame.

In the meanwhile, Arabic music and the corresponding theory, of which the most important early representative was Al-Farabi (872–950), had been transmitted to the southern, Islamic territories of the Iberian Peninsula. Al-Farabi was highly indebted to the Harmonic Science of the Greeks, which he used to describe the scales known to him, having recourse both to the superparticular proportions proposed by Archytas and the tone-fractions championed by Aristoxenus. He also presented a strikingly developed rhythmic theory based on the sophisticated Arabic (Persian) tradition, which was later appropriated by the composers of the Christian *Cantigas*. Only a small part of Al-Farabi's writings on music was translated into Latin in the twelfth century; in spite of some terminological misunderstanding by the translators, his Aristotelean demand that theory should conform to the observation of practice may have encouraged an empirical attitude among Western students of Musical Science[43].

Late-Medieval France

In the 1100s, St. Augustine's *De musica*, which had meanwhile continued to be recopied, began to receive some attention in Parisian academic circles; the number of copies slightly increased in the thirteenth-century[44]. Coincidence or not, it was in Paris, in the last quarter of the twelfth century, that a notational technique was developed to record proportional rhythmic flow analogous with, but completely independent from, poetic metre and even words.

---

[42] Gunilla Iversen, "The Mirror of Music: Symbol and Reality in the Text of *Clangat hodie*", and Charles Atkinson, "Music and Meaning in *Clangat hodie*", *Revista de Musicología*, XVI (1993), pp. 771–89, 790–806.

[43] Rodolphe d'Erlanger, *La musique arabe*, vols. I–II, Paris, 1930–1935; D. Randel, op. cit.; Owen Wright, "Al-Farabi", *The New Grove Dictionary of Music and Musicians*, ed. Stanley Sadie, London: Macmillan, 1980, vol. 1, pp. 251–52; Habib Hassan Touma, "Indications of Arabian Musical Influence on the Iberian Peninsula from the 8th to the 13th Century", *Symposium Alfonso X El Sabio y la Música*, Madrid: Sociedad Española de Musicología, 1987, pp. 137–50; George Dimitri Sawa, *Music Performance Practice in the Early Abbasid Era 132–320 AH/750–932 AD*, Toronto: Pontifical Institute of Mediaeval Studies, 1989; Manuel Pedro Ferreira, "Andalusian music and the *Cantigas de Santa Maria*", in Stephen Parkinson (ed.), *Cobras e Som. Papers from a Colloquium on the Text, Music and Manuscripts of the Cantigas de Santa Maria*, Oxford: Legenda, 2000, pp. 7–19.

[44] M. Huglo, "The Study of Ancient Sources", cit., pp. 167–70.

Rhythm was accepted in Parisian polyphony as an autonomous dimension of sound subjected to proportional rules and requiring the exact coordination of time intervals. Musical time was, however, still conceived of as somewhat elastic through expansion or reduction of its contents and acceleration or retention of its movement; it was only in the thirteenth-century that a fixed measuring stick was adopted, and with it, a spatial, geometric conception of time-constructs[45].

This occurred when the motet became a separate, autonomous, self-contained artifact, cultivated by scholars and intellectuals. Its internal time was measured by a single, invariable time-unit; its internal time proportions were manipulated as numbers. The rise of the motet as the most *recherché* of musical genres, the academic answer to the now declining *trobar clus* of the aristocratic elite, was the main artistic contribution of the mid-thirteenth century.

At roughly the same time, the knowledge of Euclidian geometry was raised to a new pitch. In the *Elements*, newly commented by Campano da Novara, arithmetic was subordinated to geometry, in the sense that geometry, dealing with the continuous, admits both rational and irrational proportions, while arithmetic, being concerned with the discrete, cannot attain the same level of abstraction[46]. This had important musical consequences.

On the one hand, it freed compositional procedure from the need to conceive musical time as the successive addition of an invariable, indivisible time unit; on the other hand, Euclid presents in the *Elements* an irrational proportion which would be frequently used in the French motet of the thirteenth and fourteenth centuries: the so-called "extreme and mean ratio" or "golden proportion" (the latter expression is a modern coinage), which occurs when the smaller term is to the larger term in the same way as the larger term is to the smaller plus the larger $[a : b :: b : (a+b)]$. In the Euclidian *Elements*, the extreme and mean ratio is not only defined as such but also used in connection with determination of areas and the construction of the pentagon and related solids[47].

---

[45] Manuel Pedro Ferreira, "Mesure et temporalité: vers l'Ars Nova", in *La rationalisation du temps au XIIIème siècle – Musiques et mentalités (Actes du colloque de Royaumont, 1991)*, Royaumont: Créaphis, 1998, pp. 65–120.

[46] Fabrizio Della Seta, "Proportio. Vicende di un concetto tra Scolastica e Umanesimo", *In Cantu et in Sermone – For Nino Pirrotta on His 80th Birthday*, Firenze, 1989, pp. 75–99 [79].

[47] Roger Herz-Fischler, *A Mathematical History of Division in Extreme and Mean Ratio*, Waterloo, Ontario: Wilfrid Laurier University Press, 1987. See also Robert Lawlor, *Sacred Geometry. Philosophy and Practice*, London: Thames & Hudson, 1982. The golden proportion was used in the planning of both medieval churches and books: cf. Carol Heitz; "Mathématique et Architecture", *Musica e Arte Figurativa nei secoli X–XII*, Todi: Accademia Tudertina, 1973, pp. 169–93 [187–90], and Jacques Lemaire, *Introduction à la Codicologie*, Louvain-la-Neuve: Université Catholique de Louvain, 1989, pp. 127–34.

This is the only three-term proportional division which is possible with only two terms, and is self-generating, which allows it to be given teological significance. It corresponds, arithmetically, to the numerical approximation $0,61803\ldots : 1$ (or $1 : 1,61803\ldots$); rougher approximations based on integers could be obtained from the Fibonacci series, available from the early thirteenth-century onwards. In this numerical series each number equals the sum of the two preceding ones: $1; 2; 3\,(= 1 + 2)$; $5\,(= 2 + 3)$; $8\,(= 3 + 5)$; $13\,(= 5 + 8)$; $21\,(= 13 + 8)$; $34\,(= 13 + 21)$, $55\,(= 21 + 34)$, etc.; the ratio between two consecutive numbers tends to approach the golden proportion. Thus, $5/8 = 0,6250$; $8/13 = 0,6154$; $13/21 = 0,6190$; $21/34 = 0,6176$; $34/55 = 0,6182$. Fibonacci himself knew of and used the extreme and mean ratio; it is characteristic of his work that he often calculated numerical approximations of it. Another important thirteenth-century mathematician, Campanus da Novara, expressed his admiration for the properties of this proportion when applied to the polihedric solids[48].

Turning to the musical domain: the motet, planned as a temporal whole divided into congruous parts, tried to capture in its time organization both the traditional Pythagorean proportions and an irrational one of special symbolic value, derived from geometrical, divisive reasoning. These proportions were marked by their association with important words, formal divisions, compositional modules, remarkable intervals, rhythmic dislocations or any other feature that called attention to itself.

In Fig. 1.7, a scheme corresponding to the three-part motet *Entre Copin/ Je me cuidoie/ Bele Ysabelos* (codex Montpellier, n° 256) shows the main structural divisions, which coincide with 1/2, the golden section, 2/3 and 3/4 of the composition. The first half is subdivided into three parts. The pauses in the upper voices and the melodic triangle in the lower, at the half-way point and the golden section of the motet, frame three statements of the word "God" which are also set out in a triangle. The hocket (a special kind of polyphonic texture with quick alternating voices) marks the proportions corresponding to the intervals of the fifth and the fourth. The last quarter of the motet is again subdivided into three[49].

It must be said in passing that harmonic practice did not evolve much after 1200. Harmonic theory kept its Pythagorean-Platonic basis, although, in the twelfth century, Theired of Dover had, in glorious isolation, resuscitated the 4:5 proportion for the ditone (instead of $9/8 \times 9/8 = 81/64$). The ratios of intervals larger than the fifth, conspicuous in polyphony, were easily calculated by subtracting from the octave their reverses: octave minus major third equals minor sixth (in modern terms, $2/1 \times 64/81 = 128/81$); octave minus minor third equals major sixth ($2/1 \times 27/32 = 54/32$, equivalent to adding a tone to the fifth, $3/2 \times 9/8 = 27/16$); octave minus tone equals minor seventh ($2/1 \times 8/9 = 16/9$); octave minus semitone equals major

48 R. Herz-Fischler, op. cit., pp. 136–44, 171.

49 M.P. Ferreira, "Mesure et temporalité", pp. 104–5.

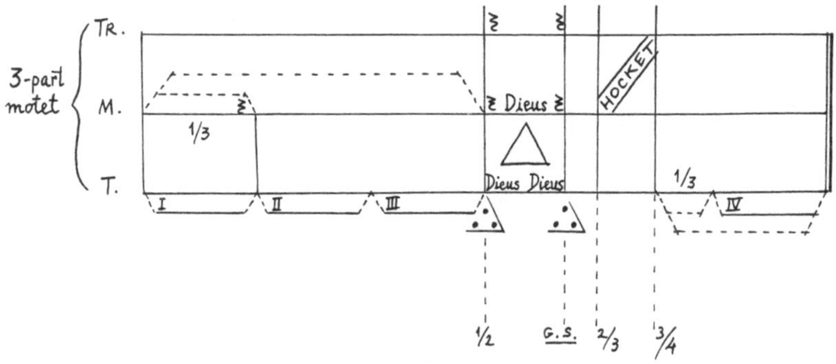

Fig. 1.7.

seventh ($2/1 \times 243/256 = 486/256$, equivalent to adding a ditone to the fifth, $3/2 \times 81/64 = 243/128$).

In general, theorists, following the lead of Johannes de Garlandia (mainly known for his innovations in the notational domain), attempted to refine the vocabulary which defined the degree of concordance, proposing intermediary categories between the obvious consonance of the octave and the obvious dissonance of the minor second; the thirds were admitted as imperfect consonances, while the status of minor and major sixth remained controversial. The fourth, which practice had definitely downgraded, retained for most authors its conventional status of consonance[50].

After Garlandia, the next turning-point was Philippe de Vitry, the indirect author of a treatise called *Ars Nova.* Vitry was a theorist-composer who, in the early-fourteenth century, reformulated musical notation and compositional technique to expand proportional manipulation of time divisions in polyphony. Praised by his contemporaries (including Petrarch), he was in touch with first-class astronomers and mathematicians like Johannes de Muris and Levi ben Gerson (Leo Hebraeus)[51]. Had he not paid more attention to royal politics, public administration and an ecclesiastical career than to

[50] Cf. John L. Snyder, "The *De legitimis ordinibus pentachordorum et tetrachordorum* of Theinred of Dover" (PhD diss., Indiana University, 1982); Erich Reimer, *Johannes de Garlandia: De mensurabili musica*, Wiesbaden: F. Steiner, 1972; Serge Gut, "La notion de consonance chez les théoriciens du Moyen Age", *Acta musicologica*, 48 (1976), pp. 20–44; Michel Huglo, "La notation franconienne. Antécédents et devenir", *Cahiers de Civilisation Médiévale*, XXXI (1988), pp. 123–132 [125–26].

[51] Eric Werner, "The Mathematical Foundation of Philippe de Vitri's *Ars Nova*", *Journal of the American Musicological Society*, 9 (1956), pp. 128–32; *Ars Nova Magistri Philippi de Vitriaco*, ed. G. Reaney et al., n. p.: American Institute of Musicology, 1964 [CSM, 8]; Ernest H. Sanders, "Vitry, Philippe de", *The New Grove Dictionary of Music and Musicians*, ed. Stanley Sadie, London: Macmillan, 1980, vol. 20, pp. 22–28.

the copying of his work, he would be nowadays as famous as the younger Guillaume de Machaut.

In Vitry's motet *Firmissime/Adesto/Alleluia*, about the Holy Trinity, the golden section falls, significantly, at "personis tribus". The lower voice (*Tenor*) is constructed in such a way that the first statement of the melody (*color* 1) occupies triple the time of the second statement (*color* 2), the same proportion 3:1 applying between their respective rhythmical divisions (for each *color* is divided into eight equivalent sections)[52].

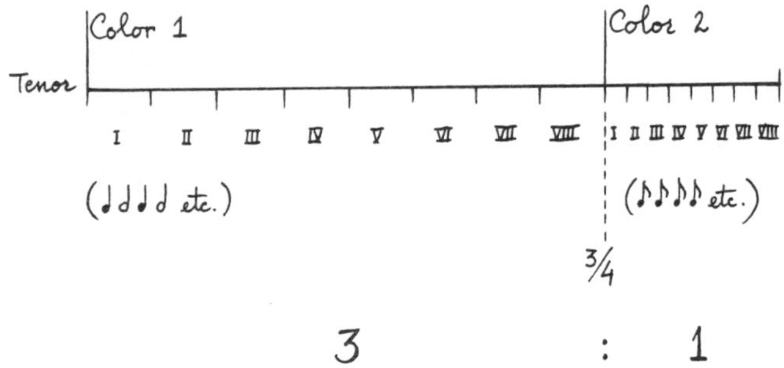

Fig. 1.8.

In the later thirteenth-century, divisive, geometric reasoning had competed with additive, numerical methodology in giving rhythmic shape to the motet not only at the overall planning stage, but also at the level of defining the shortest possible sounds to be used. By the early fourteenth-century, the geometric conception was expanded and systematized by Philippe de Vitry and his teaching first shaped into a treatise by the young Johannes de Muris, who, following Aristotle, insisted on the continuous nature of time[53].

The new notational system thus developed for mensural polyphony, the *Ars Nova* (in the strict sense of the expression) started from basically four rhythmic categories, from larger to shorter sounds: the *longa*, the *brevis*, the *semibrevis* and the *minima*. Three different levels of mensuration were distinguished: *modus*, the proportional relationship between the long and the breve; *tempus*, between the breve and the semibreve; and *prolatio*, between the semibreve and the minim. At each of these levels, the longer note could be divided into two or three shorter sounds; the whole organization had, therefore, a divisive character.

52 Manuel Pedro Ferreira, "Compositional Calculation in Philippe de Vitry", paper presented to the XVIIIth Medieval and Renaissance Music Conference (London, 6–9 July 1990).

53 Fabrizio Della Seta, "*Utrum musica tempore mensuretur continuo, an discreto.* Premesse filosofiche ad una controversia del gusto musicale", *Studi musicali*, XIII (1984), pp. 169–219.

Mensuration signs were created to indicate the use of binary or ternary divisions. At the highest mensuration level, the *modus*, the most common signs consisted of two or three horizontal or vertical strokes, often drawn inside rectangular or quadrangular boxes, but rest signs normally sufficed, serving as indirect mensuration signs. The *tempus* was ternary when a full ("perfect") circle was drawn, binary when a broken ("imperfect") circle took its place. The *prolatio* was ternary when three dots, representing minims, were put inside the circle, and binary when there were only two dots (later, the three dots were replaced by a single one, two by none). The binary and ternary divisions of *brevis* and *semibrevis* combined to give four metres, each with its time-signature. All this can be seen in Fig 1.9 below.

To enhance rhythmic variety, new procedures were found to modify the proportions between time-durations; composers used colored or void notes for metric changes within a prevailing mensuration, and introduced proportional signatures, which originally replaced mensuration signs, to jump metrical levels by diminishing (or, seldom, increasing) the value of notes in certain arithmetical ratios.

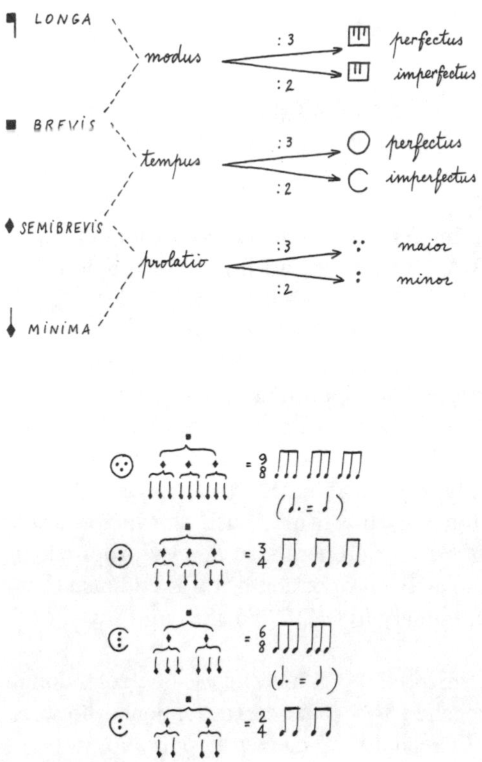

Fig. 1.9.

The built-in capabilities of the notation were not, however, exhausted at the time of Vitry and Machaut. At first the minim could be divided into (only) two semiminims; diminution (later indicated by special signs) allowed for the sudden halving of the rhythmical values or their reduction by two-thirds. This implied eight, twelve, eighteen or twenty-seven semiminims to a breve, but up to the late fourteenth-century, the semiminim was not universally adopted and the number nine remained a practical limit for proportional changes of mensuration (see Fig. 1.10).

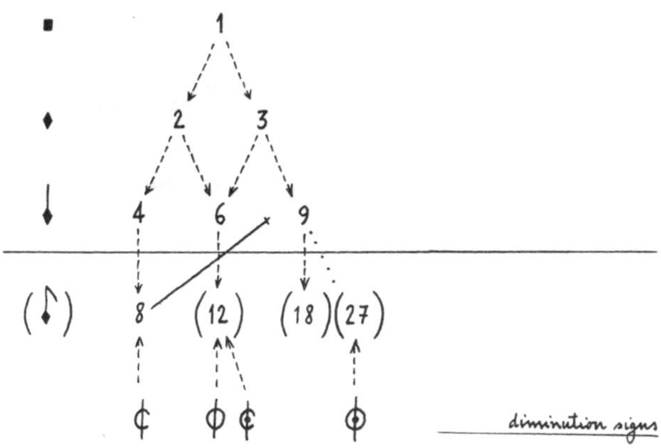

Fig. 1.10.

The proportions acknowledged in musical notation were therefore only those which involved numbers $1, 2, 3, 4, 6, 8$ or 9 and had a harmonic counterpart: the Pythagorean consonances 1:2 (octave), 1:3 (twelfth), 1:4 (double octave), 2:3 (fifth) and 3:4 (fourth); the Ptolemaic consonance 3:8 (eleventh); the Pythagorean tone (8:9) and its Ptolemaic equivalent, the ninth (4:9); or, which is the same, the proportions corresponding to tone, fourth, fifth, octave and their duplicates at the octave above. All proportions which implied a contrast in excess of 1:4 (corresponding to the larger harmonic interval, the double octave) like the theoretically conceivable 1:6, 1:8, 1:9 or 2:9, were ignored. Only at the end of the fourteenth-century, with the introduction of special note shapes and later, of Arabic numerals to denote proportions, would the mensural system allow the latter proportional ratios and also those involving hitherto excluded primes, namely five (2:5, 3:5, 4:5) and seven (2:7, 3:7, 6:7).

This polyrhythmic expansion, effected by a new class of professional, courtly avant-garde musicians, breached the close correspondence between the harmonic and the rhythmic dimensions of mensural polyphony. Once the step was taken to fully explore the divisive potential of the notation in order to represent a wider range of rhythmic proportions, the development of proportional signs took a path of its own; around 1430, however, a shift in aesthetic tendencies towards rhythmic clarity and fluidity led to the divorce of this development from practical musical needs[54].

The Decline of Proportional Thinking

Although the basic organization of the *Ars nova* mensural system was and largely remained divisive, additive ways of thinking crept in from the very beginning: for practical musicians, it was convenient to reckon rhythm from an invariable short unit of time, like the minim, while the long was too protracted to allow for easy empirical manipulation. Besides, the idea, defended by conservative theorists, that measured music was based upon number, that musical time was discreet and not continuous, did not disappear in the fourteenth century[55]. The tendency in late medieval mathematics to think of geometry through arithmetical concepts[56] may also have contributed to the slow erosion of the *Ars nova* theoretical edifice after 1400.

The fifteenth century was in various ways a transitional period, marked by the emergence of chordal sonorities and the work of innovative musical theorists like Bartolomé Ramos de Pareja, who revived the harmonic proportions 4:5 (major third) and 5:6 (minor third) and their complements 5:8 (minor sixth) and 3:5 (major sixth)[57]. The intrincacies of time proportions and the overall planning of the whole work as a step for serious composition managed to survive well into the following century, but the increasing influence of keyboard extemporization and the new intellectual trends, which encouraged dazzling, flowing polyphony and after 1540, valued clear text delivery and emotional effect, eventually won the day. The subject of proportions, once the crux of compositional thought, receeded back to the realm of prosody and instrumental tuning. A new, secularized mental world – in a sense, still our world – was about to be born.

---

[54] Willi Apel, *The Notation of Polyphonic Music, 900–1600*, 5th ed., Cambridge, Mass.: The Mediaeval Academy of America, 1953, pp. 400–435, 451–53; Richard Rastall, *The Notation of Western Music. An Introduction*, London: J.M. Dent, 1983, pp. 79–105; Anna Maria Busse Berger, *Mensuration and Proportion Signs. Origins and Evolution*, Oxford: Clarendon Press, 1993; Jason Stoessel, "Symbolic Innovation: The Notation of Jacob de Senleches", *Acta musicologica*, 71 (1999), pp. 136–64.

[55] *Quatuor principalia musicae*, cit. in A.B. Berger, op. cit., p. 45.

[56] John E. Murdoch, "The Medieval Language of Proportions: Elements of the Interaction with Greek Foundations and the Development of New Mathematical Techniques", *Scientific Change. Symposium on the History of Science*, London, 1964, pp. 247–71 [270].

[57] Carl Dahlhaus, *Studies on the Origin of Harmonic Tonality*, Princeton: Princeton University Press, 1990; Oliver Strunk, *Source Reading in Music History. The Renaissance*, New York: W.W. Norton, 1965, pp. 10–14.

2 The Sounding Algebra: Relations Between Combinatorics and Music from Mersenne to Euler

Eberhard Knobloch

Introduction

The baroque conception of music was based on a rational, mathematical foundation. Musicologists and composers of that period cited again and again the biblical verse that God had ordered ("disposuisti") the whole world through measure, number, and weight (Wisdom 11, 20). Beauty, harmony consist in order, numbers play a crucial role in this regard.

The variety of harmony stems from the composition, combination, and arrangement of its parts [8, p. 258]. They constitute the universal harmony.

This applied especially to music: "To compose" was equivalent with "to combine", "to arrange". Hence Lullism and its combinatorial art were very influential on that score. It aimed at

- musical education
- Christianization
- God's glorification
- Creativity
- Optimization.

The combinatorial approach enabled even the ignorants to learn and to practice music within one hour. This approach helped to address and to convert the infidels. It can be concluded, that God Himself was the first to apply this art. We can assume, that He was the first 'combinatorialist', when He created the world. The Creator combined the single parts of the world.

Men must imitate God in order to be creative. This especially applies to music. By means of the combinatorial art, a song can be optimized on condition that the number of songs is not too great. Their beauty consists in their variety. This variety is demonstrated by an explicit enumeration. Every possibility is evaluated according to certain musical principles.

As a consequence, I would like to speak about:

1. Mersenne (1635/1636)
2. Kircher (1650)
3. Leibniz (1666)
4. The later development in the 18th century (Euler, Mozart)

2.1 Mersenne (1635/36)

Marin Mersenne was one of the most important adherents of Lullism of the 17th century. For him, the combinatorial art was a fundamental, universal, general art. The combinatorial studies of this famous propagator of science and friend of René Descartes are set out in six publications [7]:

- *Quaestiones celeberrimae in Genesim (1623)*
- *La vérité des sciences (1625)*
- *Harmonicorum libri (1635/36) (= HL)*
- *Harmonie universelle (1636) (= HU)*
- *Cogitata physico-mathematica (1644)*
- *Novae observationes physico-mathematicae (1647)*

The first book concerns theology, the second philosophy of science, especially of mathematics, the next two deal with music theory, the last two with natural science and mathematics. By and large, combinatorial contributions of the first two publications repeat Christoph Clavius's results [3, pp. 17–19] without mentioning their source. The last two publications contain only some additional remarks with regard to the two music-theoretical books, which comprehend Mersenne's most important contributions to combinatorics. Hence I would like to concentrate on these two voluminous monographs. The complete title of the Latin written work reads as follows: "Books about harmony dealing with the nature, causes and effects of the notes, with the consonances, dissonances, proportions, keys, modes, songs, composition and harmonical instruments of the whole world. Dedicated to Henry Montmort. A book useful for grammarians, orators, philosophers, legal advisors, physicians, mathematicians, and theologians."

Mersenne cited psalm 150: Praise God with cymbals pleasing to the ear, praise Him with jubilant cymbals, every spirit should praise the Lord.

The French written work is entitled: "Universal harmony containing the theory and practice of music dealing with the nature of tones and of motions, of consonances, of dissonances, of keys, modes, composition, voice, of songs and of all sorts of harmonical instruments".

The emblem of the title page consists of four pictures illustrating the Fourth Commandment: We have to honour our parents in order to live a long time. At the upper left the emblem shows the son of Tobias healing his father by means of a fish. At the upper right one can see Aeneas leaving the burning Troja with his father Anchises on his shoulders and his son at his hand. At the lower left it is illustrated that Ruth does not leave her mother-in-law, Naemi, after the death of her husband, the father-in-law. At the lower right of the illustration Pero feeds her father Cimon (Mycon) in prison with her milk.

HARMONIE
VNIVERSELLE,
CONTENANT LA THEORIE
ET LA PRATIQVE
DE LA MVSIQVE,
Où il eſt traité de la Nature des Sons, & des Mouuemens, des Conſonances, des Diſſonances, des Genres, des Modes, de la Compoſition, de la Voix, des Chants, & de toutes ſortes d'Inſtrumens Harmoniques.
Par F. MARIN MERSENNE de l'Ordre des Minimes.
A PARIS,
Chez SEBASTIEN CRAMOISY, Imprimeur ordinaire du Roy, ruë S. Iacques, aux Cicognes.
M. DC. XXXVI.
Auec Priuilege du Roy, & Approbation des Docteurs.

Fig. 2.1.

Mersenne considered counting problems, that is a number of different arrangements or collections of objects of a particular kind:

- Permutations,
- combinations, or unordered selections, and
- arrangements or ordered selections.

2.1.1 Permutations

The number of permutations without repetition $P(n, n)$ is calculated by means of the recursion rule:

$$P(n, n) = n \cdot P(n, n-1) = n \cdot (n-1)! = n!$$

Mersenne continued the table of such permutations up to 64!, which is a number of ninety digits, an example of examplary industriousness.

33452526613163807108334022053440751647352000000000 XLI
1405006117752879898550028926244511569188784000000000 XLII
60415263063373835637651243828513997475117712000000000 XLIII
2658271574788448768056654728454615888905179328000000000 XLIV
1076599987789321741062945165124119435006597627840000000000 XLV
49523599438308800088895477595709494010303490880640000000000 XLVI
2327609173600513604178087446998346218484264071390080000000000 XLVII
111725240332824653000548197455920618487244675426713840000000000 XLVIII
5474553677630840799702686167534011030587498909590946816000000000 XLIX
127372683881542039985134308376700551529374945479547340800000000000 L
1396006857958644039241849727211728127998122219456914380800000000000 LI
72592356613899490040556185815009863655902355411759547801600000000000 LII
3847394900536672972149477848195522773762784836823256033484800000000000 LIII
207759324628980340496071803802558229783190381188455825809179200000000000 LIV
11426762854593918727283949209140702638075470965365070419504856000000000000 LV
1246943914205331060001234413005477290188301806190157746145519360000000000000 LVI
71075803098303870420070361541312205540733203952838992530294603520000000000000 LVII
4122396579711624484364080969396107921362525729364661676757087004160000000000000 LVIII
243221398202986844577480777194380367360389018032515028928668133245440000000000000 LIX
14593283892179210674648846631662822041623341081950901735720087994726400000000000000 LX
885190317422931851153579644531432144539023805999005005878925367678310400000000000000 LXI
54881799680221774771521937960948792961419475971938310364493372796055244800000000000000 LXII
3457563379853871810605882091539773956569426986232113552963082486151480422400000000000000000 XLIII
221284059310647795878786453858545533220443527118855467387637279113594747033600000000000000000 LXIV.

PROPOSITIO IV.

Quatuor vocum Tetrachordi ſeu Diateſſaron, VT, RE, MI, FA, Combinationes ſeu varietates notis vulgaribus exprimere.

Cùm ex dictis conſtet res 4, 24 vicibus variari poſſe, placet hanc varietatem ſequentibus quatuor Tetrachordi, ſeu Quartæ primæ ſpeciei notis explicare; ſi nempe canantur notæ per primam clauem in infima linea poſitam; ſi verò vtaris ſecunda claue in ſecunda linea ſcripta, canentur hæ quatuor notæ, RE, MI, FA, SOL, quibus ſecunda ſpecies Quartæ ſeu Diateſſaron conſtituitur, iterumque viginti quatuor vicibus variabuntur: denique 24 aliis vicibus mutabuntur, ſi per tertiam clauem in ſecunda regula, ſeu linea cum *b molli* ſcriptam canantur hæ quatuor notæ, MI, FA, SOL, LA: vnde conſtat hoc exemplum 72 varietates complecti, hoc eſt ter 24, vt videre eſt in notis ſequentibus:

Varietas, ſeu Combinatio quatuor notarum.

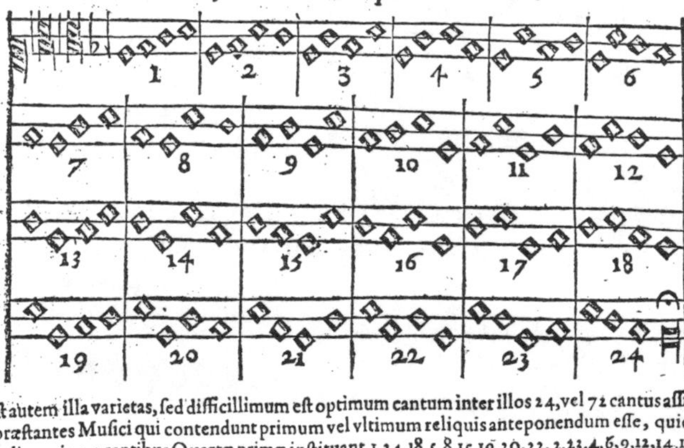

Facilis eſt autem illa varietas, ſed difficillimum eſt optimum cantum inter illos 24, vel 72 cantus aſſignare, ſunt enim præſtantes Muſici qui contendunt primum vel vltimum reliquis anteponendum eſſe, quique ſequentem ordinem in 24 cantibus Quartæ primæ inſtituant, 1, 24, 18, 5, 8, 15, 19, 20, 22, 2, 23, 4, 6, 9, 12, 14, 16 & 21: neque deſunt qui primum & 18 cæteris præferant: ſed ratione demonſtrare cantum omnium optimum, & ordinem naturalem inter reliquos cantus ſtatuere, hoc opus hic labor eſt, quemadmodum & vniuſcuiuſque temperiem, conſtitutionem, & animum ex cantibus prædictis qui magè placeant, explorare.

Porrò ſi clauis *F, vt, fa* cum ♯ ſcribatur in infima linea, fiunt iterum 24 Cantilenæ à præcedentibus differentes, quarum vnaquæque Tritono FA SOL, RE, MI conſtabit. Vbi notandum eſt eadem penitus de Tetrachordis Chromatico, & Enharmonico, & ſingulis illorum ſpeciebus, ac de præcedentibus Diatonicis intelligi poſſe. Ne verò quis laboret in illis varietatibus cantuum exhibendis, & vt promptè, & abſque vlla confuſione per ordinem diſponantur, artem Combinationis ſequenti propoſitione declaramus.

P iii

Fig. 2.2.

To my knowledge, no other author ever calculated – without a computer, of course – a greater factorial. Mersenne listed the twenty-four different songs with four notes, the seventy-two different songs of three types of songs, consisting of four different notes, respectively, the sevenhundred-twenty different songs with six notes. Considering fourths like ut, re, mi, fa or re, mi, fa, sol or mi, fa, sol, la he enumerated the possible permutations in order to discuss which song might be considered as better or more pleasant or more agreeable. The decision depends on justifiable criteria, which differ from author to author.

He systematically studied all different types of repetitions regarding a certain number n. In his case, $n = 9$. Presumably this is a reminiscence of Lull, because Lull selected nine fundamental notions for his language theory (HU, p. 116):

Table des Chants qui se peuvent faire de 9 notes	
Toutes différentes	*362880*
2 semblables	*181440*
3	*60480*
4 & 2, 2 & 3	*15120*
5	*3024*
...	
2 + 7	*36*
Toutes semblabes	*1*

In modern terms, he is looking for the multinomial numbers

$$\binom{n}{n_1, n_2, \dots, n_p}$$

where $n_1 + n_2 + \ldots + n_p = n$, $n_i \geq 1$, $i = 1, \dots, p$.

If $p = n$, all n_i must be equal to 1:

$$\binom{n}{n_1, n_2, \dots, n_p} = \frac{n!}{n_1! n_2! \dots n_p!}$$

Mersenne, however, never mentioned the identity of this expression with the coefficients of the powers of a polynom.

The equation $n_1 + n_2 + \ldots + n_p = n$ represents a partition of n. In other words, Mersenne enumerated all thirty partitions of nine in order to consider all thirty types of repetitions in the case of nine notes. Hence we have a close relation between additive number theory and permutations with repetitions, both being important subjects of combinatorics.

2.1.2 Arrangements as a Generalization of Permutations

After the permutations Mersenne deals with arrangements with and without repetitions like in modern textbooks on combinatorics [2].

If we do not take all n-elements out of n (which would lead to permutations or to $P(n,n)$), but only an ordered selection of p of the n-elements, we get $P(n,p) = n(n-1)\dots(n-p+1)$ (HL, p. 133):

Tabula generalis Combinationum

I	*22*
II	*462*
III	*9240*
...	
XXI, etc.	
XXII.	*1124000727777607680000*
Summa	*3055350753492612960484*

As always, Mersenne considered special cases like $n = 22$, $p = 1, \dots, 22$ without demonstrating his rule. The demonstration would be based on the principle of multiplication of choices:

If there are k successive choices to be made, and if the i-th choice can be made in n_i ways, for $1 \leq i \leq k$, the total number of ways of making these choices is the product $n_1 \cdot n_2 \cdot \dots \cdot n_k$.

Mersenne calculated even the sum of all twenty-two different values $P(n,1), \dots, P(n,n)$ and underlined the utility of such a table. Only in 1659, Sebastian Izquierdo published a recursion formula, in order to calculate this sum [6, p. 18f.].

The same applies to arrangements with repetitions. Every place can be occupied by one of the n elements. Hence Mersenne just said that we have to calculate the corresponding power of twenty-two or more general:

The number searched for is n^p, if we select p elements out of n (HL, p. 140):

Tabula generalissima Cantilenarum omnium possibilium

I	G	*22*
II	A	*484*
...		
IX	a	*1207269217792*
...		
XXII	ggg	*341427877364219557396646723584*
Summa		*357686347714896679177439424706*

Mersenne calculated $n^p - P(n,p)$ in order to obtain the number of those arrangements where at least one element is repeated.

2.1.3 Combinations

In order to get the number $C(n,p)$ of p-subsets of an n-set, he correctly used his results, that is tables for $P(n,p)$ and $P(n,n)$. His calculation rule reads

$$C(n,p) = \frac{P(n,p)}{P(p,p)}$$

this is in modern terms $\binom{n}{p}$ (HL, p. 137):

Tabula Methodica Conternationum, Conquaternationum, etc. utilissima

I.	*II.*	*III.*	*IV.*	*V.*
I	*36*	*1*	*36*	*35*
II	*630*	*2*	*1260*	*34*
III	*7140*	*6*	*42840*	*33*
...				

Example: $36 \cdot 35 = 1260, 1260 : 2! = 630$ or $P(36,2) = 1260,\ P(2,2) = 2,$ $C(36,2) = 630$

He used the relation $\binom{n}{p} = \binom{n}{n-p}$ without giving a reason for it, that is for the symmetry of binomial coefficients. He constructed the arithmetical triangle (HL, p. 136) by using the relation

$$\binom{n}{p} + \binom{n}{p+1} = \binom{n+1}{p+1}$$

Tabula pulcherrima et utilissima combinationis duodecim cantilenarum

I	*II*	*III*	*IV*	*V*	*VI*	*VII*	*VIII*	*IX*	*X*	*XI*	*XII*
1	1	1	1	1	1	1	1	1	1	1	1
2	3	4	5	6	7	8	9	10	11	12	13
3	6	10	15	21	28	36	45	55	66	78	91
4	10	20	35	56	84	120	165	220	286	364	455
5	15	35	70	126	210	330	495	715	1001	1365	1820
6	21	56	126	252	462	792	1287	2002	3003	4368	6188
7	28	84	210	462	924	1716	3003	5005	8008	12376	18564
8	36	120	330	792	1716	3432	6435	11440	19448	31824	50388
9	45	165	495	1287	3003	6435	12870	24310	43758	75582	125970
10	55	220	715	2002	5005	11440	24310	48620	92378	167960	293930
11	66	286	1001	3003	8008	19448	43758	92378	184756	352716	646646
12	78	364	1365	4368	12376	31824	75528	167960	352716	705432	1352078
13	91	455	1820	6188	18564	50388	125970	293930	646646	1352078	2704156
...											

He hinted at predecessors without citing their names: Cardano would have been such a predecessor. Instead of discussing combinations with repetitions in general, he considered the more difficult problem of finding combinations which represent a special type of repetitions always mainly relying on his musical examples.

Case 1

First of all Mersenne considered the case that exactly one note is repeated an arbitrary number of times. Let us represent the type of repetitions by numbers which indicate how often the single notes are repeated. The types 4 111, 2 111 mean, for example, that one note is repeated 4 times or twice, respectively, and that all other notes occur exactly once. Hence only the number of different notes matters, in our case 4.

1) He had to occupy four places by means of the given notes.
2) Every note can replace the repeated note.

Hence we have to multiply $C(n,4)$ by 4 or to be more general:
We have to calculate $p \cdot C(n,p) = p\binom{n}{p}$ or $\binom{n}{p} \cdot \dfrac{p!}{(p-1)!}$.

Mersenne's example is $n = 22$, $1 \le p \le n$ (HL, p. 138):

Tabula nova anagrammatica	
II	*462*
III	*4620*
IV	*29260*
...	
XII	*7759752*

Case 2

The problem becomes more difficult, if not one but several notes are repeated in a certain way.

type 1:	122; 112	or 1 r_1-times, 2 r_2-times
type 2:	12; 13; 34; 45	or $p, p+n$
type 3:	123; 234	or $p, p+1, p+2$

The examples of one line belong to the same type of repetitions, respectively.

1) In every line he needed the same number of different notes: 3 or 2 or 3, respectively.
2) There are equal frequencies of different numbers of repetitions of different notes: In type 1, one frequency occurs once, the other frequency occurs twice. In type 2, and type 3, every frequency occurs exactly once.

Hence Mersenne achieved the following result: Not the numbers themselves of the repetitions, but the frequencies of special numbers of repetitions matter. Type 1 leads to $\binom{n}{3} \cdot \frac{3!}{2!}$, type 2 leads to $n(n-1)$ possibilities. Mersenne does not mention the number of possibilities of type 3, namely $n(n-1)(n-2)$.

Case 3

Mersenne generalized the problem by considering all possible types of repetitions within a given number p of notes which are taken out of n notes. Hence he again wrote down the 30 partitions of 9 as in the case of permutations with repetition (HL, p. 139):

Tabella novem notarum singularis

	absque ordine	*cum ordine*
Omnes differentes	*497420*	*180503769600*
Similes 2	*2558160*	*464152550400*
2,2,1,1,1,1,1	*3581424*	*324906785280*
etc.		

The first column represents the type of repetitions, that is the partition of 9. The second column provides the number of corresponding combinations with repetitions of this type. The third column provides the number of corresponding arrangements, that is of ordered selections with repetitions of this type.

The sum of all numbers of the second column is the number of all combinations with repetitions or $\binom{n+k-1}{k} = \binom{30}{9}$ or 14307150.

Mersenne did not say this nor did he mention that the sum of all numbers of the third column is the number of all arrangements with repetitions or $n^k = 22^9$ or 1207269217792.

This result is confirmed by his table where he calculated the powers of 22, from 22^1 up to 22^{22}.

Mersenne did not explain how he calculated the numbers of the second and third column, he was only saying that he explained the method elsewhere, but unfortunately in any case not in his published works. Hence we have to reconstruct it generalizing his method of case 1 and 2:

Let p be the number of selected notes, n the number of given notes. The partition of p will be

$$\begin{aligned} p &= 1+1+\ldots+1+2+2+\ldots+2+\ldots m+m+\ldots+m \\ &= 1r_1+2r_2+\ldots+pr_p\,, \quad 1 \leq m \leq p\,, \quad 0 \leq r_i \leq p\,. \end{aligned}$$

The partition can be interpreted in the following way:

r_1 pairwise distinct notes occur exactly once etc.,
r_2 pairwise distinct notes occur twice etc.
r_i pairwise distinct notes occur i-times.

If *one* note is repeated p-times, $r_p = 1$, all other $r_i = 0$.
If *no* note at all is repeated, $r_1 = p$, all other $r_i = 0$.
There are, in fact, only $r_1 + r_2 + \ldots r_m = r$ distinct notes all in all.

Hence according to case 1, we have first of all $C(n, r)$ combinations. This number $C(n, r)$ has to be multiplied by a factor which has still to be searched for. We get a new unordered selection, a new combination of the given type of repetitions as long as a note of a certain frequency replaces another note of another frequency, for example, if two notes which occur once or twice, respectively, are exchanged. We do not get a new combination if two notes of the same frequency are exchanged, because we deal with unordered selections.

The same frequencies function as repetitions in a permutation with repetitions of a certain type we dealt with somewhat earlier. Hence we have to multiply

$$C(n,r) \text{ by } \frac{r!}{r_1! r_2! \ldots r_m!}$$

The partition 2, 2, 1, 1, 1, 1, 1 leads to $\binom{22}{7} \frac{7!}{5!2!2!} = 3581424$

This is exactly the value we find in Mersenne's table.

In a similar way we can calculate the number of ordered selections, that is of arrangements of a certain type of repetitions.

Now the order of the p-selected notes matters, but now the type of repetitions of p has to be interpreted as a permutation with repetitions. Hence the number of unordered selections has still to be multiplied by

$$\frac{p!}{(1!)^{r_1} (2!)^{r_2} \ldots (m!)^{r_m}}$$

We shall see that 30 years later Leibniz solved the same problem without knowing Mersenne's great writings on harmony and using another, more elegant method. Of course, Leibniz deduced the same results, as will be shown.

2.2 Kircher (1650)

Mersenne's combinatorial studies did not lack immediate influence on later writers. I mean the famous Lullist Athanasius Kircher and his "Musurgia universalis", published in 1650. Its unabridged, baroque title begins by promising:

> "Universal musical art, or great art of consonance and dissonance, subdivided into ten books, by which the whole doctrine and philosophy of notes and the science of theoretical and practical music are treated with the greatest versatility. The wonderful forces and effects of the consonance and dissonance in the world and hence in the whole of nature are revealed and proven by an equally new and unknown exhibition of the various examples. This is done with regard to the individual applications as well as in almost every faculty, especially in philology, mathematics, physics, mechanics, medicine, politics, metaphysics, theology."

The copperplate print illustrates Pythagoras, his famous theorem, the forging smithes who allegedly enabled him to find his consonance theory. Music is represented by a female figure with some musical instruments. The title page of the eight book is preceded by three hexameters and pentameters explaining that

> God is an "harmostes", that is somebody who puts together, who composes, that the world is His organ, that there are as many beings as metres.

Kircher cites the mythical author Hermes Trismegistos, who had said in his writing to the physician Asclepius [11, p. 77]:

> "Music is nothing else than to know the order of all things".

The eighth book is entitled "The eighth book of the great art of consonance and dissonance, that is the wonderful musical art, a new, recently invented musical-arithmetical skill, by which everyone is able to acquire in a short time a profound knowledge of composing, even if he knows nothing about music". For Kircher, too, musical composition consisted in a sequence and arrangement of consonances. The book is divided into four parts:

1) the combinatorial musical art,
2) the rhythmic or poetical musical art,
3) the practice of song-building musical numbers (musarithmi melothetici),
4) the mechanical musical art or the various transpositions of musical-arithmetical columns.

The first part explains the fundamental combinatorial operations in order to demonstrate the huge variety of possible arrangements and combinations of notes. The second part applies these rules to rhythms, that is notes and metrical sequences of syllables or metrical feet: a setting to music. The third part explains Kircher's new musical art which consists in an artistic composition of song-ordering columns. How so ever this composition is done, it will result in a new harmony. This part is by far the longest part of the eighth book. The fourth part explains the use of a musical-arithmetical box (arca musurgica).

Kircher explained the fourth part by saying that even somebody who does not know music will be taught to compose arbitrary songs by a new and easy aid that is by means of certain musical-arithmetical columns.

I would like to begin by discussing the first part. Though Kircher mentioned Mersenne several times, he did not say any word that he repeated the explanations of his French predecessor, even using the number examples: Nine notes are selected out of n notes. He left aside, however, the most difficult problem of calculating the combinations and arrangements of certain types of repetitions. Apparently, he did not always understand his source. Let us consider, for example, his table showing the number of permutations with repetitions (1650 II, 7):

Tabula II. Combinatoria ostendens numerum mutationum, rerum, in quibus non praecisa diuersitas, sed quaedam sunt similes.

1	2	3	4	5
Series rerum diversarum	Combinatio rerum omnium diversarum	Combinatio rerum in qua 2 similes	Combinatio rerum in qua 3 similes	Combinatio rerum in qua 4 similes
I	0			
II	2	0		
III	6	3	0	
...				

Mersenne had explained that there is one permutation of say nine equal notes. Kircher understood the notion of "mutatio", permutation, in the strict sense of the word and asserted that there is no permutation at all of n equal elements. He did not notice that this assertion contradicted the division rule taken over from Mersenne. The greater is the historical injustice that later on Kaspar Schott and Andreas Tacquet even spoke of Kircher's rule [6, p. 20]. The "Universal musical art" was Kircher's most important work regarding mathematical combinatorics.

As we mentioned above, part 3 deals with the practice of song-building musical numbers. The miraculous force of the combination of things can be especially found in music. Hence the principles of music are reduced to methodical tables which Kircher called "musical numbers" (musarithmi) or harmonical numbers.

He conceived three sets of tables in which several song-building tablets contain whatever is important in song-building music or in the art of composition. The first set consists of eleven tablets, the second and third sets consist of six tablets. The eleven tablets of the first set contain the whole poetical or metrical musical art. The six tablets of the second set describe the poetical or metrical musical art by flowery and ornate musical numbers. The six tablets of the third set are accommodated to the poetical as well as to the rhetorical part of musical art and help to express by conveying the charme of music.

What else is necessary? Kircher enumerated three necessary means for his "Melothesia musarithmetica":

1) a "palimpsestum phonotacticum": a writing material which can be used and wiped off. It describes the four voices: soprano, alto, tenor, and bass;
2) a "mensa tonographica": a table where the notes are written down by means of numbers;
3) the knowledge of the values of notes and time measures.

Then Kircher described the different tablets (pinaces) of the three sets (syntagmata).

The first set explains, for example, the simple contrapuntal composition for four voices and a certain poetical rhythm marked by the figures of the thorough-bass:

Gaudia Mundi

Adonia

55655	*66666*	*55545*	*55455*
88878	*88282*	*22322*	*32123*
33423	*44434*	*77867*	*55678*
88451	*44262*	*55125*	*87651*
...			

As an example, Kircher added the composition by Bernardino Roccio who had elaborated his music according to Kircher's rules.

The fourth part explains the construction of the "musical-arithmetical box" (arca musarithmica), that is of the musical-arithmetical columns and how they have to be combined. He used the expression "Abacus musurgicus" or "Abacus melotheticus" which he presumably took from Robert Fludd.

Kircher's text example consisted of five parts:

Cantate Domino	VI	short penultimate
canticum novum	V	short penultimate
laus eius	III	long penultimate
in Ecclesia	V	short penultimate
sanctorum	III	long penultimate

The Roman figures denote the number of syllables of the single parts. Every part is represented by a tablet, the five tablets are united by a certain combination. Several different combinations result in a song (melothesia). The similarity between his tablets and John Neper's small rods let Kircher speak of a "Rabdologia musurgica" ([5, vol. 2, p. 190]; [17, p. 204–212]).

His pupil, Kaspar Schott, explained Kircher's mechanical composition theory in the work "Mathematical organ" (Organum mathematicum) which appeared in 1661 [14].

ICONISMUS. I.
pag. 55.
Operculum Organi
MUSICA sive MELOPOEIA
GNOMONICA sive Horographia scioterica
COMPUTUS ECCLESIASTICUS
FORTIFICATIO
GEOMETRIA
ARITHMETICA
Organum Mathematicum Serenissimo Archiduci CAROLO IOSEPHO humillime D. D. D. P. Athanasius Kircherus è Soc. Jesu Romae 1661

Fig. 2.3.

There we find again the tablets

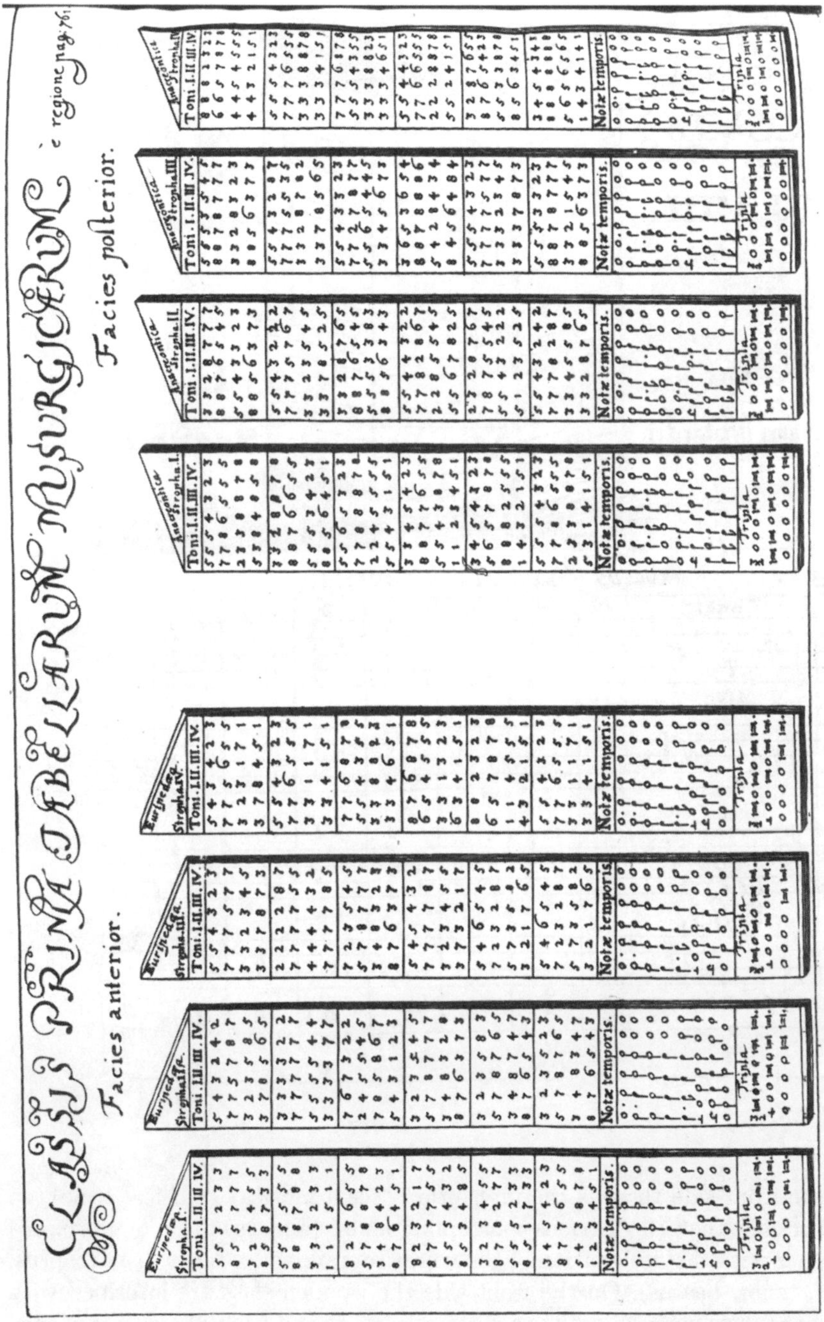

Fig. 2.4.

and Kircher's "Heptaedron Musurgicum":

Fig. 2.5.

On the front side there is the inscription "scala musica" (musical scale), on the backside the title "Abacus contrapunctionis" (abacus of the counterpoint) together with the designation of the voices soprano, alto, tenor, bass (cantus, altus, tenor, bassus). On the front side the wooden slats are inscribed with number tables corresponding to certain keys, on the backside with rhythms. By means of composition tables inscribed on the cover we first can construct the bass according to certain rules and then the pertinent trebles which form triads.

2.3 Leibniz (1666)

When in 1666 Gottfried Wilhelm Leibniz elaborated his "Dissertation on the combinatorial art" (Dissertatio de arte combinatoria), he still adhered to Lullism, wanted to reorganize logic, especially logic of discovery. He pinned his hopes on Kircher, because he knew that Kircher was writing his "Great art of knowledge" (Ars magna sciendi). Later on he frankly admitted his disappointment about it [6, p. 20].

Young Leibniz only referred to Clavius. He was neither acquainted with the extensive Lullistic literature, nor with the combinatorial achievements of Fermat and Pascal. Mathematical progress is not linear. Unlike his Lullistic predecessors, he at least demonstrated some of his solutions of combinatorial problems, though by far not all. He did not reach the mathematical difficulty or generality of Mersenne's problems. His sixth problem dealt with a number of arrangements of a certain type of repetitions, that is with Mersenne's most complicated problem without that he knew Mersenne's "Books of Harmony". Hence he developed his own solution of this problem which differed from that of Mersenne. His presuppositions were simpler than that of Mersenne:

(1) He selected up to six notes out of six notes, while Mersenne selected nine notes out of 22 notes.

(2) He enumerated nine instead of the eleven possible types of repetitions. He adhered to the Italian designations of notes (ut re mi fa sol la) without introducing – like Mersenne – the language of number-theoretical partitions. He erroneously omitted the types of repetitions 6, 51.

Indeed, there are eleven partitions of six, while he only took into account nine.

(3) He multiplied the number of possible combinations or selections of a certain type of repetitions by the number of permutations with repetitions of this type of repetitions, as did Mersenne. Unfortunately, his rule of calculating the number of permutations with repetitions was false. Insofar, he knew less than his predecessors Mersenne and Kircher. His mistake demonstrated that he was not acquainted with their results:

Let there be given the following set with repeated elements:

$$a_1^{r_1}, a_2^{r_2}, \ldots, a_k^{r_k} \qquad r_1 + r_2 + \ldots + r_k = n\,.$$

Allegedly, the number of its permutations is:

$$k(n-1)!$$

Hence Leibniz's numerical results cannot coincide with Mersenne's.

(4) But if we leave aside this factor and if we only consider the number of unordered selections or combinations, we indeed get Mersenne's result in a new way.

Let us consider his example:

ut ut re re mi fa

If we write it as a partition of six, we get

2 2 1 1

Leibniz argued in the following way:

There are $\binom{6}{2} = 15$ possibilities to select two elements. Each of them is repeated twice.

There are $\binom{4}{2} = 6$ possibilities to select two elements out of the remaining four.

Each of them occurs exactly once. Hence there are $\binom{6}{2}\binom{4}{2} = 15 \cdot 6 = 90$ combinations of the type of repetitions $1122 = 1r_1 + 2r_2 = 1.\underline{2} + 2.\underline{2}$.

Leibniz's method of arguing can easily be generalized.

Suppose $p = 1r_1 + 2r_2 + \ldots + pr_p$ is the partition of p, the whole number of selected elements. The number of combinations of this type of repetitions is

$$\binom{n}{r_1}\binom{n-r_1}{r_2}\cdots\binom{n-r_1-r_2-\ldots-r_{p-1}}{r_p}$$

$$= \frac{n(n-1)\ldots(n-r_1+1)}{r1!}\,\frac{(n-r_1)\ldots(n-r_1-r_2+1)}{r2!}\cdots$$

$$\frac{(n-r_1-r_2-\ldots-r_{p-1})\ldots(n-r_1-r_2-\ldots-r_p+1)}{rp!}.$$

(5) This product is identical with Mersenne's expression because $r_1+r_2+\ldots,+r_p = r$ or (or the expression can be reduced by $r!$)

$$M = \frac{n(n-1)\ldots(n-r+1)}{r_1!\ldots r_p!}$$

2.4 The Later Developments in the 18th Century (Euler, Mozart)

Still in the 18th century, authors took an interest in the relationship between combinatorics and musical composition, among others Euler and Mozart. Leonhard Euler's mathematical notebooks ranging from 1725 up to the end of his life contain many interesting considerations dealing with musical problems. They never appeared in his published works [9]. Hence they are worth considering, at least those which are directly connected with our combinatorial issues and which were written down between 1725 and 1727.

(1) There are 51 sequences of chords for compositions for four voices written by means of numeral figures. One of them is written by means of figures of the thorough-bass [9, p. 67]. Obviously Euler took this notation from Kircher's "Musurgia universalis".

(2) Combinatorial aspects played a role when he constructed a triad over the bass voice and its two inversions (Mathematical notebook 129, f. 53v), for example:

$$\begin{array}{ccc} 1 & 5 & 3 \\ 5 & 3 & 1 \\ 3 & 1 & 5 \\ 1 & 1 & 1 \end{array}$$

(3) Euler permuted 28 sequences of notes according to certain restricting rules, for example a sequence consisting of a crotchet, a quaver, and two pairs of semiquavers. The pairs must not be separated. Hence such a sequence represents the type of repetitions *abcc* and admits twelve permutations.

Fig. 2.6.

Ten years after Euler's death and two years after Mozart's death, that is in 1793, J.J. Hummel published a "Musical game of dice" (Musikalisches Würfelspiel) attributed to Wolfgang Amadeus Mozart (1756–1791). It seems to be possible that Mozart himself elaborated this game. There are $2 \cdot 88 = 176$ numbered waltz bars. The dice-players produce a bipartite waltz. The order of the eight bars of its two parts is determined by throwing two dice and by means of two matrices. Their Roman figures in numbering the eight columns denote the cast. The Arabic figures $2, 3, \ldots, 12$ ascribed to the rows represent the possible outcomes of the casts. If the n-th cast results in the outcome m, the element $a_{m,n}$ of the matrix denotes the number of the next waltz bar, which is to be found among the 176 written down waltz bars.

Zahlentafel

1. Walzerteil

	I	*II*	*III*	*IV*	*V*	*VI*	*VII*	*VIII*
2	96	22	141	41	105	122	11	30
3	32	6	128	63	146	46	134	81
4	69	95	158	13	153	55	110	24
5	40	17	113	85	161	2	159	100
6	148	74	163	45	80	97	36	107
7	104	157	27	167	154	68	118	91
8	152	60	171	53	99	133	21	127
9	119	84	114	50	140	86	169	94
10	98	142	42	156	75	129	62	123
11	3	87	165	61	135	47	147	33
12	54	130	10	103	28	37	106	5

2. Walzerteil

	I	*II*	*III*	*IV*	*V*	*VI*	*VII*	*VIII*
2	70	121	26	9	112	49	109	14
3	117	39	126	56	174	18	116	83
4	66	139	15	132	73	58	145	79
5	90	176	7	34	67	160	52	170
6	25	143	64	125	76	136	1	93
7	138	71	150	29	101	162	23	151
8	16	155	57	175	43	168	89	172
9	120	88	48	166	51	115	72	111
10	65	77	19	82	137	38	149	8
11	102	4	31	164	144	59	173	78
12	35	20	108	92	12	124	44	131

Fig. 2.7.

The dice replace the subjective choice of a certain bar, though the waltz could be composed without dice, too, that is by subjective choices.

Epilogue

Since the middle of the 18th century, Kircher's mechanical compositions were denounced as being "sounding algebra" [4, p. 368], the genius of an artist did not depend on mathematics any longer. Yet, they remind of John Cage who tried to eliminate every subjective influence in composing. In 1960 Hans Otte composed "tropisms". Their 93 single bars can be freely combined with one another: the composition is never finished, there is no definitive product.

The French poet Paul Valéry (1871–1945) put it in the following way: "The secret of choice is no less important than the secret of invention" (Taubert in [15, p. 8]).

References

1. Assayag, Gérard: A matemática, o número e o computador. Colóquio/Ciências, Revista de cultura científica **24**, 25–38 (1999)
2. Berman, Gerald, Fryer, K.D.: *Introduction to Combinatorics.* New York, London: Academic Press 1972
3. Clavius, Christoph: *In sphaeram Ioannis de Sacrobosco commentarius.* Rome. I cite its last edition: Opera mathematica, vol. II, first part. Mainz 1611. (Reprint, together with a preface and name index by E. Knobloch. Hildesheim, Zurich, New York 1999). 1570
4. Kaul, Oskar: Athanasius Kircher als Musikgelehrter. In: *Aus der Vergangenheit der Universität Würzburg, Festschrift zum 350-jährigen Bestehen der Universität Würzburg.* S. 363–370 Berlin 1932
5. Kircher, Athanasius: *Musurgia universalis.* 2 vols. Rome: Franciscus Corbellettus 1650 (Reprint Hildesheim, New York: Olms 1970)
6. Knobloch, Eberhard: *Die mathematischen Studien von G.W. Leibniz zur Kombinatorik.* Wiesbaden: Steiner 1973
7. Knobloch, Eberhard: Marin Mersennes Beiträge zur Kombinatorik. Sudhoffs Archiv **58**, 356–379 (1974)
8. Knobloch, Eberhard: Musurgia universalis: Unknown combinatorial studies in the age of baroque absolutism. History of Science **17**, 258–275 (1979). Italian version: Musurgia universalis, Ignoti studi combinatori nell' epoca dell' Assolutismo barocco. In: *La musica nella Rivoluzione Scientifica del Seicento,* a cura di Paolo Gozza, pp. 11–25. Bologna 1989
9. Knobloch, Eberhard: Musiktheorie in Eulers Notizbüchern. NTM-Schriftenreihe für Geschichte der Naturwissenschaft, Technik, Medizin **24**, 63–76 (1987)
10. Knobloch, Eberhard: Rapports historiques entre musique, mathématique et cosmologie. In: *Quadrivium, Musiques et Sciences,* Colloque conçu par D. Lustgarten, Joubert, Cl.-H., Pahaut, S., Salazar, M., pp. 123–167. Paris: édition ipmc 1992
11. Knobloch, Eberhard: Harmony and cosmos: mathematics serving a teleological understanding of the world. Physis **32**, Nuova Serie, 55–89 (1995)
12. Mersenne, Marin: *Harmonicorum libri.* Paris: Guillaume Baudry 1635. (The same edition appeared also in 1636)
13. Mersenne, Marin: *Harmonie universelle.* Paris: Sébastien Cramoisy 1636

14. Miniati, Mara: *Les Cistae mathematicae et l'organisation des connaissances du XVIIe siècle.* Studies in the History of Scientific Instruments, Papers presented at the 7th Symposium of the Scientific Instruments Commission of the Union Internationale d'Histoire et de Philosophie des Sciences, Paris 15–19 September 1987, pp. 43–51. Paris: Rogers Turner Books 1989
15. Mozart, Wolfgang Amadeus: *Musikalisches Würfelspiel, Eine Anleitung "Walzer oder Schleifer mit zwei Würfeln zu componieren ohne Musikalisch zu seyn, noch von der Composition etwas zu verstehen"*. Hrsg. von Karl Heinz Taubert. Mainz etc.: Schott 1956
16. Rodrigues, José Francisco: A matemática e a música. Colóquio/Ciências, Revista de cultura científica **23** (1998)
17. Scharlau, Ulf: *Athanasius Kircher (1601–1680) als Musikschriftsteller. Ein Beitrag zur Musikanschauung des Barocks.* Marburg: Görich & Weiershäuser 1969

3 The Use of Mechanical Devices and Numerical Algorithms in the 18th Century for the Equal Temperament of the Musical Scale

Benedetto Scimemi

Difficile est, nisi docto homini, tot tendere chordas.
Alciat. Embl. 2. lib. 1 ([11, p. 141])

An important subject in which music must reckon with mathematics is *temperament of the musical scale.* Let's briefly run through the terms of the problem so that any reader can grasp it. Every musical *note* has its own precise frequency; a musical instrument (especially one with a keyboard) can produce a finite set of discrete notes: what frequencies must an instrument maker choose so that the instrument can be both used and enjoyed by its player? The maker of the instrument runs into two incompatible facts.

On one hand, when several sounds are produced simultaneously and are superimposed on one another, a good musical ear likes these sounds to be *consonant.* For this, the ratios of their frequencies must be simple fractions, i.e. ratios of small integers. For example, if four sounds with the frequencies

$$3f \qquad 4f \qquad 5f \qquad 6f$$

are played together, they produce a *chord* (*accordo* also means *agreement* in Italian), and it happens that this chord is universally pleasing to the ear. Thus, an instrument that produces, say, the sound $4f$ should also be able to produce the other three. The discovery of this relationship between consonance and *simple rational numbers* is attributed to Pythagoras, and it constitutes the very basis of the close relationship between music and mathematics in classical culture.

On the other hand, in the actual practice of music, other requirements exist which have to do with *transposition* and *modulation.* It is a fact (this too is a universal experience) that a musical message remains substantially unchanged if all of its sound frequencies are multiplied by a given factor. This suggests that – in terms of frequency ratios – there should not be any privileged notes or, in other words, every sound should be able to fill the role of any other sound. Such a requirement becomes apparent, for example, when a singer who wants to use his or her voice to the very best advantage asks for the music to be *one tone higher.* But the situation mainly arises in more evolved music from the composer's need to be able to *modulate* or *transpose* within the same musical piece, an entire phrase from one *tonality* to another for the sake of expressiveness.

To give a concrete example, it should be possible to produce the *same* chord described above (i.e., the frequency *ratios* 3:4:5:6) by putting the second sound in place of the first, for which one would need the frequencies $(16/3)f$, $(20/3)f$, $8f$. Similarly, replacement of the third note by the first would require the presence of $(9/5)f$, $(12/5)f$, $(18/5)f$, etc. Carried further, this process would give rise to the necessary creation of a huge number of different notes, which would obviously make things impossible for the builder of the instrument as well as for the interpreter of the music.

The conflict is not solvable, and this explains how compromises have come about both in theory and in the construction of instruments: *equal temperament of the scale* is the best compromise that our civilization has been able to come up with to remedy the damage the incompatibility leads to. Nevertheless, describing temperament theoretically and achieving it in practical terms have, since ancient times, given rise to not a few difficulties. In modern language, the *equal* solution consists in this: the only rational numbers which are adopted as *intervals* (i.e. frequency ratios) are the *powers of 2* ($\dots 1/4, 1/2, 2, 4, \dots$; no one will refute that these consonances, called *octaves* or *diapason* in Greek, remain the leading requirement); every octave is then divided into twelve *equal* intervals using as ratios the irrational numbers

$$2^{1/12} \quad 2^{2/12} \quad 2^{3/12} \quad 2^{4/12} \quad \dots \quad \dots \quad 2^{11/12}$$

Doing this, a few simple ratios get an excellent approximation (e.g., $2^{7/12}$ is fortunately very close to $3/2$), while the result is hardly acceptable for others (e.g., $5/4 = 1.25 < 1.2599\dots = 2^{4/12}$, and a 1% error definitely rings off-key for a good musician). The good news, however, is that, in terms of transposing, the system fully satisfies the requirement: the frequency ratio of any two consecutive notes is rigorously the same – $2^{1/12}$ – and therefore any note (from the interval point of view) can be substituted by any other. Therefore, as an example, on a modern keyboard any two keys that are seven keys apart produce an almost perfect *just fifth* ($= 3/2$ interval).

In the past, though, this compromise solution, which performers could tolerate even if they didn't really like it, was downright abhorrent to theoreticians. First of all it meant forsaking the Pythagorean discovery, and all that this implied philosophically; but second, it meant that one had to be comfortable with *irrational numbers*, and these weren't yet welcome in numerical mathematics. (The notion of *incommensurable lengths* had been introduced in geometry, yes, but there it was a matter of *proportions*, and it did not lead to arithmetical operations like adding, multiplying, etc.)

But even from a practical point of view, there was no lack of obstacles. In order to produce equal intervals, makers of musical instruments could not just trust their ears, as we have seen, and rely on what sounded consonant. If they could have, they would certainly have obtained, with good precision, *pure* intervals like $5/4$ or $3/2$. They therefore had to devise empirical systems, which were not based on the ear, in order to get as close as possible to an equal scale. (Still today a piano tuner – the Italian *accordatore* could be translated *fashioner of agreements* – has to have recourse to clever stratagems to trick his instinct, and one of these is the *beat phenomenon.*) For this reason, both theoretical treatises and manuals for instrument making include descriptions of *gadgets* or graphical and mechanical devices to solve the problem. They deal in particular with the extraction of square and cubic roots, which are the necessary and sufficient (irrational) operations one needs to arrive, starting with 2, at all $2^{r/12}$-type ratios.

Gioseffo Zarlino

The impossibility of obtaining an equal scale via arithmetic was well known in times past. The theoretical difficulties can clearly be seen, for example, in musical-mathematical treatises of the Renaissance, in which approximate arithmetic alternates with geometrical-mechanical approaches.

The best known of these books is perhaps Gioseffo Zarlino's *Le Istitutioni harmonicae* [11]. It is here that one finds the famous codification of the *seven-note* (or *white-key*) scale, defined by the ratios:

1	*	9/8	*	5/4	4/3	*	3/2	*	5/3	*	15/8	(2)
do		re		mi	fa		sol		la		si	(do)

Zarlino was not out to solve equality problems. But one could wish to make his scale more symmetrical, or more suitable for transposition, and this can be achieved by adding five notes (*black keys*), which still correspond to reasonably simple ratios. For example, one could choose a sequence like this:

1	19/18	9/8	6/5	5/4	4/3	25/18	3/2	8/5	5/3	9/5	15/8	(2)
0	−6.4	3.9	15.6	−13.6	−1.9	−17.4	1.9	13.7	−15.6	17.6	−11.7	(0)

This subdivision of the octave yields fairly similar intervals between two consecutive notes and allows several pure intervals. To enable the reader to judge the degree to which it lends itself to transposing, in the second line we have shown the *deviations* from the equal scale, measured in *cents.* A *cent* is 1200 times the logarithm, in base 2, of the frequency ratio: thus in the *equal* scale the octave, equal to 1200 cents, is subdivided into 12 semitones of 100 cents each. A deviation of five cents, or one-twentieth of a semitone, corresponding to a frequency variation of about 0.3%, is clearly perceptible to a good musician. Only a few of the preceding notes are thus satisfactory, in terms of transposing.

Theoreticians' dissatisfaction have led them throughout the centuries to suggest other diapason subdivisions, and even very complicated ones. In Zarlino's treatise, which bears on Greek musical tradition, a number of intermediate intervals are mentioned which seem to have been *discovered* by different Greek musicians. For example ([11, p. 119]), he lists some sequences of numbers of four digits like

$$\begin{array}{ccccccccccc} \dots & 4608 & 4491 & 4374 & 4104 & 4096 & \dots\ 3992 & 3888 & 3648 & 3456 & \dots \\ & =2^8 3^2 & & =2^1 3^7 & & =2^{12} & & =2^4 3^5 & & =2^7 3^3 & \end{array}$$

And indeed, one has to turn to big numbers if one wants to describe very close ratios by integers. Here the reader can see the predominance of prime factors 2 and 3, which no doubt derive from the need to be able to reproduce the *just fourth* and *fifth* intervals (4/3 and 3/2) in various positions. At any rate, Zarlino himself says that he is skeptical about the significance of such precision when dealing with musical sounds.

In other sections of the book, however, the necessity of being exact does appear. One chapter is called, *"How to divide any given musical interval into two equal parts"* (*"In qual modo si possa divider qual si voglia intervallo musicale in due parti eguali"*), and it contains the description of a mechanical contrivance, the *ortogonio*, which is capable of *obtaining* the geometric mean between two lengths, theoretically quite precisely. This device consists of a frame, or loom, in the shape of a semicircle whose diameter is the hypotenuse of a right triangle inscribed in it. The vertex which is variable on the arc, has an orthogonal projection on the diameter that divides it into two segments whose ratio is r/s; then Euclid's theorem states that the corresponding height h is their geometric mean: $h^2 = rs$, from which $r/h = h/s = (r/s)^{1/2}$.

It follows that if in a lute (or harp or harpsichord) one sets three strings of lengths r, h, s respectively (obviously of the same section and subject to the same tensions), one will obtain equal intervals between two consecutive strings. The mechanical details of the device are not described by Zarlino, but one could imagine the segments being formed by strings attached to sliders with weights on them to keep them taught. Naturally, theoretical exactness cannot prevent gross experimental errors.

Further on, another chapter is entitled, *"Another way of dividing any musical consonance or interval into two or more equal parts"* (*"Un altro modo di divider qual si voglia Consonanza, overo Intervallo musicale, in due, overo in più parti equali"*). Here the author looks in particular at how to divide an octave into three equal parts (i.e., the $2^{1/3}$ interval, or equal *major third*), and to do this he uses another mechanical contraption, the *mesolabio*. This instrument, the invention of which is attributed to Eratosthenes, serves to obtain a double geometric mean: starting with two lengths r, s, it allows one to obtain two new lengths h, k such that

$$h^2 = kr\,;\ k^2 = hs \qquad \text{from which} \qquad r/h = h/k = k/s = (r/s)^{1/3}\,.$$

This time the device is made of three equal rectangular frames that can be partially superimposed on one another by sliding on a rail at their base (Fig. 3.1). Each frame has a fixed string stretched over the diagonal. If one looks for good displacements of the rectangles, one obtains three homothetic triangles that have the required proportions. The alignment of the homothetic points is ensured by a string that is stretched between mobile pegs.

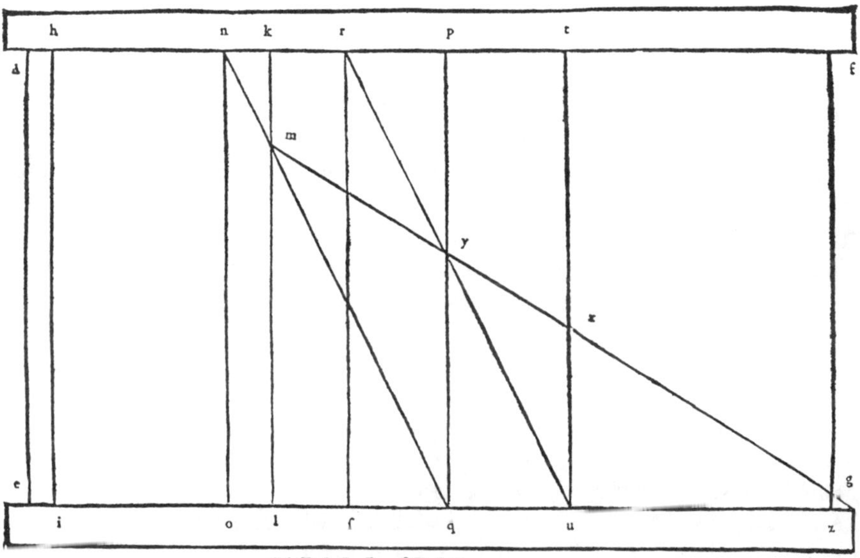

Fig. 3.1. *Mesolabio*, from Zarlino's *Le Institutioni harmonicae*, 1558. In this figure only two mobile frames are shown, yielding simple geometric means (square roots). A third frame must be introduced in order to produce double means (cubic roots)

From documents of the period, it appears that scientists in the 16th century frequently used mechanical contrivances like this, which at times they ordered from the same artisans who built musical instruments. One would even say that this mixture of technology, art and science – so beautifully embodied in the mathematics-music binomial – was one of Renaissance intellectuals' favorite pastimes.

As we said earlier, apart from geometric methods, theoreticians continued here and there to put forth more or less refined numerical proposals for approximating equal temperament, but these were inevitably rather far removed from the simple natural ratios. For example, at about the same time as Zarlino, another great musical theorist, Vincenzo Galilei, Galileo's father, proposed a simple way of dividing the octave "almost" equally, based on the fact that the rational number $18/17$ is an excellent approximation (-1 cent) of the equal $2^{1/12}$ *semitone.*

In the following two centuries, the search for a reasonable compromise continued. On the theoretical level, musical treatises continued to set forth more or less complicated proposals regarding approximation to the numbers

$2^{r/12}$. Then other, clever non-equal temperament systems were devised (called *semitonal, Werckmeister, Vallotti*, etc.): in substance, this involves carefully choosing relatively simple rational numbers that produce many combinations of pure intervals and that are thus suitable for most musical execution. J.S. Bach composed the *Well-tempered Clavier* with a view to testing some of these proposals systematically.

Giuseppe Tartini

In the 17th and 18th centuries, progress in scientific knowledge – in both mathematics and musical acoustics – caused the two disciplines to separate so that mathematical-musical treatises were no longer published in Renaissance tradition. Still, a number of mathematicians of great renown (from Descartes to Euler), even if they did not play music, delved deeply into problems of math and physics suggested by musical phenomena.

Inversely, it was rare for a musician, whether performer or composer, to deal with strictly mathematical problems without having a scientific basis as well. And yet, such is the case of Giuseppe Tartini, who lived in the second half of the 18th century, mostly in Padua. His fame as a violin virtuoso had earned him the title of *Master of Nations* in Europe. Before becoming famous, Tartini had shown himself to have unusual talent for experimentation. He was the first to observe a curious acoustic phenomenon that is produced by two simultaneous sounds; it was called the *third sound* and it is explained today as an effect of non-linearity. Tartini became very enthusiastic about this discovery, and a few years later he conceived an entire musical and philosophical theory based on the phenomenon. His *Trattato di Musica* ([8]) was at first rather flatteringly received in European scientific circles. This good fortune did not last, however, as to a reader with a scientific background, many of his arguments appeared ingenuous if not obscure[1]. At any rate, Tartini did not then apply his mathematical procedures in any significant way for music. The first chapter of the *Trattato* deals entirely with a purely mathematical subject, the approximate calculation of geometric means of integers, starting with $\sqrt{2}$. The author begins by examining the triad $(5, 7, 10)$. He observes that it gives $5^*10 = 50$; $7^*7 = 49$, and he affirms that

> *"7 cannot be a geometric mean, either by definition or by common sense ... the respective geometric mean is an irrational quantity;*

[1] The opportunity of reading Tartini was offered to me in 1977 when the *Accademia Tartiniana di Padova* printed a voluminous, as yet unpublished, manuscript [10], which contained other complicated calculations that were not all adequately explained. To my knowledge, the only specific study of Tartini's mathematics is [9]. I have recently learned from Prof. Palisca of Yale (whom I want to thank for reading this paper) that prompt confutations of Tartini's *Trattato* were published by two contemporary mathematicians: Eximeno of Spain and Stillingfleet of England.

> *and yet ... purely attributable, and demonstrable by line. The above conclusions are in harmony with the precepts of Geometric Science; it now remains to be seen to what extent they are in agreement with Harmonic Science, to which the subject belongs ...* "

What exactly *harmonic science* means is unclear, but as he uses it most frequently, it should be what we call today *diophantine approximation*, that is, a search for relatively simple rational numbers that come as close as possible to irrational numbers. For example, from the $(5, 7, 10)$ triad, Tartini constructs other triads $(12, 17, 24)$, $(29, 41, 58)$ etc. according to the following scheme:

	5	7		$5^*10 - 7^*7 = 1$	
	7	10			$5 + 7 = 12$
35	49	50	70		$7 + 10 = 17$
	12	17		$12^*24 - 17^*17 = -1$	
	17	24			$12 + 17 = 29$
184	288	289	368		$17 + 24 = 41$ ...

In each new triad the middle term comes close to the geometric mean of the extremes, which maintain a constant ratio of 2. This scheme of calculation very closely resembles the way *continued fractions* work, a subject which had been settled theoretically in the preceding century but was actually a very ancient method of calculation. A few pages further on, however, Tartini describes a second scheme in which, starting with the same $(5, 7, 10)$ triad, he derives the triads $(70, 99, 140)$, $(13860, 19601, 27520)$, etc., like this:

	5	7		$5^*10 - 7^*7 = 1$	$7^*10 = 70$
	7	10			$49 + 50 = 99$
35	49	50	70		
	70	99		$70^*140 - 99^*99 = -1$	$99^*140 = 13860$
	99	140			$9800 + 9801 = 19601$
6930	9800	9801	13860		...

As one can see, the approximation is immediately much better. In the lines thereafter, Tartini doesn't hesitate to carry out calculations with numbers of eight or more decimals (Fig. 3.2)! To be sure, here, as throughout the *Trattato di Musica*, mathematics are dealt with only through examples; there are no algebraic formulas nor even the slightest attempt to *demonstrate* (in mathematical terms) that the algorithms work.

Perchè volendoſi per eſempio aſſegnare le radici della ragione ſubſeſquialtera 2, 3, ridotta la ragione a proporzione geometrica diſcreta in 20, 24, 25, 30, ſarà 24 : $\frac{1}{2}$ il mezzo aritmetico tra i due mezzi 24, 25. Duplicati eſtremi, e mezzo in 40, 49, 60, ſaranno 40, 49; ed egualmente 49, 60, radici della ragione 2, 3, con l'eccesſo della unità nel prodotto di 49. Perchè

$$\begin{array}{r} 40 \\ 40 \\ \hline 1600 \end{array} \qquad \begin{array}{r} 49 \\ 49 \\ \hline 2401 \end{array}$$

Ma ſottratta la unità da 2401, reſta 2400, e 1600, 2400, eguali a 2, 3. Dunque ec. Se ſi vuole minorar la differenza, ſi moltiplichino i tre termini

$$\begin{array}{ccc} 40 & \times & 49 \\ 49 & \times & 60 \\ \hline \end{array}$$

1960, 2400, 2401, 2940. Aſſegnato il mezzo aritmetico 2490 : $\frac{1}{2}$ tra i due 2400, 2401, duplicati eſtremi, e mezzo in 3920, 4801, 5880, ſaranno radici molto più proſſime della ragione 2, 3, così 3920, 4801, come 4801, 5880. Perchè

$$\begin{array}{r} 3920 \\ 3920 \\ \hline 15366400 \end{array} \qquad \begin{array}{r} 4801 \\ 4401 \\ \hline 23049601 \end{array}$$

Ma ſottratta la unità da 23049601, reſta 23049600, e però 15366400, 23049600 ragione eguale a 2, 3. Dunque ec.

Fig. 3.2. Page 2 of Tartini's *Trattato di Musica*, 1754

In modern algebraic terms (not then in use), we can give the following explanation for both algorithms. In the first scheme,

$$\begin{array}{cccc} & x & y & \\ & y & 2x & \\ \hline\hline xy & y^2 & 2x^2 & 2xy \end{array} \qquad \begin{array}{cccc} & x+y & 2x+y & \\ & 2x+y & 2x+2y & \\ \hline\hline \dots & (2x+y)^2 & (2x+y)^2 & \dots \end{array}$$

beginning with an $(x, y, 2x)$ triad, which fulfills $y^2 - 2x^2 = \pm 1$, one obtains a triad $(x+y, 2x+y, 2x+2y)$ that does not satisfy precisely the same condition, but rather its twin with the opposite sign: $(2x+y)^2 - 2(x+y)^2 = 2x^2 - y^2 = -(\pm 1)$. The second scheme works like this:

$$\begin{array}{cccc} & x & y & \\ & y & 2x & \\ \hline\hline xy & y^2 & 2x^2 & 2xy \end{array} \qquad \begin{array}{cccc} & 2xy & y^2+2x^2 & \\ & y^2+2x^2 & 4xy & \\ \hline\hline \dots & (y^2+2x^2)^2 & (8x^2y^2)^2 & \dots \end{array}$$

and gives $(y^2+2x^2)^2 - 2(2xy)^2 = (y^2-2x^2)^2 = 1$. Whoever has studied any theory of numbers will not fail to see that here we are dealing with the famous *Pell–Fermat equations*: $y^2 - 2x^2 = 1$ and $y^2 - 2x^2 = -1$. It is known (see, for example, [2, p. 210]) that all their solutions can be obtained by developing $\sqrt{2}$

in continued fractions, and one sees immediately that this algorithm is equivalent to Tartini's first algorithm. The second algorithm produces a square and therefore only involves the *even* equation $y^2 - 2x^2 = 1$. Both procedures can also be explained as follows: if a indicates an integer, then an element $x+y\sqrt{a}$ of the ring $Z[\sqrt{a}]$ is *a unit* if and only if $y^2 - ax^2 = \pm 1$. It is also known that units form a *cyclic group*; now it is easy to check that, if u is a *generator*, then Tartini's two algorithms come out respectively to the sequences

$$\begin{array}{llllllllll} u^3 & u^4 & u^5 & u^6 & u^7 & u^8 \ldots & & & & \\ u^3 & & & u^6 & & & & u^{12} & & u^{24} \ldots \end{array}$$

and this explains the plus sign in the second algorithm (all exponents are even) as well as the faster convergence towards the geometric mean. In the next paragraphs, Tartini looks at how, with a similar method, to approximate the geometric mean of 2/3, which amounts to dividing the just fifth into equal intervals (an insignificant operation, musically speaking). The resulting diophantine equation is less popular but not less interesting: $2y^2 - 3x^2 = 2$.

It is difficult for the modern-day reader to imagine how and where Tartini may have learned these algorithms, or how he could become so impassioned with such a dreary computational activity. It is worth recalling, however, that Padua at that time knew no lack of illustrious scientists among which shine one of the Bernoulli's and the Riccati family.

Daniel Strähle

For an instrument the strings of which are plucked (lute, guitar, mandolin), temperament of the scale is established once and for all from the outset by the maker of the instrument when a series of little *frets* are inserted under the strings at fixed positions along the fingerboard. When the musician puts a finger on a fret, it is virtually as if the string were shortened, that point becoming one of its ends (the other end is fixed). Since the frequency is inversely proportional to the length of the string, in order to produce the octave, one of the frets (let's say the last one) will have to be situated at the halfway point of the fingerboard. For an equal scale eleven other frets must be placed at intermediate points, so that every time the finger moves from one fret to the next the string shortens by a ratio of $2^{-1/12}$ to produce a semitone.

Now how does one systematically establish the proper positioning of the frets? Here is how a simple lute maker, Daniel Strähle, who in 1743 published his "recipe" in a scientific Swedish journal[2], tells us how to go about it. You

[2] I learned of Strähle's method from I. Steward's book ([6, pp. 246–253]) but the rediscovery of the original article is due to J. Barbour [1]. In [6], besides the formula and the figure I have reproduced here, one finds interesting details regarding the undeserved criticisms addressed to Strähle by his contemporaries and the recent re-evaluation of his work.

build an isosceles triangle (Fig. 3.3) the base AB of which is 12 long and the other two sides AC, BC of which are 24. On AC you choose a point D distant 7 from A. Then you divide the base into 12 equal parts via the points $A = A_0, A_1, A_2, \ldots, A_{11}, A_{12} = B$. Now you place the string such that one end of it is on point B, and its halfway length on point D. (This is always possible if one chooses an opportune unit length, or else moves along the triangle, letting C be the center of a homothety). Strähle's recipe consists in placing the bridges on the points D_i where the string meets the straight line CA_i.

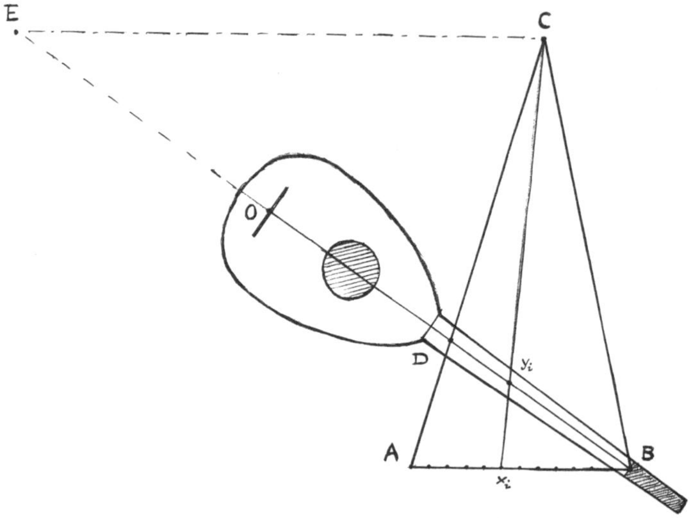

Fig. 3.3. D. Strahle's (1753) method to locate the frets positions on a the fingerboard. It may be noticed that this device has itself somehow the aspect of a musical instrument

That the result is not a precisely equal temperament but only an approximation is obvious to a present-day mathematician by the following reasoning. Introduce coordinates x on line AB and y on line BD such that the values $x = 0$, $x = 12$ correspond respectively to points A, B, and the values $y = 1$, $y = 2$ on points D, B. Since the construction is based solely on *projection* (from C) and *intersection* (with BD) operations, we have here a *projective transformation* and therefore the coordinate y_i of fret D_i is a rational function $y_i = d(x_i)$ of the coordinate x_i of point A_i. We can therefore write $d(x) = (\alpha x + \beta)/(\gamma x + \delta)$ and impose the initial conditions $d(0) = 2$, $d(12) = 1$ on $\alpha, \beta, \gamma, \delta$. Moreover, in a *projectivity* an infinitely distant point on the line AB has a corresponding point on line BD that is the former's intersection E with the line which passes through C and is parallel to AB. One can quickly calculate, by similarity of triangles ABD,

CED, that the value for E is $y = -10/7$. This third condition $d(\infty) = -10/7$ yields

$$d(x) = (10x - 408)/(-7x - 204)$$

Bearing in mind that the frequency is inversely proportional to the length, one can calculate the rational values $2/d(x_i)$ for $i = 1, 2, \ldots, 11$. The following table shows the intervals one obtains and the relative deviations (in cents) from the equal temperament:

1	211/199	109/97	25/21	29/23	239/179	41/29	253/169	65/41	89/53	137/77	281/149	2
0	1.36	1.92	1.84	1.30	0.46	−0.51	−1.46	−2.22	−2.62	−2.50	−1.69	0

As the reader can see, the maximum deviation is less than 3 cents, which is very small and more than acceptable musically speaking.

One can regret that this temperament consists of rather complicated fractions and that it no longer contains any of the pure intervals, apart from octaves. I have therefore tried to modify the numbers in this recipe, leaving the type of construction unchanged, i.e., remaining within projectivities. For example, one can substitute Strähle's suggested condition, $d(\infty) = -10/7$, by a condition where the just fifth is *salvaged*: $2/d(x_7) = 3/2$. The function then becomes $d(x) = (4x - 168)/(-3x - 84)$, and it too is easy to achieve in practice through the use of a different isosceles triangle, where the equal sides are 20 long (rather than 24) and point E at distance 6 (rather than 7). One arrives at the following values:

1	87/82	9/8	31/26	24/19	99/74	17/12	3/2	27/17	37/22	57/32	117/62	2
0	2.46	3.91	4.50	4.44	3.88	3.00	1.95	0.90	0.02	−0.53	−0.59	0

As we wanted, the fractions are simpler, and some natural intervals have reappeared (the 9:8 interval with respect to the first tone). The deviation, however, nearly doubles, up to 4.5 cents, and therefore Strähle's temperament is definitely better.

Another curiosity we can observe has to do with the sixth step, namely the approximation to $\sqrt{2}$. Here too Strähle's 41/29 is better than our 17/12. Indeed, these two fractions are respectively the fourth and third *convergents* in the continued-fraction development of $\sqrt{2}$. The fifth convergent, which is 99/70, will make its appearance in the next paragraph.

Christoph Gottlieb Schröter

In the same part of the 18th century, there lived in Saxony a most knowledgeable author of musical treatises ([4,5]), and no less clever maker of harpsichords, Christoph Gottlieb Schröter. Along with Händel and Telemann, he was a member of the very exclusive *Societät der Musikalischen Wissenschaften*, an academy whose very aim was precisely collaboration in the fields of music and mathematics. This society is better known under the name of the founder, Mizler, whose determination led to the nomination of J.S. Bach as the academy's fourteenth member.

In one of his works[3], Schröter describes a mathematical algorithm which, in a few and simple calculations, arrives, as we shall see, at an incredibly precise equal temperament. Here (apart from some minor, irrelevant changes we have chosen for typographical reasons) is Schröter's recipe. One writes the following twelve numbers in succession: $1, 2, 2, 2, 2, 2, 2, 2, 3, 3, 3, 3$. Having added them up, to 27, one constructs a second line of twelve numbers, carefully placing the figures under the first series; one starts with 27 and continues by adding each figure with the one above it: 27, $28(= 27 + 1)$, $30(= 28 + 2)$, $32(= 30 + 2)$, etc. Having added up the numbers of the second line, which comes to 451, one repeats the procedure with a third line. This is what it looks like:

1	2	2	2	2	2	2	2	3	3	3	3	
27	28	30	32	34	36	38	40	42	45	48	51	(54)
451	478	506	536	568	602	638	676	716	758	803	851	(902)

One can now observe that by *normalizing* the second line (i.e., dividing it by 27), one obtains a subdivision of the octave into twelve intervals, which come pretty close to the equal scale, and where some of Zarlino's ratios reappear:

$$1 \quad 28/27 \quad 10/9 \quad 32/27 \quad 34/27 \quad 4/3 \quad 38/27 \quad 40/27 \quad 14/9 \quad 5/3 \quad 16/9 \quad 17/9 \quad 2$$

By normalizing the third line (dividing the sum of the numbers by 451), one gets an extraordinarily close approximation of the equal division, as can be seen in the following table of deviations in cents:

$$0 \quad 0.6 \quad -0.7 \quad -1.0 \quad -0.6 \quad 0 \quad 0.5 \quad 0.6 \quad 0.1 \quad -1.1 \quad -1.2 \quad -0.7 \quad 0$$

One might think that the excellent results merely depend on a lucky choice of numbers in the first line. But that is not the case. One can experiment with any first line and observe that the algorithm still works, even if its convergence is slower. For example, starting with a *constant* line, here is what happens:

1	1	1	1	1	1	1	1	1	1	1	1	
12	13	14	15	16	17	18	19	20	21	22	23	(24)
210	222	235	249	264	280	297	315	334	354	375	397	(420)
0	−3.8	−5.2	−5.1	−3.8	−1.9	−0.8	1.9	3.3	4.0	3.8	2.5	0

[3] I learned of Schröter and his algorithm from conversations with the musicologist Mark Lindley, who consulted me in the 1980s to get a mathematical explanation of it. I assume that the book he had in his hands was either [4] or [5], but since then I have never been able to locate it in a library.

One sees that the deviations are more than double the preceding ones, but they are still acceptable from a musical point of view. It is interesting to find the just fourth and fifth intervals ($4/3 = 280/210$ and $3/2 = 315/210$), and, for the approximation to $\sqrt{2}$, the fifth convergent $99/70 (= 297/210)$. One could go on to verify that a third application of the procedure leads to numbers of about 5000 with a less than 0.1 cent deviation. For the starting line one can do a bit better than Schröter, but his choice is already very good; for example, among lines in which only the numbers $1, 2, 3$ are used, a computer has found that the one that leads to the closest approximation is

2	2	2	2	2	3	2	3	3	3	3	3	
30	32	34	36	38	40	43	45	48	51	54	57	(60)
508	538	570	604	640	678	718	761	806	854	905	959	(1016)
0	−0.6	−0.6	−0.3	−0.1	−0.2	−1.0	−0.3	−0.8	−0.7	−0.2	0.0	0

Using linear-algebra language, it isn't difficult at present to convince oneself that the algorithm leads to the desired result. In fact, let's take the lines of the preceding table as (column) vectors $\boldsymbol{u}$. Then the procedure described is like multiplying $\boldsymbol{u}$ (on the left) by the 12×12 matrix :

$$A = \begin{pmatrix} 1 & 1 & 1 & \dots\dots\dots & 1 & 1 \\ 2 & 1 & 1 & \dots\dots\dots & 1 & 1 \\ 2 & 2 & 1 & \dots\dots\dots & 1 & 1 \\ \dots & \dots & \dots & \dots\dots\dots & \dots & \dots \\ 2 & 2 & 2 & \dots\dots\dots & 2 & 1 \end{pmatrix}$$

This real positive matrix allows for a real positive *eigenvalue* which is *dominant*, that is, greater in absolute value than all of the others ($\lambda_0 > |\lambda_i|, i = 1, 2, \dots 12$), and this produces a *real eigenvector*. This can be verified by calculating the *characteristic equation* with standard procedures, and the result is $(1+\lambda)^{12} = 2\lambda^{12}$. The twelve complex (simple) zeros are the numbers $\lambda_i = \left(2^{1/12}\varepsilon^i - 1\right)^{-1}$, where $\varepsilon = \mathrm{e}^{2\pi \mathrm{i}/12}$ is a primitive twelfth root of 1, and the dominant eigenvalue is $\lambda_0 = \left(2^{1/12} - 1\right)^{-1}$. Then, letting $\mu = 2^{1/12}$, it is confirmed that $\boldsymbol{u}_0 = \left(1, \mu, \mu^2, \dots, \mu^{11}\right)$ is an eigenvector of λ_0. Indeed, from

$$\lambda_0 = 1 + \mu + \mu^2 + \dots + \mu^{11} = \left(\mu^{12} - 1\right) / (\mu - 1) = 1/(\mu - 1)$$

it follows that

$$\begin{aligned} & 2\left(1 + \mu + \dots + \mu^{i-1}\right) + \left(\mu^i + \mu^{i+1} \dots + \mu^{11}\right) \\ & = \left(1 + \mu + \dots + \mu^{i-1}\right) + 1/(\mu - 1) = \mu^i/(\mu - 1) = \lambda_0 \mu^i \end{aligned}$$

for every index i, and that reads, precisely, $A\boldsymbol{u}_0 = \lambda_0 \boldsymbol{u}_0$.

For the approximate calculation of $\boldsymbol{u}_0$, since there is a dominant eigenvalue, one can apply what has now become the classical *power* method (e.g., cf. [3, p. 84]) which is as follows. Let us write the generic vector $\boldsymbol{v}$ as a linear combination of the eigenvectors $\boldsymbol{v} = c_0\boldsymbol{u}_0 + c_1\boldsymbol{u}_1 + \ldots + c_{11}\boldsymbol{u}_{11}$, and let us calculate $A^k\boldsymbol{v} = \lambda_0^k[c_0\boldsymbol{u}_0 + \sum_{i>0} c_i(\lambda_i/\lambda_0)^k\boldsymbol{u}_i]$. Since it is $|\lambda_i|/\lambda_0 < 1$ for every $i > 0$, it is clear that, as k increases, the direction of the vector $A^k\boldsymbol{v}$ tends towards that of $\boldsymbol{u}_0$, so that in the end $A^k\boldsymbol{v}$ tends to an eigenvector of λ_0. Thus we explain the fact that the choice of the initial line (= vector) is not critical, and the considerable speed of convergence is guaranteed by the fact that in our example $|\lambda_i|/\lambda_0 < 0.12$.

We cannot explain how Schröter ever conceived of such a fast and precise algorithm. Again, we cannot exclude that a professional mathematician might have suggested the procedure to him, but we are more inclined to think that this is an empirical recipe that Schröter fashioned with patience in the manner of a craftsman.

Conclusion

The few examples we have given are sufficient demonstration, we believe, that temperament of the scale has provided great opportunities for collaboration between mathematics and music in the course of history. The impossibility of precisely solving the problem arithmetically led, as we have seen, to the idea that approximate solutions be sought, through either interpolation (projectivity) or recurring algorithms (continued fractions, matrix powers). It is almost certain that, initially at least, all these methods were fine-tuned by trial and error, without their authors' realizing the far-reaching consequences they would have, but not without a certain abundance, either, of ingeniousness – perhaps the same ingeniousness which inspired the fashioning in Italy in the 18th century of those incomparable violins. It is difficult to decide whether, in the course of history, music gained more from mathematics or the contrary. What we certainly must not do, at any rate, is to place the two disciplines in two separate and incommunicable worlds. History teaches us that the boundaries of art, craftsmanship and science are often shaded and movable.

References

1. Barbour, J.M.: A geometrical approximation to the roots of numbers. Am. Math. Monthly **64**, 1–9 (1957)
2. Hardy, G.H., Wright, E.M.: *An introduction to the theory of numbers.* Clarendon Press 1970
3. Johnson, L.W., Riess, R.D.: *Numerical Analysis.* Addison Wesley 1982
4. Schröter, C.G.: *Der musikalischen Intervallen Anzahl und Sitz.* 1752
5. Schröter, C.G.: *Letzte Beschäftigung mit musikalischen Dingen, nebst 6 Temperaturplänen und einer Notentafel.* Nordhausen 1782
6. Steward, I.: *Another fine math you've got me into ... .* New York: W.H. Freeman and Co. 1992

7. Strähle, D.: *Proceedings of the Swedish Academy*, 1743
8. Tartini, G.: *Trattato di musica secondo la vera scienza dell'armonia.* Padova 1754. Padova: Cedam 1973 (Reprint)
9. Tartini, G.: *Traktat* ... . German translation of [8], with comments by Alfred Rubeli. *Orpheus Schriftenreihe zu Grundfragen der Musik*, Band 6, Düsseldorf 1966
10. Tartini, G.: *Scienza platonica fondata nel cerchio.* Padova: Cedam 1977
11. Zarlino, G.: *Le Istitutioni harmonicae.* Venezia 1558

4 Lagrange, "Working Mathematician" on Music Considered as a Source for Science

Jean Dhombres

Permanent secretary to the First class of the French National Institute, which was then a revolutionary replacement for the Academy of Science, Jean-Baptiste Delambre has left an interesting and rather personal portrait of Joseph-Louis Lagrange (1736–1813). In the tradition of academic life, the purpose of such an account written in 1813, was less to explain the scientific achievements of a man, than to portray what a great scientist should be[1]. Delambre then goes as far as telling us a socially bad story about Lagrange. And it concerns Lagrange's taste for music, or better said his misuse of music.

Delambre quotes an opinion told to him in confidence by Lagrange himself: "I like music, claims Lagrange, because it leaves me alone; I am listening to the first three bars; but I already hear nothing of the fourth. I am then left to my own reflections; nothing interrupts my thinking. And that is the way I solved more than one difficult problem"[2]. There is no way of rhetorically escaping from what Lagrange claims: music avoids him the burden of any conversation with others. And in many respects, Lagrange justifies what has often been said about mathematicians: they prefer their own music to any other. Contrary to philosophers, who so often crowded salons of the Parisian life in the 18th century, and this happened to be more or less the same in Lisbon or Vienna, all cities which are linked with the present video-

[1] Academic life has its own social life, with differences between actual behaviour of academicians and what their behaviours should be. It was through "history", or anecdotes told by official *éloges*, etc., that an Academy had the possibility of telling what a scientist should be. We cannot really consider such portraits as myths: it has to be recalled that the portrait was officially read in front of the peers at the Academy, and they had a direct knowledge of the habits of the dead scientist who was celebrated. The interest of the *éloges* was the way "true" facts were selected and organised: such a function to compose an éloges was reserved to the permanent secretary only, but he had the right to ask around him for help. See J. Dhombres, *Le portrait du bon savant* in J.-N. Bouilly, *René Descartes, trait historique*, Palomar Athenaeum, Bari, 1996, pp. 115–134.

[2] "Je l'aime parce qu'elle m'isole; j'en écoute les trois premières mesures, à la quatrième je ne distingue plus rien, je me livre à mes réflexions, rien ne m'interrompt, et c'est ainsi que j'ai résolu plus d'un problème", Notice sur la vie et les ouvrages de M. le Comte J.-L. Lagrange, *Hist. et Mém. Institut National des Sciences et des Arts*, Paris, 1813, reproduit dans les *Œuvres de Lagrange*, publiées par les soins de J.-A. Serret (éd.), t. 1, Paris, Gauthier-Villars, 1867, p. XLVIII. We will refer to this volume by just OL.

Fig. 4.1.

conference, Lagrange had no shame to express what he wished to look as a mathematician.

It is thus quite normal, that a glory of such salons, another mathematician, Jean Le Rond d' Alembert, complained about Lagrange to Voltaire: "C'est un homme peu amusant, mais un très-grand Géomètre" (a rather boring person but what a great Geometer!). Is there not a wish, among so many mathematicians doing music, or listening to music, to avoid being described as boring characters[3]?

[3] Historians of science have sometimes described some moments of thought against mathematics, and therefore against mathematicians. Very few have tried to analyse a sort of social reaction against mathematicians, as boring persons, more or less like theologians, who were no longer authorised to speak in any salons of the 18th century. In his biography of Buffon, J. Roger tries to philosophically explain why Buffon, who had begun to work in mathematics by translating from the English version Newton's *Method of Fluxions*, gave up all sorts of mathematical thinking. He claims that Buffon rejected the unrealistic propension of mathematics to create abstract entities having no physical or natural existence (J. Roger, *Buffon*, transl. into English, Cornell Univ., 1994). There is a different explanation, and Buffon made a fool of himself around 1747, proving his lack of mathematical awareness, by dogmatically attacking a proposed change in Newton's attraction law (J. Dhombres, The mathematics implied in the laws of nature and realism,

However, in his twenties, that is around 1760, Lagrange very carefully studied texts on music. He wrote a sort of history about the role mathematics had played in music, and even spent some time explaining the theory of wind instruments. I am wrong in saying that he explained such a theory: in fact, he confronted the mathematical theory he had of wind vibrations with musicians' practice, and what could be described as the theories provided by musicologists of the day. Isn't strange this work on music for a man who escapes listening to music by immediately thinking to his own mathematical activities?

I would like to explain such a paradoxical attitude, and to begin by the description of an ideological factor directly related to what was the position of a scientist. For the last decades of the 18th century, and particularly in France, the mood was that a scientist should entirely be devoted to his subject. A scientist was prescribed as a sort of a priest, completely occupied with his vocation, which could be summarised by one commandment: to do science. The requirement for such a behaviour can easily be read when one browses over discourses which were pronounced by the three successive permanent secretaries of the Academy of Sciences in Paris. At the beginning of the 18th century, science was then presented as a pleasant hobby and geniuses like Leibniz and Newton seemed able to do it so easily. "What has been obtained through work only is not equal to what Nature freely provides" even wrote Fontenelle, while explaining the career of Guillaume de I'Hôpital, the man who introduced Leibniz infinitesimal calculus to a general audience[4]. But with the Enlightenment, things changed. The last secretary, the philosopher, mathematician and future politician Condorcet, could even reproach a scientist to loose his talent by having a social life.

Recall that an academician of science had no teaching duties, but received money from the State, on a regular basis. The word to denote the highest class of academicians, "pensionnaire" precisely refers to the salary or "pension ". The duty of an Academician became science as a professional duty, whereas the regulations written down in 1699 insisted more on curiosity, even if it was an organised curiosity. By explaining how little he was interested in music, Lagrange thus exhibited his disdain of social life, in order to better show how committed was his activity as a mathematician.

It also meant that he thus associated music with leisure, and perhaps with aristocratic life. The French Revolution accentuated this representation, or what may be described as a bourgeois process in the Academy, at least a deep sense of why the meaning of work may improve the representation of

or the role of functions around 1759, to appear in Proceedings of the Arcidosso conference).

4 "Ce que l'on n'obtient que par le travail n'égale point les faveurs gratuites de la nature". In Fontenelle, Éloge de M. de I'Hôpital, *Hist. Acad. Sc.*, 17.

science among the general public[5]: scientists of a high level became sort of civil servants, paid by the State to promote new ideas, new tools. Therefore scientists had to justify their position as "professionals" or as "experts". They were experts in innovation.

If they really were to become the priests of the new Society, as Saint Simon and some others wished them to be in the early decades of the nineteenth century, then they also had vows to make: you shall adore one God only, Science. They had to prove their devotion, not really by extraordinary results, and priests are not obliged to do miracles, but by a proof of their activity at work, what used to be called being zealous. Science was no longer a natural intellectual activity. For mathematicians, a social answer was found for this new requirement: teaching activities, either in the active form of teaching classes to bring new knowledge, or in the task of writing text-books to change older habits. The kind of "distinction", which Pierre Bourdieu with good reasons, tries to uncover in every intellectual activity, took the form of "professionalism" around 1800. Lagrange's confession reveals how the representation of science evolved in the habitus mentis of the time, and the futility of music helped to exemplify the seriousness of a scientist's work.

This idea is certainly not against an idea common among musicians, or more generally among artists. Because for those who practice arts, there is always associated with it the need of a regular, somewhat painful, but necessary work. However, the purpose of an artist is generally to hide this tedious work while performing his art. On the contrary, in the early nineteenth century, the practice of a science like mathematics required to exhibit a regular and painful work. This became part of the new distinction between art and mathematics. It is noticeable how Stendhal, in his autobiography,

[5] It is quite certain that the values generally attributed to work also changed during the century, and thus scientist adapted to the new kind of ideas. We may see this around the question sometimes raised about the possibility for a man of aristocratic ascent to be a professional scientist. Such a question is raised in a very witty way by Fontenelle, but the idea of work is not really used. Fontenelle is bitterly complaining the lack of thinking of high aristocrats belonging to the army: "Car il faut avouer que la nation française, aussi polie qu'aucune nation, est encore dans cette espèce de barbarie, qu'elle doute que les sciences poussées à une certaine perfection ne dérogent point, et s'il n'est point plus noble de ne rien savoir" *(op. cit.)*. For mathematicians, as well as for musicians, to practice an instrument or their voice, there has always been a tension between the requirement on regular and tedious work and the other requirement of being natural and fresh with no apparent effort. Descartes, for example, clearly wishes to show that his mathematics, for example his solution of Pappus' problem, is almost obvious, and just required the task of thinking in the right way. Newton is not the opposite, and in his published works, for instance in his *Principia*, hides the rather technical and sometimes particularly arguable parts, like those on the theory of tides. Such examples, and many others, are useful to understand that the way a mathematician represents his work also depends on a historical circumstances of his own society.

tends to show how natural came his taste for art, and in a solitary way, and how be had to work in mathematics around 1797, with the imposition of a collective work in a school[6] (with exams, where he first failed, and had to take special courses with a small number of pupils, etc.).

Such ideas about the work of a scientist, or perhaps should we say the ethos which became common among scientists, were not prominent during the sixties of the eighteenth century. That is at the time when Lagrange gave his *Recherches sur la nature et la propagation du son* to the first issue[7] of *Miscellanea Taurinensia*, that is to what will become the Proceedings of the Academy of Turin, a city where he was born in January 1736. But at this earlier time, there very well might have been another sort of an opposition to Music.

Music was being presented by some intellectuals as too subtle a domain to be invaded or explained by mathematics. In the second part of the 18th century, mathematics was not generally viewed with due consideration. Lagrange, at the very beginning of his paper, admits that Mathematics has not yet been too far in the explanation of Nature. And, for him, musical instruments are belonging to Nature, that is to civilised Nature.

> In spite of the fact that the Science of Calculus was in recent times brought to its highest degree of perfection, however it does not appear that one has moved a lot in the application of this Science to phenomena of nature[8].

This is a rather rhetorical sentence, and it has been written to better prepare the last sentence of the introduction to Lagrange's paper.

> The convenience between my results and experiments will perhaps be of some help to destroy prejudices among those who seem to despair that Mathematics can even bring true light into Physics[9].

6 Stendhal, *Vie de Henry Brulard*, V. del Litto (ed.), *Œuvres intimes*, t. II, La Pléiade, Gallimard, Paris, 1982. See J. Dhombres, Un seul côté des objets, Stendhal et l'académisme mathématique, *Etudes littéraires*, to appear in 2002.

7 "Recherches sur la nature et la propagation du son", *Miscellanea Taurinensia*, t. 1; Nouvelles Recherches sur la nature et la propagation du son, *Misc. Taur.*, t. 2; addition aux premières recherches sur la nature et la propagation du son, *Misc. Taur.* All those papers are collected in OL, respectively pp. 39–150, pp. 150–318, and pp. 319–334. I have not tried here to go back to manuscripts, or to check reactions among mathematicians about Lagrange's ideas. From Lagrange's papers on music being put into a mathematical form, I am trying to build the image of mathematics European society was having during the second half of the eighteenth century.

8 "Quoique la science du Calcul ait été portée dans ces derniers temps au plus haut degré de perfection, il ne paraît cependant pas qu'on se soit beaucoup avancé dans l'application de cette science aux phénomènes de la Nature", OL.

9 "L'accord de mes résultats avec l'expérience servira peut-être à détruire les préjugés de ceux qui semblent désespérer que les Mathématiques ne puissent jamais porter de vraies lumières dans la physique", OL, p. 46.

Lagrange seems to have to justify mathematics. Was not music a good choice? Only thirteen years earlier, another scientist had also noticed how mathematics was wrongly used in physics. And remarkably used this word of Physics as well, contrary to what should have been the tradition, that is to speak of natural philosophy. And as Lagrange, he sought to modify this situation by proposing a successful mathematical theory for the stability of boats. In addition, Pierre Bouguer, in his *Traité du Navire, de sa construction, et de ses mouvemens*[10], published in 1746, explained the reason of the misuse, or the disappointment about carrying mathematics into Physics or in the management of professional techniques. For him, the main reasons were that physical properties of objects were not sufficiently observed or analysed, so that the mathematics used was too rough and irrelevant. In other words, mathematics was dogmatically applied to Physics, not stemming from Physics. Physics was not correctly done.

> Some people obtained the solution for some questions, and so they have enriched us, in the sense they enlarged the domain of Science; but as those questions are of a very limited nature, and as they cannot leave the realm of purely hypothetical truths, they really have no application to Marine, which cannot be satisfied with hypotheses or so simple suppositions. The same kind of inconvenience is to be found in too many other circumstances; and even it is too common a situation, in as much we have seen the use of a saying, which could be so strange if it was correctly established, that a proposition could theoretically be true and practically be wrong[11].

Member of the French Academy of science, having spent a long time in Peru in order to measure the length of an arc of the meridian, Pierre Bouguer was not considered as a pure mathematician[12]. His work on light, with the foundation of photometry, is certainly his most remarkable achievement; but he naturally thought that there is no possible distance between a mathematical thinking and its use for any explanation in physics. In other words, mathematics could not only give a correct representation for some

[10] P. Bouger, *Traité du Navire, et de ses mouvemens*, Paris, Jombert, 1746.

[11] "Ils sont parvenus à la solution de nouvelles questions qui nous enrichissent, il est vrai, en augmentant le domaine des Sciences: mais comme ces questions sont trop limitées, et qu'elles ne sortent pas des termes des vérités purement hypothétiques, elles n'ont réellement aucune application dans la Marine, qui ne se satisfait pas d'hypothèses ou de simples suppositions. Le même inconvénient n'a que trop lieu dans d'autres cas; ce n'est même que parce qu'il est trop ordinaire, qu'on a vu s'introduire cete distinction qui serait si étrange, si elle était bien fondée, qu'une proposition peut être vraie dan la théorie, et fausse en même temps dans la pratique", *Traité du Navire, op. cit.*, préface, pp. xij–xiij.

[12] A special issue of *Sciences et Techniques en Perspective* (2e série, 3, fasc. 2, 1999) has been devoted to the kind of science Bouguer developed, and contain the proceedings of a symposium *Eprouver la science, le premier XVIIIe* siècle, held in 1997.

phenomena, but there was nothing else to be known in such phenomena than what could be provided by mathematics. This is a strong and new position in terms of philosophy of nature, and it can be understood as the turning point for its transformation into physics. I do not wish here to speak of mathematical physics, and wait for using the terms physico-mathematics, which became to be used in the *Encyclopédie* for example in 1751. Bouguer is very explicit about the fact that his new point of view does not depend on the new Calculus. On the contrary, so he says, in spite of what has been claimed, up to now, Calculus had not been used correctly.

> There is almost no question where physics is involved, which can be rigorously solved, in spite of what is too often said in recent times about modem analysis[13].

We should go further and conceive that by providing mathematical theories to Marine or to Music, Bouguer and Lagrange were trying to check Calculus. Not that they were having doubts about its correctness and its ability in terms of computations; they were challenging the specific analysis provided by Calculus to represent physical phenomena without forgetting some physical properties, without over simplifications. In other words, they were trying to find physical ontological bases for Calculus, and so questioning Calculus as a reality, and not as a model. Lagrange, in his work on sounds and music, gave a definition for the integral as a limit process; Bouguer in his work on stability of ships also had a numerical definition of the integral, without any intervention of differentials.

The purpose of Bouguer, and the one of Lagrange as well for music, was certainly not to propose experimental physics only: for both of them, experiments were mainly to verify the quality of theories developed. That is to verify that the physical analysis of phenomena had correctly been made. There was no question about the quality of a mathematical theory, as it was a logical analysis once the correct physical properties had been provided.

Among those experiments, which Lagrange also sought to explain, there were musical experiments, even quite practical ones. Lagrange had carefully read the explanation given by so well known a musician as Giuseppe Tartini. And the best is to leave to Lagrange the task of expressing how he recovered Tartini' s practice about the third sound, that is the sensation that one has in the presence of two different sounds.

> It takes a particularly fine ear in order to perceive those compounded sounds; thus only some of the most gifted artists had been able to distinguish such sounds. M. Tartini is the first one, as far as I know, who

[13] "A peine est-il en effet une seule question de celles qui sont mêlées de Physique, qui soit résolue en toute rigueur, malgré les fréquentes applications qu'on a faites, comme à l'envi, dans ces derniers temps de l'analyse moderne", *Traité du Navire, op. cit.*, préface, p. xiij.

> undertook the care of examining such sounds, as can be seen in his Treatise on Music printed in Padua the year 1754. The famous author tells us that by bringing together two sounds either from the same instrument, in so far as it is able to maintain a sound, such like a violin, a trumpet, etc., or from two instruments somewhat distant one from another, a third sound is being heard. It is the more audible as one comes closer to the middle point of the given interval[14]

For those who may immediately reproach a triumph or even a domination of mathematics over the arts, and particularly over music, the best is to quote once more Lagrange. In the way, and this time he had to listen to music, he explains that music had the advantage to oblige him to make progress in Mathematics.

> I quite far from believe that my *Dissertation* might contain a complete theory about the nature of sound and its propagation; it is something in itself to have contributed to the advancement of the physico-mathematical Sciences by being able through analytic computation to prove many truths up to now unexplained in Nature[15].

A quotation from Henri Bouasse writing in 1926, may confirm that Lagrange really wished by studying music to do mathematics, and to enrich mathematics by trying to understand the causes of phenomena, and even of recent phenomena discovered by musicians

> Lagrange's analysis is remarkable. It is not useful to add that Lagrange explained so many ideas that it takes a great deal of patience in order to extract from what he says, propositions which in themselves have an absolute evidence. But, mathematicians seem to have as their main objective to be unreadable[16].

[14] Il faut une extrême finesse d'oreille pour percevoir ces sons composés; aussi n'y-a-t-il que quelques-uns des plus habiles artistes qui les aient reconnus. M. Tartini est le premier, que je sache qui se soit attaché à les examiner avec soin, comme on peut le voir dans son *Traité de Musique* imprimé à Padoue l'année 1754. Ce célèbre Auteur nous apprend qu'en tirant d'un même instrument capable de tenue, comme les violons, les trompettes, etc., deux sons à la fois, ou bien en les tirant de deux instruments éloignes l'un de l'autre de quelques pas, on en entend un troisième, qui est d'autant plus sensible qu'on se rapproche plus du point milieu de l'intervalle donné. (OL, p. 142–143)

[15] "Je suis bien éloigné de croire que [ma *Dissertation*] contienne une théorie complète sur la nature et la propagation du son; mais ce sera du moins avoir contribué à l'avancement des Sciences physico-mathématiques, que d'avoir démontré par le calcul plusieurs vérités qui avaient jusqu'ici paru inexplicables dans la nature", OL, p. 143.

[16] H. Bouasse, *Acoustique. Cordes et Membranes*, Paris, 1926; nouveau tirage, A. Blanchard. Paris, 1987, p. 242. ("Cette analyse de Lagrange est très remarquable.

Such a conclusion, here given by a professor of physics of the early XXth century, has been heard very often, today and as well during the 18th century. In spite of the fact that the mathematics Bouguer discovered, proved technically very useful for naval architecture – up to the point of modifying the forms of vessels, there was a reaction among professionals. They had understood that more mathematics in their art provided a way to have more standards, and even to give to administrators the power to judge about the performances of a ship, just using drawings. They also understood that a state policy was to organise an education of engineers via mathematics, and so to govern the entrance to the body of Marine officers who will be responsible for shipbuilding. In reaction, judgements were pronounced about the fallacies of mathematics. Then came an interesting compromise, mathematics having its own requirements, and uses of mathematics some others. It can be read for example in the advertisement which was written for a new French edition of *Théorie complète de la construction et de la manœuvre des vaisseaux, mise à la portée de ceux qui s'appliquent à la navigation*, in 1776. In fact, Euler had published in 1749 a *Scientia Navalis*, containing in particular the stability theory that Bouguer had explained some years earlier. It was considered to be far too difficult from a mathematical point of view, and it is true that Bouguer had a better explanation, without presupposing some knowledge of Calculus, and providing what was necessary. Let us read what the editor added in 1776.

> ago a book for specialists in geometry; this one is entirely conceived for sailors. Not only he gave away questions, which are depending on too uncertain physical principles, with purely hypothetical coverage, but he too avoided too complicated problems, which cannot be solved but using long computations or a too difficult kind of analysis, or which would be of just pure curiosity. Here, everything is certain, useful and simple[17].

The idea of certainty is essential, and a psysico-mathematical science is a warrant; usefulness and simplicity are other requirements. It is noticeable that the word "evidence" is not used. Professionals as well as scientists knew

---

Inutile d'ajouter que Lagrange l'enveloppe de considérations telles qu'il faut une patience à toute épreuve pour en extraire les propositions précédentes, dont l'évidence est absolue. Mais les mathématiciens semblent avoir pour principal objectif d'être illisibles").

[17] "L'illustre auteur de ces Elémens avait donné il y a longtemps, sur la Science de la Marine, un grand Ouvrage destiné aux Géomètres: celui-ci l'est surtout aux Marins. Il en a écarté, non seulement les questions qui dépendent de principes physiques trop peu certains et purement hypothétiques, mais encore les problèmes trop compliqués qui n'auraient pu entre résolus que par de longs calculs ou par une analyse trop difficile, enfin ce qui ne serait que de pure curiosité. Ici tout est certain, utile et simple." Léonard Euler, *Théorie completète de la construction et de la manœuvre des vaisseaux, mise à la portée de ceux qui s'appliquent à la navigation*, Paris, Claude-Antoine Jombert, 1776, avertissement.

that both mathematics and physics were preparing a new evidence, but it was not already an evidence.

It is also because of the existence of such judgements that a man like Lagrange felt obliged to claim that he was not really listening to music, but in any concert was thinking about his own mathematics. The evidence, of which Henri Bouasse spoke, has not always been so: specifically, what Lagrange tried to do, was to show why it was legitimate to consider that vibrations in the air, making sounds, whether musical or not, were to be treated as just vibrating chords. In other words, Lagrange invented the now common idea that there is a common phenomena in the vibrations of the chords of a violin, or the vibrations of air which make the sound of a violin, or even the same as the vibrations of air in a trumpet. We describe this achievement as a reduction, and thus as a reduction of music to vibration theory. I prefer to use the word "invented" to describe Lagrange's achievement in his *Researches about the nature and propagation of sound.* When we do now use Lagrange's technique, we no longer invent, but use the evidence he has provided.

To invent, for a mathematician, is the same as to prove. It is because he had proven something that Lagrange could claim that he had found the "nature" of sound. And it is because his mathematical results led him to explain the most recent discoveries of musicians, like Tartini and even further, that Lagrange considered that he could explain to others his own music, his own discovery. It will rather quickly become evidence for the layman.

And in Lagrange's case, which is certainly not the case of all mathematicians, in order to prove something he had to find it in himself, outside any music. At a higher level than just a refusal of social relations, Lagrange's refusal to hear music was his way of creation. One may read the way he reduced his program of research – and I insist on this verb, implying a simplification of the physical world. I just translate from the original text which appeared in 1761.

> Given an indeterminate number of elastic particles which are ordered along a straight line, and are kept in equilibrium because of their mutual forces of repulsion, find the moves such particles have to follow once they have been disturbed in any way, without the particles being able to get out of the straight line[18].

For a working mathematician of the 18th century, it was rather easy to perform such a program, which was also a model for sound propagation; but it was not trivial to have reduced the problem of sound to such a program. In fact, it proved enough to study infinitesimal moves of the particle, that

[18] "Étant donné un nombre indéfini de particules élastiques rangées en ligne droite, qui se soutiennent en équilibre en vertu de leurs forces mutuelles de répulsion, déterminer les mouvements que ces particules doivent suivre dans le cas qu'elles aient été, comme que ce soit, dérangées, sans sortir de la même droite" (OL. p. 44).

is to obtain as many differential equations as there are particles. This way of introducing time in physical processes is typical of the 18th century. Once such differential equations were obtained, Lagrange was able to show they were identical to differential equations obtained from vibrating chords. He had in fact another model for such musical chords: he saw such a chord as a chain with suspended, regularly disposed masses. It is important to see that the two models are completely different: in the chord case, masses are not at all on the same straight line.

But it is because the differential equations are the same in the two different models that Lagrange was able to perform one act more. He considered that in the case of chords he could go to infinity, that is he might add more and more masses smaller and smaller. There exists – it is not at all obvious – a limit, and in the mathematical vocabulary it just means that there exists an integral. Lagrange then discovered that this integral could be greatly simplified because of properties of sines and cosines in the way they are multiplied and integrated. Thus is he discovered what we now call the orthogonal relations of sines and cosines. He then got the general solution for the vibrating chord, and as well, the general solution of his problem about sound propagation.

A first consequence is for vibrating chords: whatever might be the form initially provided to the chord, or whatever is the way one acts on the violin, its oscillations are periodic. This solves the case of vibrating chords.

A second consequence is about the speed of sound, which proved independent of the force of the first move; it goes as well to a theory about echoes, and about resonance and natural moves of harmonic chords to the sound of the principal.

Lagrange is willing to show what has been gained in this rather long process, which involved a lot of computation, dozens of pages in Lagrange's original text. Well, for echoes for example, he showed that to understand it as reflection on a surface is correct, but for the quality of the sound, for its harmonic analysis should we better say today, the physical qualities of the surface were unimportant: the surface had just to be understood as an obstacle, which is to say as a boundary condition. Mathematics was there understood in its task of analysing and reducing phenomena.

It is clear that it is not enough just to say that Lagrange had finally achieved a mathematical theory of sound, or that he put music into mathematics, at least in the sense when music is compared of sounds of given frequencies. One has to qualify the kind of mathematics here used. In the classical sense, there was no geometry in the process; and number theory played almost no part. What Lagrange achieved was to provide a representation, pot an object but a canonical and possible computation on an object. The name of this object, which we nowadays would use, is a periodic function, and the representation would be called Fourier series. But this does not do justice to what Lagrange did, to his personal music. As he cannot be claimed to have understood Fourier analysis. Outside the analysis of what

100 RECHERCHES SUR LA NATURE

$$-\frac{3\varpi H dx}{aT} P_3 \sin\frac{3\varpi x}{2a} \sin\frac{3\varpi H t}{2T}$$
$$-\ldots\ldots\ldots\ldots\ldots\ldots$$
$$+\frac{2dx}{a} Q_1 \sin\frac{\varpi x}{2a} \cos\frac{\varpi H t}{2T}$$
$$+\frac{2dx}{a} Q_2 \sin\frac{2\varpi x}{2a} \cos\frac{2\varpi H t}{2T}$$
$$+\frac{2dx}{a} Q_3 \sin\frac{3\varpi x}{2a} \cos\frac{3\varpi H t}{2T}$$
$$+\ldots\ldots\ldots\ldots\ldots\ldots$$

37. Présentement il faut substituer dans ces formules les expressions des quantités $P_1, P_2, P_3, \ldots, Q_1, Q_2, Q_3, \ldots$, d'où, en ordonnant les termes par les quantités connues $Y_1, Y_2, Y_3, \ldots, V_1, V_2, V_3, \ldots$, on trouvera autant de suites infinies, dont chacune sera multipliée par une de ces quantités.

Soit $\frac{X}{a}$ la raison générale des indices des Y et des V au nombre m, X dénotera la partie de l'axe qui leur est correspondante dans le premier état du système; donc, si l'on emploie le signe intégral $\int$ pour exprimer la somme de toutes ces suites, on aura

$$\begin{aligned}y = \frac{2}{a}\int dx Y \Big(&\sin\frac{\varpi X}{2a}\sin\frac{\varpi x}{2a}\cos\frac{\varpi H t}{2T} + \sin\frac{2\varpi X}{2a}\sin\frac{2\varpi x}{2a}\cos\frac{2\varpi H t}{2T}\\ &+ \sin\frac{3\varpi X}{2a}\sin\frac{3\varpi x}{2a}\cos\frac{3\varpi H t}{2T} + \ldots\Big)\\ + \frac{4T}{\varpi H a}\int dx V \Big(&\sin\frac{\varpi X}{2a}\sin\frac{\varpi x}{2a}\sin\frac{\varpi H t}{2T} + \frac{1}{2}\sin\frac{2\varpi X}{2a}\sin\frac{2\varpi x}{2a}\sin\frac{2\varpi H t}{2T}\\ &+ \frac{1}{3}\sin\frac{3\varpi X}{2a}\sin\frac{3\varpi x}{2a}\sin\frac{3\varpi H t}{2T} + \ldots\Big),\end{aligned}$$

et de même pour u

$$\begin{aligned}u = -\frac{\varpi H}{aT}\int dx Y \Big(&\sin\frac{\varpi X}{2a}\sin\frac{\varpi x}{2a}\sin\frac{\varpi H t}{2T} + 2\sin\frac{2\varpi X}{2a}\sin\frac{2\varpi x}{2a}\sin\frac{2\varpi H t}{2T}\\ &+ 3\sin\frac{3\varpi X}{2a}\sin\frac{3\varpi x}{2a}\sin\frac{3\varpi H t}{2T} + \ldots\Big)\\ + \frac{2}{a}\int dx V \Big(&\sin\frac{\varpi X}{2a}\sin\frac{\varpi x}{2a}\cos\frac{\varpi H t}{2T} + \sin\frac{2\varpi X}{2a}\sin\frac{2\varpi x}{2a}\cos\frac{2\varpi H t}{2T}\\ &+ \sin\frac{3\varpi X}{2a}\sin\frac{3\varpi x}{2a}\cos\frac{3\varpi H t}{2T} + \ldots\Big),\end{aligned}$$

Fig. 4.2. A glimpse on a page of Lagrange's paper (1759) on "Recherches sur la nature et la propagation du son" (OL, p. 100)

heat propagation is, which fundamentally is a work in physico-mathematics, Fourier's contribution is to have transformed what was just a tool for Lagrange, integrals in Fourier's coefficients, into the proper modes of a large series of physical phenomena, which are governed by some second order linear differential equations[19]. He has shown that the mathematical explanation for musical sounds was a universal explanation; music, if one may say so, or harmony, was far more general a process than believed, and this could be proved.

Now, I should really enter in my subject, at least if I were to speak to mathematicians only. That is I would have to study carefully the kind of solutions provided by Lagrange, his definition of the integral for example, and how he was new compared to other solutions provided by Euler, or Bernoulli, and how he could not discover Fourier' analysis. That is I will do a professional kind of history of mathematics, which cannot be understood by others than mathematician. Speaking then to musicians, I should enter into the detailed way Lagrange adapted his computations to musical instruments. And for this, it is not enough to show what he did, but I would have as well to explain how Lagrange's theories were modified. Clearly, I would have two more papers to write[20]. But I chose one only, and to show that the link of music and mathematics could also be discussed in terms of the vision of the world a mathematician may have.

And to conclude I prefer to use Lagrange's own conclusion to the first of his paper on the subject of sound, at the end of a discussion about consonant sounds and dissonant ones, which is always the test of a possible imperialism of mathematics towards music[21].

> I believe that, whichever the system of music one can imagine, one shall never elude such difficulties without using common sense and common taste, upon which common habits and prejudices have somewhat more power than is ordinarily thought[22].

And Lagrange to claim that his paper was not the right place to discuss such things. He however was referring to a long paper entitled "Fondamental" in the *Encyclopédie*, or *Dictionaire raisonné des arts et des métiers.* It had been written by d'Alembert. It is really part of my story to add that thirty years later, mathematicians seemed less concerned by music as the paper *Fondamental* was not reproduced in the three volumes devoted to mathematics in the *Encyclopédie méthodique.*

---

[19] Jean Dhombres, Jean-Bernard Robert, *Fourier, créateur de la physique mathématique*, Paris, Belin, 1998.

[20] I thought that at least a glimpse on Lagrange's way could be obtained by just reproducing as illustrations some of his pages. I chose this reproduction form the original paper published in Torino.

[21] Lagrange explicitly criticised Rameau's claim to have based musical harmony in a natural way.

[22] OL, p. 148. The French original can be read as an illustration.

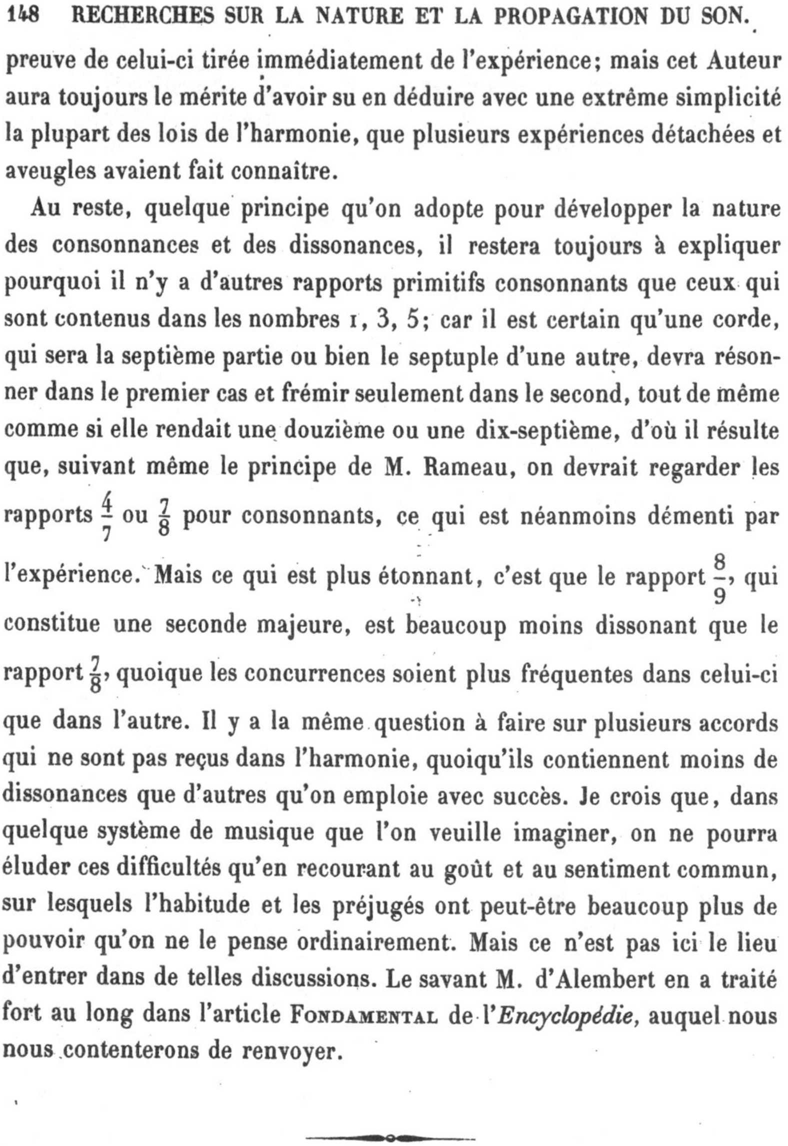

Fig. 4.3. The conclusion of Lagrange's memoire of 1759 in Miscellana Taurinensia (OL, p. 148)

5 Musical Patterns

Wilfrid Hodges and Robin J. Wilson

In this article we look at several musical pieces that illustrate mathematical devices used by composers in their writing. These devices include canon, expansion, retrograde motion and inversion.

Canon

The first of these is *canon*, where the main theme is translated from voice to voice. Many canons are very familiar, such as *Three blind mice*, *Frère Jacques*,

Fig. 5.1. Extract 1 – Sumer is icumen in

and *Row, row, row the boat.* One of the most remarkable canons ever written dates from about 1300 – *Sumer is icumen in.* Here the two lowest parts are in canon, and in addition the four high parts have their own canon above. The result when we put them all together: is shown in Fig 5.1.

In one of the most famous canons, the 16th-century *Tallis's canon*, the tenors sing exactly the same notes as the sopranos, but four beats later. This canon is shown in Fig 5.2.

Let's see what's happening mathematically. It's just the usual mathematical idea of translation, $y = x + c$. This is just one of several mathematical transformations that we can find in music – others include retrograde symmetry, where the music is the same backwards and forwards, and inversion, where we can turn it upside down, or combinations of these.

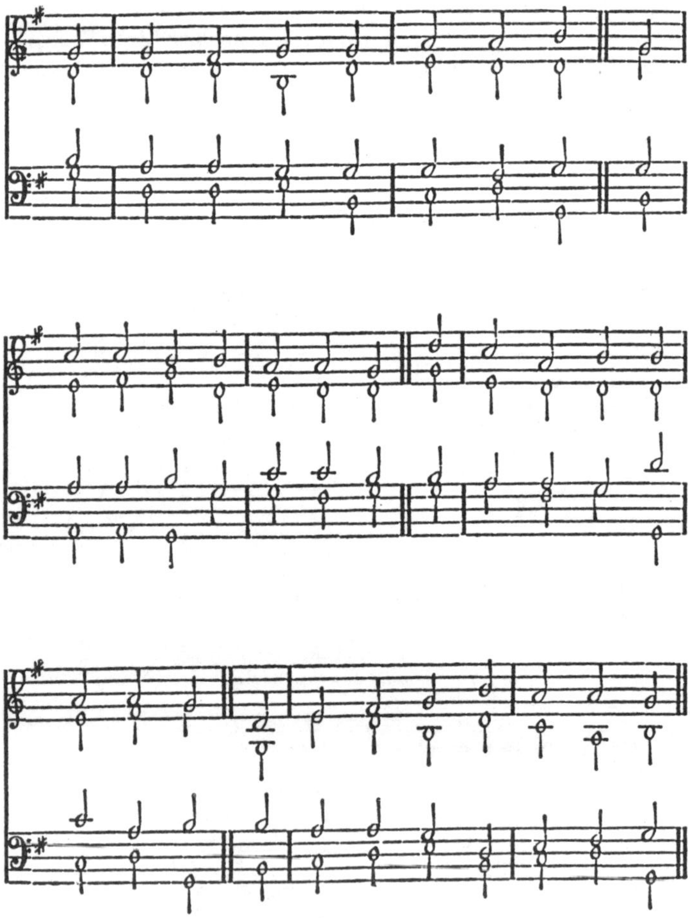

Fig. 5.2. Extract 2 – Tallis's canon

From the earliest times, composers seemed to enjoy setting themselves challenges, seeing whether they could compose acceptable music subject to given mathematical restrictions. There is even an unusual Tudor four-part canon in which three parts start in close canon, while the other part, a *cantus firmus*, sings a four-note phrase over and over again – but first with 8 beats on each note, then 7, then 6, and so on, down to 1 and then $\frac{1}{2}$.

Modified canon

Another challenge is to compose music in which the parts are in canon, but at different speeds – or $y = mx + c$, where $m \neq 1$. A brief but effective example of this occurs at an emotional high point of Brahms's *Requiem*. Here the sopranos sing 'I will see you again', which indeed we do with the tenors at half speed.

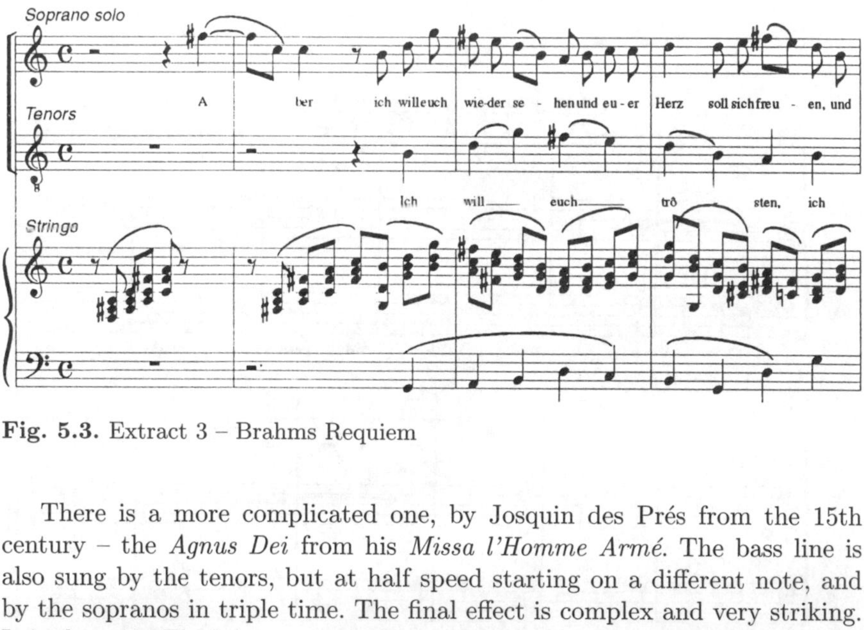

Fig. 5.3. Extract 3 – Brahms Requiem

There is a more complicated one, by Josquin des Prés from the 15th century – the *Agnus Dei* from his *Missa l'Homme Armé*. The bass line is also sung by the tenors, but at half speed starting on a different note, and by the sopranos in triple time. The final effect is complex and very striking. It is shown in Fig 5.4.

An even more complicated example is by the American composer Conlon Nancarrow, who died in 1998. In one of his canons, four voices play the same theme at speeds in the ratios 17:18:19:20, all entering at different times in such a way that most of the piece seems completely chaotic, but at the climax of the work they all come suddenly and miraculously into time with each other.

One of the most beautiful canons is the Chaconne from Henry Purcell's masque *Dioclesian*. Here the two recorder players are in exact canon throughout, the second just two bars after the first, while underlying this is a *ground*

Fig. 5.4. Extract 4 – Josquin Agnus Dei

bass – a six-bar phrase continually repeated throughout the piece. It is shown in Fig 5.5.

Fig. 5.5. Extract 5 – Purcell's Chaconne

Retrograde motion

Another mathematical device used by composers is that of retrograde motion. This yields pieces that can be played forwards or backwards, and are sometimes known as *palindromes*, or *crab canons* – here $y = -x$. Here is a palindrome from J.S. Bach's *Musical Offering* – reading the first line backwards gives the second line, and *vice versa.*

Fig. 5.6. Extract 6 – Bach's Crab canon

Another example is a delightful minuet by Joseph Haydn from his *Piano Sonata 41*, which the composer liked so much that he re-used it in his *47th Symphony*. Here's the beginning of it.

Fig. 5.7. Extract 7 – Haydn Piano Sonata 41

The earliest known piece of this type is by Guillaume de Machaut in the 14th century, who used this device in setting the words '*My end is my beginning*', and a recent example is by the American composer George Crumb, whose madrigal '*Why was I surrounded by mirrors*' is full of patterns that run the same way backwards as forwards. Such reflections were also used by Alban Berg in his opera *Wozzeck* – where large sections are the same backwards as forwards – and there's even a whole opera by Paul Hindemith which one can listen to either way round.

Inversion

The device of turning music upside down – or inversion – has been used effectively by several composers. The Hungarian composer Béla Bartók used it in his *Mikrokosmos*, in a piece called *Subject and reflection.*

Fig. 5.8. Extract 8 – Bartók's Mikrokosmos

Several of J.S. Bach's compositions are full of mathematical symmetries. Here's an ingenious example of inversion: each of the following fugues is the inversion of the other.

Fig. 5.9. Extract 9 – two Bach fugues

An example that combines canon, reflection and inversion is Haydn's *Canon cancrizans*, where each of six sections is a three-part canon on the words, *Thy voice, O Harmony, is divine.* Each part in Section 2 is the same as one of the parts in Section 1 backwards, and the same thing happens also in later sections when the music is turned upside down! He wrote it on receiving an honorary degree from Oxford University in 1791.

Another interesting example of symmetry is Paul Hindemith's piano work *Ludus Tonalis.* Here the whole of the last movement is obtained by taking the first movement and playing it upside down and backwards. Apart from the last chord, rotating the whole movement through 180° does not change it at all.

Fig. 5.10. Extract 10 – Ludus tonalis

The twelve-tone system

Many of these ideas are combined in twelve-tone music. In Schönberg's *Piano Suite* the twelve notes of the octave appear in some order, then transposed up six semitones, then inverted, and then the transposed version inverted, and they can also be played backwards. Figure 5.11 shows the result, with six forms of the tone row – the tones can appear in any octave.

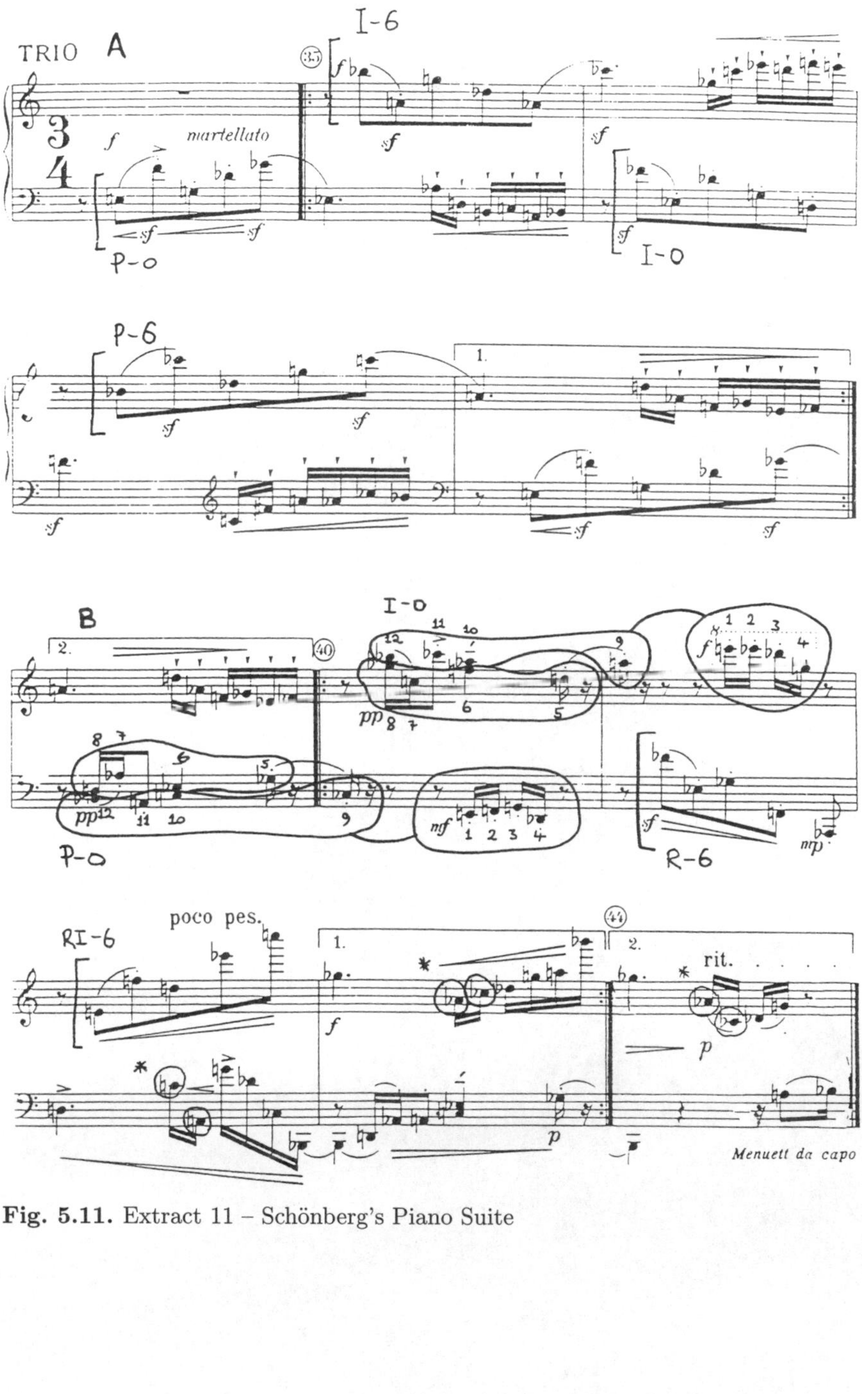

Fig. 5.11. Extract 11 – Schönberg's Piano Suite

6 Questions of Logic: Writing, Dialectics and Musical Strategies

F. Nicolas

Introduction

What is logic in music? Is there any practice in music that can be called logic? Is the nature of this possible logic in music musical or mathematical?

ISSUES

These questions entail today a considerable degree of subjectivity.

Today we are witnessing a proliferation of the power of calculation applied to music, witness Ircam. What might be called *musical reasoning* risks to lose some of its rights: the right to direct and to canalise this surplus power in calculation that is extraneous to music and invading the terrain of well-established operations, traditionally conducted with pen and paper. The increased capacity of calculation can no longer be ignored by composers. Their task is to identify the lines of force capable of alimenting their own thought.

Reasoning in music, what use could and should it make of this power of calculation which arrives unasked? How can it be made to serve music, without itself serving as a foil for new techniques that invade and assail it?

Musical thought cannot, in this situation, avoid reflecting on the relations between reason and calculation, and logic is necessarily involved in such an examination.

Twentieth century mathematical logic shows that there is a twofold excess in the heart of these relations:

1) Rationality exceeds calculation, particularly in cases of the undecidable where rationality is not calculable
2) Conversely, calculation exceeds reason: for mathematical model theory[1] shows that a coherent theory generates *ipso facto* the existence of a countable model whose nature is pathological (completely foreign with respect to the "natural" model of the theory).

A twofold excess therefore, capable of reorganising in the musical universe a double excess in relations between music's written and audible dimensions:

[1] See Löwenheim–Skolem theorem, to be discussed later.

1) On one hand musical reasoning goes beyond what is written, beyond literal formalisation. Not everything in music can be written, that is literalised (usually the impossibility of writing everything is affected in music to the dimension of timbre). It follows that the impossibility in question is not a technical impossibility *in abstracto* (timbre can always be digitised by projection onto the calculable) but an impossibility *in the conditions themselves* of the musical act, within the frame of musical writing itself.
2) On the other hand, the coherence of writing, the consistency of a literal order has often been taken (particularly by serialism) as a guarantee of a sensitive musicality, of a musical meaning as it is heard. If the eye is coherent, the ear will be obliged, and able, to follow it[2].

In other words, with the end of thematism, of tonality and of metrical structure, a gulf has been created between two systems of music – I mean "serious" or so-called "contemporary music" whose sensitive power is established on the basis of writing. It is a gulf which separates its sensitive and its literal order, writing and perception and, more generally, score and hearing.

The point of my initial questions is this: how are we conceive music if its universe is irremediably divided into two orders that will never be fully complementary? How are we to write music intended to be heard if it is impossible either to write something listened to or to listen to something written?

CAN MUSICAL LOGIC BE DEFINED?

To answer these questions, do we have to start with a definition of what musical logic is? I do not think so. It may be argued, instead, that for a musician there is no more meaning in defining musical logic than there is in defining music itself. Indeed, for a musician there is no good definition of music. Furthermore, such a definition *cannot exist* for him, no more than a satisfactory definition of mathematics can exist for a mathematician. To be more explicit: I do not claim that in general no satisfactory definition of music or mathematics can be arrived at. The encyclopaedias, for example, tell us that music is the art of sound; and so it is? They also state that mathematics is the science of figures and numbers – which is less convincing!–. It is possible, I believe, to frame definition of mathematics that would be completely satisfactory (satisfactory for me, perhaps because I am not a mathematician

[2] It did not imply that the ear should follow the coherence of the eye, just like there is no semantic transcription of the syntactic model in mathematical logic; the only purpose was to show that the ear can find its own way to follow what has been organised by the eye according to a coherence that remains singular to it.

in the strict sense of the word, but only a friend of mathematics). The definition I take from Alain Badiou's philosophy: *"mathematics are ontology"*[3]; therefore: mathematics are everything that can be said about being, about being as such is for me an excellent definition, but I am convinced that for a mathematician, especially for what the Anglo-Saxons call a *"working mathematician"*, this definition will be inefficient. He might feel gratified to find himself raised to the dignity of *ontologist*, but this will not make him a subject in his own enterprise. In the same way art might be defined philosophically as "the truth of the sensitive" and music as "the truth of the audible", or, even better, "the truth of listening". But, as definitions, these statements (whose philosophical origin is the same as the above definition of mathematics) will be unefficient within musical thought.

Thus, for a musician, one does not define music, even though the word itself remains capital for him, a word he cannot do without.

A musician spends his time asking himself in anguish if, by interpreting such or such work, by composing such or such other, the happy outcome will really be music. And one musician is all too ready to disagree with another: "This is truly music! This is certainly not!".

Eventually, for a musician, the word *music* is a name, precisely a proper name that acts without any need of a definition, just like the woman you love, to your eyes, has a name that is enough to enliven you without any need to be defined.

If *music* cannot be defined, could *musical logic* be? Surely not, again, because of the too strong proximity between the *names.*

Just a very simple demonstration scheme *ad absurdum*: if one could define *musically* what musical logic is, one would be so able to define *musically* what music is. From the impossibility to do this (see the previous axiom, or *axiom of proper name*) follows directly the impossibility to define a musical logic. QED

If one cannot define musical logic, what is there left to be done? We have to identify where the logical dimension operates in music. We have to distinguish the logical proceedings of music.

In a way, we not so much have to speak about "the" logical music or "a" logical music, but more about "logics in music", that is what could be named *logical* in music by setting out to locate its proper job: its specific field, its singular operations, its particular effects.

Which are therefore in music the properly logical proceedings and how are they working?

---

[3] Cf. *L'Être et l'événement* (Seuil).

IS THE LOGIC OF MUSIC A MUSICAL LOGIC?

To define these logical proceedings, musical thought must face the following question: strictly speaking, is the logic of music a musical logic?

The most common answer is that it is not. Received opinion (the contemporary *doxa*) has it that the logic of music is not a musical logic, i.e. it is over-determined by different logic from other domains of thought.

There are essentially three ways of placing the logic of music under the tutelage of another domain of thought:

Mathematical tutelage (arithmetic)

The first places the logic of music under the tutelage of mathematics, mere arithmetic for the most part and, commonly, a combinatory of the first integer numbers. An exemplary formulation of this approach is this statement which underlies St. Thomas d'Aquin's *Summa theologica* : "Musica credit principia sibi tradita ab arithmetico" (music submits itself to principles which it derives from arithmetic).

Physical tutelage (acoustics)

Others propose tutelage of a physical, more precisely of an acoustical, nature. One might mention here Aristoxenes of Tarento, or Rameau, but it is an orientation which even today seems quite natural.

In this case, music would be structured by a purely physical logic.

A word of warning: this conception brings into play not only specific physical laws (a truism) but it also establishes a tutelage of a specifically logical nature.

Music would not only borrow its material from physics, physics would not only endow music of a shaping framework or furnish the material conditions of its existence (this goes without saying: obviously, there is no music without sounds), but musical thought must also base its logical principles of inference on acoustics, which is quite a different matter. The idea is that musical logic must derive its norms from the physical-acoustical logic of sound itself.

Psychological (and physiological) tutelage

There is a third figure of tutelage, that of psychology. Music would derive its norms externally from a psyche of the feelings and a physiology of sensations. This psychological approach does not go back very far in time, although from the time of the Greeks it is constantly present, and is still present in the modern age. As in the two previous approaches, this tutelage posits a musical coherence based on an exogenous logic: to paraphrase St. Thomas, music should submit itself to psycho-physiological principles.

Musical autonomy

I will defend a quite different thesis: that music as thought is capable of self-normalisation (which by no means signifies self-entrenchment or self-definition, and even less self-demonstration of its own coherence: musical autonomy is neither autarky nor, strictly speaking, a nomology[4]). What is claimed is that the logic of music is a musical logic. It is not a mathematical-arithmetical, physical-acoustical or psycho-physiological logic.

*

In terms of this (hypo-)thesis, how are we to recognize logical operations in music?

I will proceed in two stages. To begin with, I will analyse these questions by means of variations. I will then, more synthetically, distinguish different logical statements in music and try to articulate them.

Variations on the Logical in Music

I shall begin by offering a series of variations involving the logical in music. By so doing, I shall provide my discourse with musical, rather than mathematical, form, since, if mathematics means demonstration, to "make music" means variation.

There are two great modes of variation in music:

The first one, the most common, consists in developing an object in such a way that it changes throughout the discourse. This is the case with Beethoven's variations.

The second, less usual, consists in varying the context of presentation of an object which itself remains unchanged. The simplest way of doing this is to change the lighting, swivelling a projector around the still object, so that each new disposition will reveal a renewed profile. Henry Pousseur has drawn our attention to this kind of variation in the case of Schubert. These two methods have a point in common: they move from the same to the other, more precisely, from the same to various others. Both set out from the statement of an identity (for example our theme would be in this paper the *logical in music*) to generate otherness: in the first case, generating other objects; in the second case, showing other facets of a same object. I prefer to work on a completely different kind of variation that, unlike the previous ones, moves from the others to the same. The qualifying issue of this third kind of variation

[4] Even if the logic of music is a musical one, musical thought cannot demonstrate, infer or even define it, as we have noticed above. This musical logic is a data with which music makes and creates. It is something like an axiom which musical thought decides, without demonstrating it of course, and then puts to the tests of its consequences. From this point of view, musical logic is no more defined than the concept of set or the concept of membership in set theory.

is not the alteration of an initial identity but the extrication of a common trait within a dispersed diversity. The point, in some sense, is to bring close together separate elements which stand apart, without obvious relations, in order to *recognise* the underlying, we might say incognito operation of one and the same figure within the initial diversity. Certainly, this third kind of variation assumes a particular interest when it is not the pure and simple reversal of the first two types, i.e. when it does not lead up to the presentation in conclusion of the theme which has the same nature as the one presented from the outset by the other types of variations. It is not a question of proceeding as did Liszt in his *Fantaisia "Ad nos"* or as does Franck in his First Choral for organ, where at the very end we are given the initial theme that establishes the retrospective key of the work.

If I call my first two types *alteration* and my third *recognition*, I am highlighting the fact that *recognition* produces an object which does not readily lend itself to the generation of alterations. *Recognition* is not a retrograded *alteration*. While the latter is related to a deduction, the former cannot be linked to an induction.

Here my method will come within these *variations-recognition* and I will submit the term of *logic* to different instances in such a way as to delimit what at first appears as a black hole, soaking up all light rather than enlightening the mind.

These variations, some negatively delimiting, others positively comprehending and interrelating, will aim at recognising, under a simple word, a name.

VARIATION 1: HISTORICAL EMERGENCE

Let us briefly examine the historical moment when the category of *musical logic* first appeared.

It is a category which seems to emerge at the end of the XVIIIth Century, in a very specific context. I shall quote at length the musicologist Carl Dahlhaus:

"[...] The forces in compositional technique that made possible an *autonomisation* of instrumental music may be brought together under the concept of *musical logic* – a concept closely allied to the idea of music *as speech*. [...] In his "Fourth Grove of Criticism" of 1769, Johann Gottfried Herder could still speak of *logic* in music with unconcealed disdain. [...] Herder dismisses as a merely secondary force the *logic* that in music lies in the context of the chords [...] Herder, it seems, was the first to use the term, but only with Johann Nikolaus Forkel, two decades later, did the concept of musical logic gain aesthetic respect. "Language is the garment of thought, just as melody is the garment of harmony. In this respect, one may call harmony a logic of music, because it stands in approximately the same relation to melody as logic in language stands to expression." [...] Where

Herder contrasted, Forkel mediated, calling the harmonic regulation of tonal relations *musical logic*"[5].

All in all, according to Dahlhaus, the emergence of the category of *musical logic* should be characterised by the following elements:

1) The theme of a musical logic appeared when music tried to conceive its own autonomy with respect to natural language, when music had to apprehend its *en-soi* and its *pour-soi.*
2) The theme of musical logic was at once correlated with the model of language, music tending to reflect on its autonomy through the modality of a specific musical language.
3) Musical logic had as its immediate test field its ability to link harmony and melody, in an historical period in which counterpoint and polyphony (that had until then regulated both the horizontal and the vertical dimensions of musical discourse) had given way to the accompanied melody.
4) The question of logic, defined as the possibility of regulating new relations between sounds, was placed at the heart of the musical discourse, not in its periphery: musical logic was the harmonic centre animating the melodic surface.

These are the four initial features:

1) The question of musical logic is posed when music attempts to think of itself as an autonomous universe and to reflect on what it is that guarantees its inner consistence.
2) Musical logic is originally conceived in correlation with the category of language.
3) Musical logic works on the unity of a divided field (divided between melody and harmony).
4) Musical logic is a centre that animates a peripheral appearance.

VARIATION 2: FUGUE (BACH)

The begin of Johann Sebastian Bach's fugues introduces a very simple matrix that can be represented as follows: the theme, classically called the *subject*, continues in a *counter-subject* at the same times as it is repeated in an altered form, in what is called a *reponse.* We thus have the association of an altering repetition (the reponse repeats the theme by transposing it and changing it – with possible transformations) and an extension into another melodico-rhythmic figure (the counter-subject). Hence: the one of the subject splits into two (into a counter-subject and a reponse) in the movement of its self assertion.

[5] *The idea of absolute music* (translated by Roger Lustig), Chicago and London: The University of Chicago Press, 1989 (pp.104–105).

$$\text{Subject} \longrightarrow \frac{\text{Counter-subject}}{\text{Reponse}}$$

Fig. 6.1.

I believe that this "logic" can be compared to the *principle of non-contradiction* that finds its emblematic expression in Aristotle's *Metaphysics*, where the principle is asserted as the first and irreducible foundation of all coherence of logos[6].

To this principle of non-contradiction can be oppose a musical principle that I shall call the *principle of forced negation*: any musical object, once *asserted*, must *come to terms* with its opposite, compose itself in becoming. In our short example, the theme exists by becoming other through a *split* [*scission*] according to a twofold alteration: that of the counter-subject and that of the reponse. This is the first feature that contradicts the idea of any parallelism between classical logic and musical logic, posing indeed the idea of an *anti-symmetry* (or of an orthogonality) between these two logics.

VARIATION 3 : ILLOGICALITY 1 (XENAKIS)

We now have a case of a negative variation (variation of delimitation). Let us give an example from the composer who has made himself the bard of parallelism between mathematics and music: Iannis Xenakis.

The first page of *Herma*, a work for piano dating from 1961, presents a material whose pitch structuring process is stated to be *stochastic*. This seems plausible considering the erratic character of the material, except that – and it is no small matter – we find from first bars a pure and simple twelve-tone series, which is not the outcome of any probabilistic draw. A mistake in calculation, we might ask? Or liberating gesture on the part of the composer, who defies the laws of his own calculation in order to effect a musical ratio emancipated from the mechanical enchainment? An examination of the score as a whole tends to invalidate this hypothesis, since this twelve-tone gesture is not followed up: we find no altered reiteration nor any influence upon the dominant stochastic gesture that will go on and on producing tornadoes of

[6] Aristotle asserts it as follows:
"The same cannot simultaneously belong and not belong to the same, according to the same" (3, 1005 b 19–20 – translated to French by Barbara Cassin in "La décision du sens. Le livre *Gamma* de la *Métaphysique* d'Aristote", Paris, Vrin, 1989, p. 125).
Then "No one can assert that the same is and is not." (3, 1005 b 23–24; ibid. p. 125).
And finally "The most widely accepted opinion is that two contradictory statements cannot be simultaneously true." (6, 1011 b 13–14; ibid. p. 153).

notes. Here the word *illogical* imposes itself thanks to the musical principle asserted above (the *principle of forced negation*) which a minima demands that an assertion (in this case, that of a twelve-tone series) should entail some consequences and not remain without a continuation, a dusty axiom lying unused in the corner of a theory. Worse than an unnoticed axiom: a useless axiom![7]

VARIATION 4: THEMATIC LOGIC (MOZART)

Let us return to composers of a different dimension, say Mozart, and examine this short extract of the development in the Piano Concerto number 25 (in C major K. 503) (see Fig. 6.2).

We are faced with a chemically pure example of a thematic development in which the theme asserts itself as self-consciousness, that is, as a capacity to norm its inner alteration. In fact, we have here a sequence in which the theme is repeated three times, first transposed from initial C major to F major, then to G major and, finally, to A minor. It can easily be linked to the pitches forming the head of the theme, and deduce that the theme has shifted in accord with a macroscopic path that is isomorphic to the microscopic structure of its outset.

We are dealing with a fragment of development that exemplifies what could be called *thematic logic*[8].

VARIATION 5: REPETITION (HAYDN)

In his symphonies and string quartets, Haydn plays a game of surprise that he loves to repeat. He likes to surprise us at once and then to repeat this surprise a second time with a knowing wink, creating an expectation to which we shall become dependent: does not this repetition simply mean that a third occurrence is on the way?

Haydn is addicted to this "*good* luck comes in three" game, in the course of which he sometimes deludes our expectations, and sometimes satisfies them, surprising us by insisting on continuing three times. It is an example which seems to me to show how music contravenes a fundamental logical principle, that of *identity* (something doubly asserted is always the same, independently from its different instances). The principle of musical logic, anti-symmetrical from this identity principle, might be called a *principle of differentiation* and defined as follows: any musical term which is doubly asserted undergoes an

[7] See my article against Xenakis: *Le monde de l'art n'est pas le monde du pardon* (Entretemps, n° 5, February 1988): http://www.entretemps.asso.fr/Nicolas/textesNic/Xenakis.html

[8] See my essay in the concept of theme: *Cela s'appelle un thème* (Cf. Analyse musicale n° 13, October 1988): http://www.entretemps.asso.fr/Nicolas/TextesNic/Theme.html

25° Concerto pour piano en Do majeur K 503

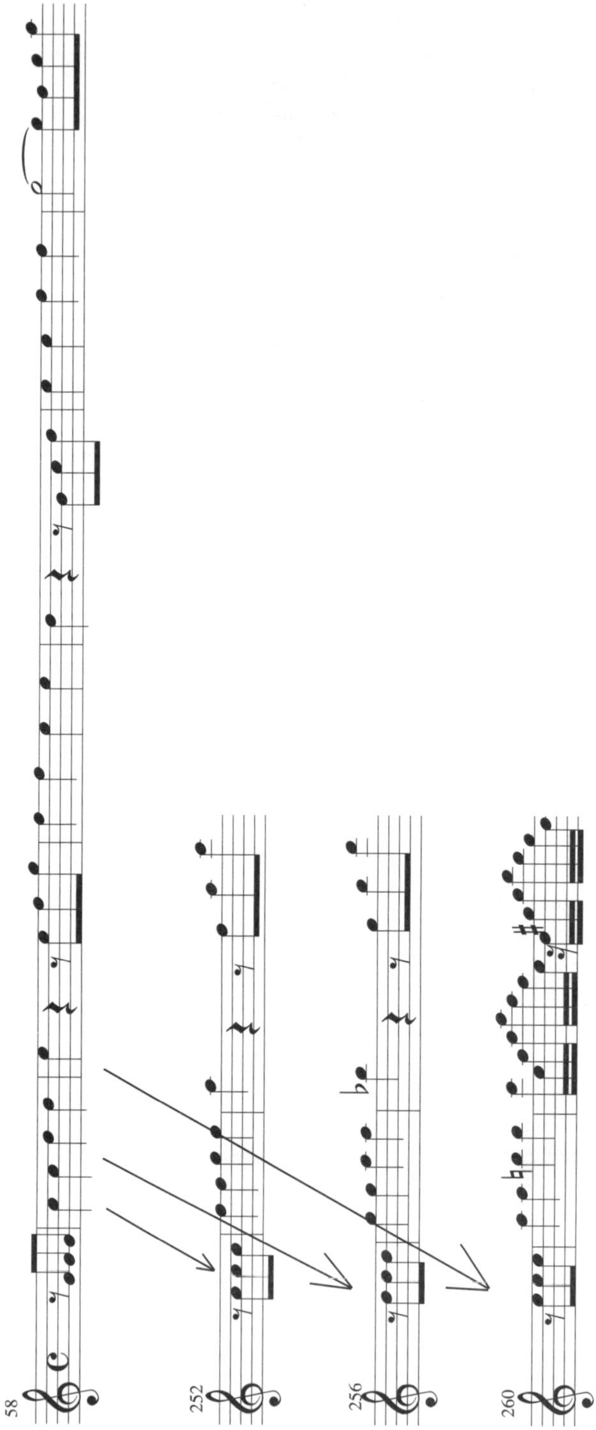

Fig. 6.2.

alteration; i.e.: no term which is repeated is identical to itself. Furthermore: in music to repeat means *ipso facto* to alter.

VARIATION 6: ILLOGICALITY 2 (SCHOENBERG)

In 1952, Boulez pointed out, with justified vigour, what he called a misinterpretation [contresens][9] in the understanding of the twelve-tone system by its inventor. He refers to those cases in Schoenberg in which the twelve-tone series logically structures the melody (following the laws of its own order) whereas the harmonic accompaniment of the melody is governed by a principle of distribution of the remainder: to construct the chords, use is made of the pitches of which the melody has not made use and they are grouped in small packages and somehow or other associated with the proper horizontal order.

Boulez rightly saw in this a failure of serialism to structure a material that remains subject to the outdated rule of the accompanied melody.

If one could index the success of tonal harmony (as it appears in Forkel) to the existence of a musical logic, then Schoenberg's failure effect this in terms of the twelve-tone principles has to be considered as an instance of illogicality.[10]

VARIATION 7: TAUTOLOGY (LIGETI)

The next negative example highlights what I propose to call musical tautology. Take, as an example, Ligeti's *Coulée*, a score for organ dating from 1969. One need only listen to the opening of the piece to have a look and then look at the score to understand how the relations between ear and eye are purely functional and musically redundant. In other words, the proper order of writing is, in this case, brought back to its bare nucleus: univocal codification and, as a consequence, the dialectic between writing and perception, where the former feeds the latter, has been reduced to a truism.

VARIATION 8: LOGIC OF LISTENING (FERNEYHOUGH)

Last example but one, drawn from Brian Ferneyhough's *La Chute d'Icare*. At the end of the piece, after the cadence of the clarinet and at the beginning of the coda, something surprising happens: three instruments (the piccolo, the violin and the cello) successively state a little regular pulsation in a context of writing that leaves very little space for this type of regularity, the prerogative of music in previous centuries. What is singular here, what raises an unusual

[9] *Relevés d'apprenti*, p. 268.

[10] Boulez opposes this treatment to his own, which he describes as follows: "[Complexes] are derived one from the other in a strictly functional way, they obey a logical coherent structure" (*Penser la musique aujourd'hui*, p. 41).

"logical" problem, is the fact that this intervention occurs so late in the work that it annihilates all possibility of consecution and imposes a retrospective examination, that acts as a revaluation of what preceded rather than as an opening for what follows (the work is almost finished). Hence the impression that this short moment is a *logical* problem in listening that might be said to be of the inductive order[11].

VARIATION 9: MAGNETIC FIELD (MONTEVERDI)

The last example is taken from Monteverdi's Madrigal *Hor ch'el ciel e la terra* (see Fig. 6.3). The initial accord in the tonic is repeated so intensely that, despite its function of rest (according to the tonal logic), it inevitably acquires an increasing degree of tension. As a result, it will be the arrival of the long retarded dominant chord that will act as a relaxation, precisely where, in tonal logic, the dominant chord ought to create tension and call for further resolution.

This example shows how a work assumes a musical logic of its own (in this case a tonal logic) which is not to be submitted to codified and operationalised chords progression (as happens with a logical law of inference, as the *modus ponens*) but in order to bring into play lines of force and energy fluids that the work will be free to distort and to change.

From this point of view, tonal logic must appear as the construction of a magnetic field that can be traversed in any direction, provided one has enough energy to deviate the trajectories traced out by the field.

CODA

To summarise briefly, these variations raise the following points:

1) Mathematical logic and musical logic are not so much parallel as anti-symmetrical. One could systematise this anti-symmetry by opposing to the three main logical principles of Aristotle three elements characteristic of musical dialectic:
   - the *principle of differentiation* versus the *identity principle* (see variation 5);
   - the *principle of forced negation* versus the *principle of non-contradiction* (see variation 2);
   - finally, where Aristotelian logic prescribes the *principle of excluded middle* (there is no middle position between A and not-A, hence I must choose between one or the other), musical composition would suggest a *principle of obligatory middle*: any musical term must entail another

[11] For a more detailed discussion on this point, see *Une écoute à l'œuvre: D'un moment favori dans* La chute d'Icare *(de Brian Ferneyhough)* – Compositeurs d'aujourd'hui: Brian Ferneyhough (éd. Ircam–L'Harmattan, 1999).

Fig. 6.3.

term which is different from the evolving negation of the previous term.

It is a kind of neutral[12] term, for it is "neither the one nor the other". In conclusion, these three principles would suggest that musical thought should benefit from a confrontation with stoic logic[13] rather than with Aristotle.

2) Let us call *musical tautology* any correlation between two orders that is merely a univocal and mechanisible functionality.
3) In music, logic would act upon the juncture of two dimensions: e.g. the two dimensions of the horizontal melodic and the vertical harmonic, or again the macroscopic and microscopic dimensions where we find that logic is the link.
4) Ferneyough's example poses logical questions that have less to do with the work's musical structure than with its singular dynamics: how does a given work deal with the logical principles of a musical nature that it inherits from the musical situation in which it is set.

Interlude: Mathematics, Music and Philosophy

It will perhaps have been noticed that my mode of demonstration, by variations which aim to define an object by the arrows that point it is intimately related to the basic idea of topos theory, for which the whole network of arrow-relations is more important than the object itself. The existing common point between a field of mathematics and a musical necessity is by no means a matter of chance, as we shall see.

As far as logic – as well as other subjects – is concerned, I believe that there exists no direct link between mathematics and music and that all attempt to relate them passes (that is *has to pass*) via philosophy. Any attempt to link mathematics directly to music[14] can only be effected within what I shall call an engineering problematic, that is in the mode of an application of mathematics to music. This relation based on the applicability is completely

[12] In an etymological sense: *ne-utrum.*

[13] One could refer to Claude Imbert's philosophical works. For example: *Pour une histoire de la logique* (PUF, 1999).

[14] Unfortunately, I am not aware of any attempts to work the other way round, from music to mathematics. [Note of the translator: an interesting counter-example is given by the problem of construction of musical rhythmic canons, as formalised by the Rumanian mathematician Dan Tudor Vuza. It leads him naturally to non trivial results in the domain of factorisation of cyclic groups. In particular Vuza provides a method of constructing all factorisations of a given cyclic group into non-periodic subsets, by clarifying the properties of the so-called non-Hajós groups. Rhythmic canons associated with such a factorisation have the very fascinating property to "tile" the musical space (that is, no superposition between different voices or holes). Vuza's algorithme has been recently implemented in *OpenMusic* by IRCAM's *Equipe des Représentations Musicales* (http://www.ircam.fr/equipes/repmus/)].

independent from the content of mathematical thought and uses only results susceptible of formulisation, in other words, the classical situation in which mathematical rationality takes the shape of a pure calculation equivalence between the two sides of a sign *equals* ("="). The most dominant approach today is unfortunately that of reducing mathematics to a collection of formulas that can be applied to, or transposed as, music. Xenakis has built his reputation on operations of this kind.

My thesis is that any assumption of a relation between music and mathematics must proceed by way of philosophy, not through a compendium of calculations. If there is a question of contemporaneous thought between mathematics and music, not of vassalage or of application, then it is philosophy to which we must delegate the setting up of a conceptual space capable of containing it. This is so because musical thought is not scientific but artistic, so that direct links between, for example, mathematics and physics, have no counterpart in the case of mathematics and music: in the former case, such relations are rendered valid if one assumes the ontological character of mathematics (for, everything that makes sense for being as such [*l'être en tant qu'être*] makes sense *ipso facto* for any being [étant]). But, more precisely, music is not a science, and musical logic is not an acoustical logic ...

Musical Proceedings of Logic

After having explored some logical issues which relate to music, let us attempt a more synthetic elaboration by offering what mathematicians would call a "compendium of results".

In our chosen field of music, I propose to understand by *logic* everything that in formal terms conditions possibilities of existence.

Not all conditions of existence are logical; logical are those that entail possibilities of existence in *formal terms.*

To give an elementary example, the so called logical rule of *modus ponens* (if A $\rightarrow$ B and if A, then B) takes into account the validation of B by assuming A and A $\rightarrow$ B, independently of whether A really exists nor if the implication A $\rightarrow$ B really subsists.

Thus logic does not concern itself with things that really exist. It is concerned only with the prescription of a coherence of the possible, without taking into account effective realization.

As Leibniz has it, logic establishes a configuration of possible worlds but it delegates to God the task of determining the unique world that really exists.

With this delimitation, I suggest we distinguish three proceedings of logic in music:

- the writing of music;
- the dialectic of musical pieces;
- the specific strategy of a musical work.

MUSIC AND WORK

A central point in musical logic concerns the difference between a structural level of music and the concrete and singular level of a work.

A tonal logic may exist, for instance, but no one work would ever display it as such.

Only a treatise of harmony would be capable of accounting for it. From the point of view of the work, which is what most interests us, music is influenced only superficially by this logical prescription (in the case, of course, of tonal music), without being completely subject to its law.

The work, on the one hand, is endowed with a "need-to-say" [*devoir-dire*] which is in fact a prescription involving its being – i.e., the necessity of asserting its unity as a musical *being* [*étant*] –, what is usually called a "piece" of music. At the same time, the work takes upon itself a kind of strategic prescription.

Hence, the general process of inference of a work – or the consistency of its "need-to-say" – needs to be separated from its strategy – or insistence of a "want-to-say" [*vouloir-dire*] –, for which a singular process is operating.

In what follows, we will describe as a *piece* of music this first element of the musical opus (the general process of inference or consistency of its have-to-say), and use the term *work* for the second element (the strategy and its singular process of inference, or insistence of the want-to-say). The *piece* of music is the level at which the opus establishes itself as a being, existing in a situation. The musical *work* is the level at which the opus takes the shape of a project, of a musical subject.

There are three proceedings:

- Writing influences in a formal way the coherence of a possible world of music: it represents the logical proceeding of music as a universe.
- Dialectic influences in a formal way the consistence of a piece and the possibility of its unity: it is the logical proceeding of a piece of music.
- Strategy means the logical proceeding of a musical work seen as a subjective singularity. Its influence formally concerns the insistence of the work, that is the possibility to sustain a musical project throughout the dimension of the piece.

We will now proceed to clarify one by one each of these proceedings by disengaging their specifically logical aspect.

WRITING

Musical writing can be thematicised as a logical dimension through two reflections:

Writing and sound material: topos theory

The first point concerns how musical writing takes into account the sound dimension. The answer to this question may be logically clarified through topos theory. In mathematical category theory (or topos theory), logic appears as a logic of universes. The collection of logical operations may be characterised as the relationships between the totality of objects in the universe and the so called *subobject classifier.* The validity of any logical connection in the universe is given by a particular and well-determined point of this universe. Therefore, following Alain Badiou's philosophical interpretation of mathematical category theory[15], it can be stated that a logical operation corresponds to a *centration* of the universe[16].

This new approach to logic leads us, from a philosophical point of view, to differentiate being in situation or being-there [être-là] and appearing.

This could clarify the problem of a musical logic for the following reasons:

1) The relation between the score and listening to it may be conceived as the relation between the being-there of music and its appearing to the senses.
2) The logical centration in music may be characterised as a centration on the writing: musical writing is what takes into account the musical dimension of the sound created. By distinguishing what in music does exist and what does not, writing states the validity of appearing[17]. It also validates *in a musical situation* the real existence of any appearance of sound; for it would remain a mirage without the effectiveness provided by the writing itself. Who, after having heard a piece of music, has not been led to compare the consistency of what he seemed to have heard with the score? By so doing, you show you are familiar with the transcendental use of writing, even if it remains an intuitive use that is not analysed as such. From this point of view, musical writing takes into account the passage from a sound level to a musical level by structuring the musical logic through the sound situation.

Writing and listening: model theory

My second point is: how does musical writing relate not only perception and hearing, but – more specifically – the musical listening?

Relating writing to the singularity which is the musical listening implies a specifically logical dimension that can be expressed as follows: in music

[15] Cf. *Court Traité d'ontologie transitoire* (Seuil, 1998).

[16] For reasons that I will not develop here, this function can be philosophically named the concept of *transcendental,* following the very precise meaning that Alain Badiou originally introduced in his philosophical interpretation of mathematical topos theory.

[17] In philosophical terms, the writing measures the *there* of sonic *being-there*, the *da* of the *Dasein.*

writing is what calculates and demonstrates, whereas listening starts precisely at the point from which things cannot be expressed and ordered according to some strict rules of perception[18]. The *privileged moment* [*moment favori*][19] of a work, when listening takes wing, is founded on a logical condition: that some demonstrated things cannot be shown, that speaking about something really existing remains meaningful, even if it cannot be presented according to the traditional schemes of showing. Mathematics provides us with many examples of this.

Musical listening, which entails a thought linking the sensitive aspect to the intelligible, is effected when sensitivity splits off from sheer perception and gives way to a new principle of intelligibility which no longer depends on showing, but embraces a musical rationality of infinities for which no representation is possible.

How can writing take into account this new schema of sensitivity? More precisely, how can musical writing influence the possibility of such listening taking place and continuing to operate in a work?

This question implies a logical dimension which can be related to what the 20th century mathematical logic called model theory, a theory that analyses the articulation of reason and calculation.

The mathematisation of logic and, thus, its literalisation has, since the end of 19th century, led to a split between, on the one hand, the scheme of the letter and, on the other, that of its interpretation; in other words, an horizontal barrier separates a purely syntactical scheme from a semantic one. This barrier does not disappear, in model theory because its semantic interpretation of objects is not concerned with logical connectors, which are confined to the syntactic field. It is clear that logic develops, on the one hand, in an horizontal dimension (that is the propositional calculus) and, on the other, in a vertical relation between syntactical inferences and semantic interpretations (this is model theory).

This fact enables us to characterise what musical logic is concerned with: the relations between writing and musical perception are similar to those linking syntax and semantics. From this point of view, musical logic is the science of the way in which writing progressions dialectise themselves into sound consecutions.

By interpreting the mathematical duality theory/model by projection on the musical duality score/audition, it is possible to "interpret" different

---

[18] Although one might be tempted to criticise the idea of what is conceived as an adequacy between what is shown [*montré*] and proved [*démontré*], in music the category of *perception* is very close to the philosophical concept of *apperception*.

[19] Cf. *Les moments favoris : une problématique de l'écoute musicale*, Cahiers Noria n° 12 (Reims, 1997) : http://www.entretemps.asso.fr/Nicolas/TextesNic/momentsfavoris.html

logical/mathematical theorems of our century by proposing the following theses[20]:

1) *Musical listening proceeds according to some determinations that cannot be written as such.*[21]
2) *Any score is compatible with at least two radically heterogeneous listenings.*[22]
3) *Any consistent musical writing guarantees ipso facto the existence of a possible listening.*[23]

DIALECTICS

My second question concerns musical dialectics. As already pointed out, music can be organised according to dialectical principles that are anti-symmetric to those of Aristotelian logic. These principles are organised by schemas of inference that are equally logical, of the type "*If ... then ...* ". The principle has already been established in the two axioms of *forced negation* (see variation 3: "If A, then not-A") and *obliged middle.*

Four dialectical issues

In music all these inferences assume a more systemic character when we consider the strictly logical base of different compositional styles. The fact that in music *logic* means *dialectic* becomes obvious once we have observed that all musical historical situations have set up a specific dialectic issue with respect to the works they embraced.

1) In the case of the baroque fugue, the dialectic issue was that of a split [scission] of its single subject (into a counter-subject and a response: see variation 2).
2) In the case of the classical sonata form, the dialectic issue was that of a *resolution* of two deployed opposite forces[24].

[20] I will not analyse here the chain of propositions in detail. One may refer to my contribution to the Colloquium *Ars Musica* (Bruxelles – 2000): *Qu'espérer des logiques musicales mises en œuvre au XXe siècle?* (forthcoming).

[21] See Gödel's well known theorem.

[22] See later, Löwenheim-Skolem theorem.

[23] See Henkin's theorem. This third thesis tends to validate the serial statements that we have mentioned before (as: *"perception has to follow the writing"*), once one has noticed what follows: if perception has to *follow* (serialisme), the "real" model does not *follow* the logical mathematical theories (model theory), for deductions by the latter have no semantic translations in the model; consistency of the model and coherence of the theory are not isomorphic to each other. This means that musical listening does not work by *following* the writing (listening is not a perception of written structures) but by deploying according to its own rules.

[24] Cf. Charles Rosen's works, e.g. *The Classical Style* (The Viking Press, 1971).

3) In the case of the romantic opera of Wagner, the dialectic issue was that of a *transition* between the multiple entities that make it up.
4) As to Boulez' serial work, the dialectic issue was that of an inversion [renversement] of order[25].

Dialectic of the same

We return here to our previous statement: any dialectic is characterised by the fact that a musical variation (in its broad meaning) is considered as an alteration of a principal unity. We might be tempted to say that it concerns *classical* musical dialectic and we might remark that there is no reason at all to restrict musical dialectic to this classical dialectic (as if one were to confine mathematical investigations to bivalent classical logic, that of the excluded middle).

I have already remarked that there exists a possible alternative which involves my work as a composer, an alternative that I called the *recognition-variation* [*variation-reconnaissance*]. Let us attempt a brief description of the conceptual space of its realisation.

The idea is to delineate a musical dialectic which, contrary to the *classical* dialectic, goes from the others to the same, in a kind of conquest of the generic, of anybody, of the anonymous. The alterity would be a starting point, the first evidence so that what is astonishing and precious will be attached to the universality of the same, rather than to the differentiation of particularities.

Of course, as has already been pointed out, this dialectic cannot be a retrograded *alteration*, transforming deductions into inferences. It is a dialectic that must make up operations of its own that definitely cannot be a mere inversion of *classical* operations.

I suggest that we should adopt Kierkegaard's approach, in particular, the three following operations: reply [reprise], reconnaissance and reduplication.

The reply (that is coming back forwards) is a second occurrence which turns out to be the first one, whereas the reconnaissance (of an unknown[26]) is a first occurrence which turns out to be the second one.

The reduplication is a reflection that seals the one of a single gesture thanks to the how [*comment*] that reduplicates the what [*ce que*], the enunciation that validates the statement, the making [*faire*] that seals the saying [*dire*] ... [27].

These three formal operations concern two different aspects of the two: *reply* and *reconnaissance* concern an ordinal two (for they fix an order and

[25] Cf. Célestin Deliège's works, in particular *Invention musicale et idéologies* (éd. Christian Bourgois).

[26] Of an *incognito*, in Kierkegaard's terminology.

[27] Its opposite would be the Hegelian *redoublement*, when polarisation shows the two divisions of a primary unity (the redoublement concretises the two faces of a same thing).

determine what is primer and what is not) whereas reduplication – and its Hegelian counterpart redoublement – is concerned with a cardinal two (which means that it is linked to the quantity and determines how and whether 2 may be identified with $1 + 1$ or not).

These operations, I believe, could provide a logical alternative to the development of classical music. In fact, they already exist in contemporary music (see, for example, the works of Eliot Carter or Helmut Lachenman). Our task is thus to acquire consciousness of what has already been done, of casting new light on the principles that already exist, something like what happened with the *axiom of choice* at the very beginning of 20th century, an axiom that many mathematicians of the previous century had already used implicitly.

STRATEGY

We now come to my third major logical concern, centred on the specific strategy of the individual work and which leads us to distinguish two fundamental rules from the standpoint of logic.

1) The strategy of each work's must be thought of within a specific inferential framework and not just in a more or less selective deviation in relation with the broader system it has inherited.
2) The work must be brought to an end. It has to finish somehow, without arousing any suspicion of suicide.

Let us briefly analyse both points.

Inferential system

By prescribing a systematic strategy for the work we are suggesting that it must pursue insistently and even relentlessly a musical project of its own, independent of the variety of sound situations it might encounter. In order to be a real *musical subject*, the work cannot limit itself to simply noting down a punctual clinamen, without consequences. Nor can it content itself with placing some local declination in relation to a musical system that would constitute its global envelope. Such a work would imply a hysterical and unilaterally rebellious subjectivity.

The challenge is a different one: the work should create an expression, a "need-to say" based on its own force. It should create a persisting instress [*intension*] due to a systemic *parti pris* and not only to some spontaneous reaction toward a path standardised by a tonal or serial system, or even by some systemic dialectic of the same ...

I am not asserting here that this systemic character has to be formalised, nothing indicates that it could replace the well-known musical systems. It has more to do with a subjective quality of insistence than with a system that can be codified.

This singular systemic character of the work could be seen as its personal modality of inference, just as a singular mathematical theory adds its own rules of inference.

To give an elementary example, the order relation states that if $A \leq B$ and if $B \leq A$ then $A = B$, which is a new way of inferring the identity of A and B. The assumption that this strategy has to be systematic inverts Boulez' problematic System and Idea, because the principle of the work no longer consists in confronting and deviating the musical system; on the contrary, it superposes itself on the musical system.

As a radical example of my systematic presentation, I suggest the category of the *diagonal*, which derives from Cantor's mathematical ideas and owes nothing to Boulez' concept of *oblique*. As I have given an account of this method elsewhere, I shall not deal with it here[28].

When the end occurs

If this insistence orientates the desire of the work within the infinite of its situations, it is the end that has the work face the necessity of a conclusion.

The moment of its end, when the work entrusts itself to the subsequent outcome of what it enacts, within the dialogue with other works it establishes, poses a number of significant questions of logic. In order to answer them, let us have recourse to the mathematical idea of *forcing*. Here too, Schoenberg's manner of working is highly illuminating, particularly for what extends the whiles of his works[29].

The correlation of two

Hence system and conclusion are connected from a logical point of view: interruption only intervenes because a strategy is involved (in the case of a chaotic collection of events this would be impossible).

The strategy of the work involves the relation between the finite and the infinite within the work. This relation is a product of a logical approach, insofar as it is investigated, as it is here, in terms of a formal examination, i.e. an analysis based on a scheme which takes account of the conditioning of the possible. If the work is really a work, that not only *must say* but that *wants to say* (i.e. if the work is a real musical subject which is no reducible to a *piece* of music activated by the structuring situation), then its "want-to-say" *must* be part of a singular process of insistence and *must* reach the point of deciding to come to a conclusion. The fact that a singular "want-to-say" – which is not the same as a general "need-to-say" – thus generates a necessity interlinking all its own, may be seen as a specific example of what has been called, in other fields (philosophy, psychoanalysis, ...) a *logic of the subject*. This logic is the formal scheme of inference to which the subject

[28] Cf. *La singularité Schoenberg*, éditions Ircam-L'Harmattan (Paris, 1997).

[29] Cf. *La singularité Schoenberg*, op. cit.

freely submits itself, provided the freedom of the musical work means freedom of determining itself (Kierkegaard) or of considering itself accountable for its own actions (Nietzsche).

Conclusion

If we see logic as a formal scheme of conditioning and inference, then the logical prescription in music will take the shape of a triple injunction, by projection on a threefold level, that of the musical world, of the opus considered as a *piece* of music or as a *musical work*:

1) The musical world is a universe of thought, and this means it is not only capable of indent musical being [*l'être musical*] – by determining what comes about in music [*étants*] – but of controlling appearing [*les apparaîtres*], appreciating their existence, insofar as musical writing is capable of defining a central field (the score). It is this that, at one and the same time, is located at the centre of the musical world and capable of providing a centre for music itself.
2) The musical piece will be endowed effectively with unity, being countable-as-one [*comptable-pour-une*] once a specific dialectic is brought into play, involving the specific musical situation (i.e. the very exceptional status of the music universe) inside which it has been placed. This dialectic governs in formal terms a general scheme of inferences and consequences that has been incorporated by the piece.
3) The musical work will become a subjective process rather than an act of pure subjectivation, provided that it involves a strategy, i.e. an aptitude to make insistent, throughout the piece of music that identifies with it, a singular want-to-say structured by some singular principles of inference that enable it to exist as a musical project.

To put it more schematically:

- The writing provides the logical coherence of the musical world[30].
- The logical consistency of musical pieces is related to the types of musical dialectics historically established.
- The logic of insistence of a musical work takes the shape of a specific musical strategy.

In other words: in music logic acts as coherence in the writing of the world, a dialectic of consistency in the pieces and a strategic insistence along each work.

(Translated from French by Moreno Andreatta)

[30] More precisely: of the world of music that we are concerned with and which is, as we must keep in mind, just one of the many real or virtual musical worlds (there are such examples as the different worlds of oral music tradition, of popular or of improvised music ...).

7 The Formalization of Logic and the Issue of Meaning

Marie-José Durand-Richard

Introduction

This paper is not directly concerned with musical logic, because I am unable to raise pertinent issues concerning it. But, as is clear from the introductory paper by François Nicolas, some parallelism is not lacking between musical logic and formalised logic, in terms of their ontological foundations and symbolical calculations. Thus, my purpose is to focus on the questions linked to the separation of a blind formal calculus from its possible interpretations, as they engage our present queries about meaning. I shall first present the specific point of view of an historian of science, a very different matter from our present knowledge of logic. Then, I shall define the periods I have chosen to show how these questions were dealt with historically. I shall go on to analyse how the relationship of logical calculus to the question of its meaning was viewed during these different periods.

7.1 Some Indispensable and Significant Steps Forward in the Mathematicization of Logic

If we consider the development of logic from the point of view of our knowledge now at the end of the Twentieth Century, a number of especially significant breakthroughs in the mathematicization of logic claim our attention since they predetermine the present state of relations between mathematics and logic:

- the breaking away from mathematical realism, when different forms of non-Euclidean geometry merged in the 1830s. This led to a search for original sources of rigor in the logic of mathematics itself, beyond the physical world,
- the foundational crisis in mathematics, at the end of the Nineteenth Century, when several antinomies in set theory were pointed out. It dealt with such examples as the class of all classes which do not belong to themselves[1] [3, pp. 445-48], [4, pp. 238–44]. In 1918, Bertrand Russell (1872–1970) expressed such antinomies in a popular form, the "barber paradox":

[1] Letters from Cantor to Dedekind: 28th of July and 28th of August 1899.

A barber advertises that he shaves all those who do not shave themselves, and it is asked if he is going to shave himself.
All these examples draw on the concept of non-predicative class, the extension of which can not be defined, because its definition is a circular one: it deals with a class of objects which refers to the object being defined [26, pp. 1183–84]. Such a definition is made possible because of the reflexivity of natural language. As they were used to define other concepts in analysis, cases of this kind made it clear that classical analysis in mathematics contains contradictions.

- the growth of various realms of thought intended to solve these specific questions:
  * the Intuitionistic School, whose fore-runner was Leopold Kronecker (1823–91). Conceived to put order in the number system, it was further developed by Luitzen E.J. Brouwer (1881–1966). Mathematical thinking was conceived as a process of construction, founded on the perception of succession in time. It was claimed that a mathematical definition has not only to secure the existence of a mathematical object, but that it must guarantee that this object can be arrived at through a finite number of steps. This school also contested some demonstrative principles such as the indirect method of *reductio ad absurdum*, or the law of excluded middle for infinite sets.
  * the Logistic School, so named in 1904 at the International Congress of Philosophy in Geneva. Its aim was to reduce mathematics to logic, by offering a logical definition of whole numbers. Gottlob Frege (1848–1925) is the first uncompromising exponent of this undertaking. Subsequently, Bertrand Russell and Alfred North Whitehead (1861–1947) will attempt to eliminate an ultimate antinomy discovered by Russell in Frege's *Grundgesetze der Arithmetik*. The overall aim was to write logic as a calculus, to furnish it with a formal derivation from axioms, in an attempt to clarify all the logical laws on which the validity of deductive processes must ultimately rest.
  * the Formalist School of David Hilbert (1862–1943). From this viewpoint the objects of mathematical thought are the symbols themselves [24]. First put forward in 1904, at the International Congress of Mathematics at Heidelberg, the 1920's saw its effective development. It set out to clarify the axiomatic foundation of the separate branches of mathematics, constituted as they are of concepts both mathematical and logical. With this formalist philosophical approach, Hilbert hoped to give a positive answer to the three following questions, which he expounded during the International Congress of Mathematicians held in Bologna in 1928:
    1) is formal axiomatic complete[2]? in other words, can it be demonstrated that any statement can be either confirmed, or refuted?

[2] To this question, Gödel will give a negative answer in 1931.

2) is formal axiomatic consistent, or non-contradictory[3]? in other words, can it be demonstrated that it is impossible to reach a contradictory statement, such as "$2+2=5$", by a sequence of valid steps? The question becomes crucial once the system of axioms no longer refers to a realm of familiar and well defined objects. The preceding antinomies had shown that the apparent evidence of elementary notions was not sufficient to assure the consistency of the system that they helped to build.
3) is formal axiomatic decidable[4]? in other words, does there exist a method capable of deciding, before beginning the demonstration, if a mathematical statement is true or false? This question is usually called the "Entscheidungsproblem".

This school elaborated a first theory of demonstration, by which it proved the consistency of simple formal systems. It is a theory, however, in which truth and demonstrability are not clearly distinguished.

- the incompleteness theorem formulated by Kurt Gödel (1906–78) in 1931, which pointed out the impossibility of reducing mathematics to logic. He reached this conclusion after his demonstration, in 1930, of the completeness of the quantification theory of Frege. Gödel went on to show that some assertions, which can be considered as true by non formal arguments, cannot be proved in the axiomatic system chosen for the demonstration, which makes it possible to build a formalised theory of arithmetic. Such a system of axioms is thus incomplete: it is not adequate when it comes to proving all the assertions that can be formed in this system. Since Gödel's incompleteness theorem, truth and demonstrability are established as concepts, and clearly distinguished: the concept of truth which is semantic, is stronger than that of demonstrability which is syntactic, since a statement can be established without being true or false [23, pp. 592–617].

Nevertheless, history of this kind, typical of reference books, shows a recurrent grasp of the phenomena of formalising logic. What is really being attempted is a demonstration of the way in which logic has become what it nowadays is. It tends consequently to project our own conceptions on to the past, instead of attempting any real understanding of the kind of queries which produced it. This kind of history often starts with the work of Frege, who leaves logic in a state which can be identified with the present calculus of predicates, because quantification no longer bears on the subject of the proposition, as it had since Aristotle, but on one or several arguments, or variables, in a proposition analysed as a polyadic function, and where several levels may be concerned, since every function can in its turn play the role of argument for another function. Of course, logic has been qualified as formal since Aristotle, in his analysis of propositions, put letters in the place of the subject and predicate. Moreover, effectively, after Frege's theory of

[3] To this question, Gödel will give a negative answer in 1931.
[4] To this question, Turing will give a negative answer in 1936.

quantification, logicians will look for a systematic elucidation of each step of the demonstrative sequence of propositions, and of all the principles which legitimate their validity.

7.2 What could be a More Historical Investigation of the Mathematicization of Logic?

My own interest in the history of mathematics and logic is not an attempt to supply reassuring filiations; the aim is rather to question the state of any knowledge by starting from its antecedents. An event can only be seized upon – and this seems to me essential – by starting from the locus of its emergence, when it took on its first meaning, i.e. the locus which inscribes it in its own history. For this reason I wish to examine the conditions which first made possible an effective connection between mathematics and logic, after more than twenty centuries of close proximity between the grammatical analysis of language and Aristotelian logic, refined as Scholastic logic during the Middle Ages. The first effective connection took place in the work of George Boole (1815–64) who supplied, in the middle of the Nineteenth Century, an algebraic writing of a calculus of propositions, in a form which is not exactly the same as our present Boole's algebra.

It is important, from my point of view, to set out from the process of mathematicisation of logic in Boole's work, for the following reasons:

1) As far as I know[5], Boole was the first author who wrote logic completely as an independent symbolical calculus, operating on symbols deprived of meaning. He was also the first author among the logicians who explicitly questioned the relationship of this calculus to the lack of meaning of the symbols on which it works.

2) When Boole is referred to, his work is never related to that of his contemporary Charles Babbage (1791–1871), who conceived, from 1834 to his death, the drawings, plans, and mechanical notations for an engine designed as "analytical", and which corresponds to our present automatic computer with an external program. This engine isolated the different logical functions, the same functions that Von Neumann would adopt more than a century later for the classical architecture of the EDVAC computer: an input-output system for data, which might be numbers or symbols; the storage of data and results; a mill to calculate them; a command system of operations, constituted as a punched-cards system of command of the operations, derived from the Jacquard loom, which can use an external program; and a control system of the operations, governed by the movements of ring-formed barrels

[5] Leibniz have considered logic as a calculus, but not as an independent one, and Boole's work is in no sense a direct continuation of his texts, which at the time were still unpublished. Boole and his contemporaries knew only his work on infinitesimal calculus, thanks also to the way in which it had been developed on the Continent.

with studs, which may be considered as micro-coded parts of an internal program [12, pp. 26–30]. To me it is apparent that the relations between the work of both Boole et Babbage must be investigated from the same point of view if we are to understand this innovative connection between mathematics and logic.

3) When conferences or books on cognitive sciences or artificial intelligence deal with the opening up of these new realms, they tend to refer to René Descartes (1596–1650), Thomas Hobbes (1588–1679) or Gottfried Wilhelm Leibniz (1646–1716). However, when Hobbes, in his *Leviathan* and in his *De Corpore*, spoke of thought as a calculus, of ratiocination as computation, he adopted this metaphor in order to insist on the consecution which inscribes the composition of thoughts in time, in specific opposition to the uncontrolled imagination.

To begin with, he gave a number of examples: numbers, lines and figures for geometricians, consecution of words for logicians, laws and facts for jurists. He went on:

Out of all which we may define (that is to say determine) what that is, which is meant by this word Reason, when we reckon it amongst the Faculty of the mind. For REASON, in this sense, is nothing but Reckoning (that is, Adding and Subtracting) of the Consequences of all names agreed upon, for the marking and signifying of our thoughts; I say marking them, when we reckon by our selves, and signifying, when we demonstrate, or approve our reckonings to other men [25, p. 22].

Here, addition and subtraction have an exclusively arithmetical definition, describing, as they do, operations for collection or separation, for synthesis or analysis.

In his project of *Mathesis universalis*, as in his *Discours de la Méthode*, and as early as his letter to Mersenne of the 20th November 1629, Descartes laid stress on the composition of words and ideas, and by their means he characterised reason as a universal instrument. He further distinguished his projected philosophical language as universal, indeed as the only possible way of enumerating all thoughts, and of isolating simple thoughts as clear and evident. But, if Descartes separated grammar from the meaning of words, he did not say anything specific either about the nature of these combinations, or about the links between these combinations and the meaning of the words [9, pp. 164–67].

As to Leibniz, his own invention of notations for the infinitesimal calculus led him to consider calculability as an ideal norm for the art of invention. Firstly, such calculability remains within in the realm of combinatory and, secondly, the calculus depends on concepts, more precisely on characters such that what they stand for can never be forgotten. In a manuscript written in 1674, but published only at the beginning of the Twentieth Century, Leibniz wrote:

"De la méthode de l'universalité".

4) ... *"C'est la caractéristique qui donne les paroles aux langues, les lettres aux paroles, les chiffres à l'arithmétique, les notes à la musique ; c'est elle qui nous apprend à fixer comme des traces visibles sur le papier en petit volume, pour être examinées à loisir ; c'est elle enfin qui nous fait raisonner à peu de frais, en mettant des caractères à la place des choses, pour désembarrasser l'imagination."*

And Leibniz considered as his task to elucidate this "caractéristique", which he also named specious or symbolical. Obviously, all these approaches are strongly motivated by a reflection on the nature of algebraic notation, whose existence had been stabilised less than a century before.

It is to these, and to these alone, that reference is usually made, but I am led to ask if the persistent neglect of Boole and Babbage derives from ignorance or is the outcome of deliberate choice based on a desire for a precise kind of affiliation. The question is far from anecdotal, when cognitive sciences now often take the hypothesis of a naturalisation of such a logical calculus, considered as computation on symbolical representations.

5) Moreover, to understand a moment like the present when an ever more widespread use of computers in the organisation of whole sections of the economy and the simulation of phenomena of all kinds, induces us to speak of a Second Industrial Revolution, it is crucial to call to mind the fact that Boole and Babbage's work coincided with the moment in which England was endeavouring to assimilate the consequences, both institutional and conceptual, of the First Industrial Revolution.

It is in this light that I shall now attempt to define the stages that in my opinion characterise philosophical discussion of specific relationships between an independent logical calculus, the meaning of the calculus and that of the symbols it is based on. I wish to distinguish:

- the conditions which, in the England of the first half of the Nineteenth century, made it possible to effect the first inter-penetration of mathematics and logic. Here, a network of algebraist reformers played a role absolutely central, Babbage first among them, followed by Boole;
- the reasons why this connection, once established, necessarily entailed for its originators the idea that the operating properties of symbolical calculus and the meaning, or what they called the "interpretation" of the results, should be independent.
- the reasons which, on the contrary, led Frege to reaffirm a strong connection between the form and the content of logic,
- and lastly, the renunciation of any reference to an ontology of calculus, a renunciation linked to the emergence of plurivalent logic and of model theory, totally different approaches which are concerned with investigation of the mutual relationships between syntax and semantics.

7.3 The Connection Between Mathematics and Logic. The First Phase: Great-Britain (1812–1854): How is the Permanence of this New World to be expressed?

The movement which led to the mathematicization of logic goes back to the debates which took place in Great-Britain, in the first half of the Nineteenth Century. They centred on the nature of the fundamental knowledge that Universities should be expected to impart. The Universities of Cambridge and Oxford were, at the time, the only universities in England. They were both of them Anglican and the statutes on which they were based which had not been significantly modified since 1570. They were hidebound in their conservatism. The reformers reacted with impatience, the most radical of them with hostility, when they compared this conservatism to the dynamism of the new industrial towns, new centres whose interest in science was profound but strongly utilitarian. For the utilitarian approach of science, institutions should be an expression of society, efficient expressions aiming at the happiness for the greatest number. On the contrary, at the Anglican universities, mathematics at Cambridge and logic at Oxford were regarded as perennial disciplines, whose task it was to demonstrate the permanence of a world whose overall harmony derived from its divine origin. A search for the means by which the new forms of knowledge, both empirical and instrumental, could be integrated into the university curriculum became a central issue. And this at a time in which the established order in the country was threatened from within and without. From within, by the social contradictions which arose from the Industrial Revolution; from without, by the new political horizon of the French Revolution, some of whose aspects found an echo in Great Britain in Radical circles. In the initial debate, in *The Edinburgh Review*, the following questions were primary: which branch of learning, mathematics or logic, represents the true foundation of all knowledge, and what epistemological conditions must be satisfied if they are to incorporate newer forms of knowledge? The debate centred on the conditions under which the instrumentality of operating processes could be accepted as a permanent form of knowledge, whether mathematical or logical [13].

The movement for a renewal of logic that the philosopher William Hamilton (1788–1856) called "The New Analytic" was formulated in this context [21, p. 199]. The problem was to understand if the new truths produced in various branches of knowledge such as political economy, belonged to the permanence and to the universality of the world, or if they depended exclusively on the contingence of its transformations. Furthermore was this truth of theirs to be based on the data of this changing world, or exclusively on reasoning. The *Treatise of Logic*, written in Oxford and published in 1826 by Robert Whately (1787–1863), was conceived along these lines with the aim of assimilating Political Economy to traditional teaching, while adapting it to the epistemology induced by Scholastic logic [6]. Boole in 1847, in his

Mathematical Analysis of Logic, will make of this textbook his point of departure. In like fashion, the philosopher Dugald Stewart (1753–1828) will, in the footsteps of John Locke (1632–1704), makes use of mathematics to put forward the notion of demonstrative internal consistence as against that of an external truth, in any case out of reach [37, 1814, p. 141 & 329]. It is only in the case of mathematical objects that nominal and real existence are confounded; and Whately stresses the analogy between logical and mathematical operations, emphasizing in both cases the use of arbitrary symbols, "without any regard to the things signified" [43, p. 6]. Discussions between Whately and Hamilton on the nature of Logic evidence the acceptance of the former, and the refusal for the latter, of the instrumental aspect of Logic, whose automatic and mechanical character eliminates any possibility of a conscious reflection capable of controlling each step of a reasoning process, challenging too the necessity of so doing.

In Cambridge, as early as 1812, the young reformers, grouped around Babbage, refer to Bacon and further develop the conception of language Locke had expressed in his *Essay on Human Understanding* (1694). Algebraic writing was employed with the intent of arriving at a theory of invention which would break with the repetitive forms of traditional teaching, open to a spirit of investigation. These men all shared the conviction that the Cambridge mathematician Robert Woodhouse (1773–1827) had earlier expressed in terms of Locke's own language:

There can be neither paradoxes[6] *nor mysteries inherent and inexplicable in a system of characters of our own invention, and combined according to rules, the origin and extent of which we can precisely ascertain...Demonstration would be defined to be a method of showing the agreement of remote ideas by a train of intermediate ideas, each agreeing with that next it; or, in other words, a method of tracing the connection between certain principles and a conclusion, by a series of intermediate and identical propositions, each proposition being converted into its next, by changing the combination of signs that represent it, into another shewn to be equivalent to it.* [Woodhouse, 1801, 93,107]

But these young reformers went further. Their stated intention was to legitimate algebra as a science, in line with their political engagement to renew teaching in Cambridge. Such teaching had until then been based on geometry, regarded as the canon of all theoretical and permanent science, given the deductive exposition of Euclid's *Elements*, and the geometrical exposition of Newton's *Mathematical Principia of Natural Philosophy* (1687). Both these texts stood for an adherence to values which looked to the past for truth and order. The reformers first joined forces with Babbage, to have the continental Leibnizian notation of the infinitesimal calculus introduced into the university examination system, where it had been ostracised thanks

[6] These paradoxes essentially concerned the developments of functions by series, when they include calculus on impossible quantities.

to the old conflict between Newton and Leibniz as to who had first invented the infinitesimal calculus and which yet became so efficient to deepen Newton's *Principia*, particularly in Laplace's hands. In the meantime however developments, particularly at the hands of Laplace, has come to supplement Newton's own *Principia*.

Throughout the 1820s, the Cambridge reformers conducted a detailed discussion centred on inventive methods, such as the language of signs, analogy, induction and generalisation. And in the 1830s, George Peacock (1791–1858), a leading member of this group in the University of Cambridge and active in a number of new scientific societies, first introduced the idea of a radical separation between a Symbolical Algebra and its possible interpretations. He achieved thereby a very significant epistemological break-through, which consists in investigating operations, no longer in the light of their numerical results, but in terms of their properties as combination laws, seen as absolutely independent of the values of the symbols with which they worked.

7.4 Peacock's Symbolical Algebra and its Underlying Epistemology

To win acceptance for algebra as the foundation of knowledge in terms of teaching at Cambridge, it was imperative that it should be characterised as a science, that its instrumental character should be subordinated to the fundamentally deductive nature inherent in a speculative science. This prospect imposes several conditions:

- resolution of the paradoxes which in the course of the Eighteenth Century had arisen from the uncontrolled introduction of "impossible quantities" (our complex numbers) in the calculus; operations on these quantities had led to multi-valued results, not all of them systematically identified, and thus not systematically expressed,
- legitimising of algebraic results without recourse to analogy, it being common practice at this time to call upon corresponding geometrical properties or corresponding properties on whole or fractional numbers,
- refusal of an algebra based on operations whose definitions had been, until then, of an arithmetical nature.

Peacock was a very stubborn reformer, even if a moderate one and the radical reform of algebra he desired to effect was conceived in a modern perspective. He wished to integrate into his new form of algebra all the most recent results of analytics, even those arrived at by controversial processes. By so doing, he intended to furnish them with a rational foundation. To achieve these goals, Peacock abandoned any idea of founding algebra on the arithmetical definitions of operations. He replaced them with properties based on the existence of a unique operating structure, conceived as the "language of symbolical reasoning" [33, p. 1], as a "science of combinations of symbols absolutely arbitrary", i.e. "general in their form and in their value", as a science whose consistence is only assured by a "necessary connection" between the conclusions and assumed primary principles, considered as the only ultimate facts [34, pp. 194–200]. But these assumed principles do not sustain any axiomatic construction of this Symbolical Algebra, – a word that is not even pronounced. As in Locke's conception of hypothesis, these principles are first rules given *a posteriori* which serve to resume the previous constitution of abstract ideas and the comparison of their relations, obtained by experience.

It is because he aimed at such a conceptual compromise that Peacock's work gave rise to an epistemological break-through in algebra. On the one hand, he set out to harmonise all the achievements of algebraic analysis, on the other, he intended to neutralise criticism of the instrumentality of algebraic practices. His aim was to confer upon algebra the legitimacy which pertains to necessary truths, while refusing any reference to contingent truths. He chose to take into account all the results obtained by algebraic practices, even when it was difficult to authorise them as perfectly well founded on reason, in the case, for example, of operations on impossible quantities or on the differential operator, already written as d/dx and submitted to operations similar to arithmetical ones. He did though endeavour to legitimate them by processes which are exclusively operating processes, of a symbolical nature and conceived as logically pre-existent to any interpretation, though he does accept that its forms may be suggested by practice.

Moreover, in so far as he refuses to found his exposition on a constructive process as well as on an axiomatic system, the conceptual leap from inductive process to general knowledge can only be legitimated by an implicit ontological assumption. This assumption is encapsulated in his double central principle of the permanence of equivalent forms, which regulates the relations between Arithmetical Algebra and Symbolical Algebra. Peacock was who clearly distinguished between arithmetical equality, which must hold for all numerical values, and symbolical equivalence, which must hold only for the validity of operations [33, pp. 97–98]. This double principle authorises him to transfer into Symbolical Algebra all the general forms which have been arrived at in Arithmetical Algebra:

A): Whatever form is algebraically equivalent to another when expressed in general symbols, must continue to be equivalent, whatever those symbols denote.

B): Converse Proposition: Whatever equivalent form is discoverable in arithmetical algebra considered as the science of suggestion, when the symbols are general in their form, though specific in their value, will continue to be an equivalent form when the symbols are general in their nature as well as their form. [34, p. 194]

This double principle cannot be other than ambiguous if the logical existence of Symbolical Algebra is not pre-supposed. And such an implicit assumption contains the idea of a pre-established finality, since any new idea concerning operations will thus be perceived as the sign, in other words, as the mark which suggests a structure already existing

This conception of Symbolical Algebra forms part of a genetic epistemology founded on history, an epistemology which sets out to account for the heuristic process leading to a statement of principles, and the passage from inductive to deductive reasoning [11]. Like Locke in his *Essay on Human Understanding*, Peacock conceives his Symbolical Algebra as universal precisely because it is a speculative science and, because in such a science, what we call "truth" quite simply entailed the possible combinations of ideas and words, considered respectively as natural and as artificial signs. Demonstration has only to introduce, between two ideas whose relation is not immediate, a sequence of intermediate ideas named "proofs". The relationship thus established between them is such as to entail intuitive knowledge [Locke, 1694, IV.1.9]. This Lockean conception of truth, demonstration and reasoning is perfectly adequate when applied to algebraic writing and its transformations. As for Locke on the statute of language [Locke, 1694, IV.4.4; IV.6.14; Duchesneau, 973, 199–202], the principle of permanence for Peacock is founded on the implicit idea of the finality of the world conceived as an achieved creation, even if language is conceived of as an invention intended to discover it. This implicit finality is an essential part of Peacock's thought: it assumes the role of an implicit legitimisation by means of which Peacock can avoid a further epistemological break-through, one which would have led him to champion the freedom by which a mathematician is at liberty to build a formal language, from axioms and definitions.

With Peacock's work, the object of mathematical knowledge changes its nature. At the time where instrumentality was laying claim to positive values, it reinforced its links with the effecting of operations considered in themselves, without reference to the meaning of the objects involved, while maintaining an ontological link between this idea of operating and the faculties of mind whose effects they represent. Implicit recourse to the finality of this language of symbolical reasoning assumes, rather than a renunciation of any reference to meaning, the transference of this function from the treatment of objects to the treatment of operations. What is really meaningful now is the working of operations, conceived of as the process of combining ideas in the mind.

7.5 How Boole uses this Symbolical Conception of Mathematics to Algebrize Logic

This radical separation between a fundamental operating compass and one or several subordinate interpretative areas will characterize the work of this Symbolical [29] or Philological [22] School of Algebra. Thus, for the contemporaries of Babbage, his "analytical engine" represented the materialization of a symbolical calculus founded on the separation between symbols of operation and symbols corresponding to numerical or symbolical data and results. It seems to afford a perfect testimony of the radical independence of the operating functions from the values of the symbols. Babbage and L.F. Menabrea[7] identify it as a universal engine capable of calculating all values of any function, whereas they consider that the "difference engine", which Babbage conceived in the 1820s to eliminate human error in the drawing up of nautical and astronomical tables, was only a particular engine capable of effecting only a specific sequence of operations. Actually, the "difference engine" can obtain any recursive primitive function whose calculability by machine can be established [31, p. 75]. Lady Ada Lovelace will subsequently provide an accurate definition of the symbolical character of the computations carried out with the analytical engine [1, III, p. 144].

Hence, the formulation and application of the idea of a radical separation between operations, whose nature is solely symbolical, and their possible meaningful interpretations, conceived as playing no role at all in the logical legitimacy of their operational status. But they will soon lead to the renunciation of the unique character of the operating compass.

Boole follows closely in the steps of Peacock when he asserts that logic is a science, and when he first assumes the arbitrary character of symbols as a constitutive aspect of language in general. But he can define more specifically the relations between the automatic character of logical calculus and the interpretation of symbols because, since Peacock's time, D.F. Gregory (1813–44) had worked on the problem and stated the following theorem:

whatever is proved of the latter symbols, from the known laws of their combination, must be equally true of all other symbols which are subject to the same laws of combination. [20, p. 34]

Gregory defines Symbolical Algebra as "the science which deals with the combination of operations defined by the laws of combination to which they are submitted", so that this science is now concerned with the specific investigation of various classes of operations, each of them being characterised by one particular system of laws of combination [19]. And Boole refers explicitly to this theorem, and to this conception of Symbolical Algebra, both in his mathematical and in his logical works.

Boole is also attentive to the purely instrumental character of logic, and goes further, basing it directly on the operations of the mind. Rejecting any metaphysical prospect, he shows that he obtains the same analysis when his investigation sets out, on one hand, from the syntactic structure of natural language and, on the other, from the mental act of selection to which corresponds the denomination of things from a given Universe of discourse.

[7] Menabrea was an Italian military engineer from Piedmont. He subsequently became Prime minister of United Italy.

In fact, "the business of words is to direct these (precise) operations of the mind" [2, p. 42].

Proposition 1.
All the operations of Language, as an instrument of reasoning, may be conducted by a system of signs composed on the following elements, viz.:
1st. Literal symbols, as x, y, & c., representing things as subjects of our conceptions.
2nd. Signs of operations, as $+, -, \times$, standing for those operations of the mind by which the conceptions of things are combined or resolved so as to form new conceptions involving the same elements.
3rd. The sign of identity, $=$.
And these symbols of Logic are in their use subject to definite laws, partly agreeing with and partly differing from the laws of the corresponding symbols in the science of algebra. [2, p. 27]

Thus, if x alone stands for "white things" and y for "sheep", let xy stand for "white sheep". [2, pp. 28-9]

In other respects,

If x stands in for "men", and y for "women", $x + y$ stands in for "men and women".

But this statement assumes that:

These classes are perfectly distinguished, in such a way that no element of the one is element of the other. [2, pp. 32-33]

In this manner, Boole can point to the operational analogies between algebraic calculus and logical operations, but, like his contemporaries and predecessors of the Symbolical School, he carefully stipulates that, if we can perceive these properties as analogies, they manifest, most essentially, the existence of common symbolical operations. In any case, he can now write logical propositions as algebraic terms, and he can use Gregory's theorem to obtain by their means the same properties as for the algebraic calculus [2, p. 31], when the laws applied are the same.

Thus, Boole can write some of the logical properties of these laws of combination as algebraic properties, and he can use Gregory's theorem to transform them by means of algebraic equations:

$$xy = yx$$
$$x + y = y + x$$
$$x - y = -y + x$$
$$z(x + y) = zx + zy$$
$$z(x - y) = zx - zy$$

His algebra is, nevertheless, a special algebra, whose symbols represent nothing else than 0 and 1, because the 0 and 1 are the only solutions of the fundamental equation $x^2 = x$, obtained from the fact that the repetition of the operation of selection from one class gives the same class. As this fundamental equation can be written as $x(1-x)=0$, Boole can thus state that he is demonstrating what for Aristotle was a principle, but which is now a law, the law of the excluded middle.

The efficiency of these formal laws does not involve the nature of the mind, but only the phenomena which constitute the general operations by which it acts. And Boole shows that Aristotelian logic becomes a particular case of his own logic. Algebraic writing concerns first the propositions which express relations between things or facts, called by Aristotle categorical propositions, and by Boole primary propositions, whose calculus corresponds to our calculus on classes. As for our calculus of propositions, it corresponds to Aristotle's hypothetical propositions, which Boole named secondary propositions, such as: "If the sun has a total eclipse, then, the stars are visible" or "either the sun will shine or the walk will be postponed", and which he integrates to his calculus by considering the time during which each elementary proposition is true or false. In this way, Boole can assert that these two calculus formally maintain such a tight and remarkable analogy that they constitute one and the same calculus from an operating point of view, and that subsequent interpretation will suffice to distinguish between the two. The Aristotelian syllogism treatment is then reduced to the elimination of one unknown between two equations, and this method is of considerable interest since it can be generalised in any number of equations.

As Gregory had done in mathematics, Boole places much emphasis on the separation between this symbolical calculus and the interpretation of symbols, and on the fact that the calculus works on writings for which the interpretation of the symbols is not involved in the symbolical calculus.

Whatever our a priori *anticipations might be, it is an unquestionable fact that the validity of a conclusion arrived at by any symbolical process of reasoning, does not depend upon our ability to interpret the formal results which have presented themselves in the different stages of the investigation. There exist, in fact, certain general principles relating to the use of symbolical methods, which, as pertaining to the particular subject of Logic, I shall first state. ...*
§ 4 The conditions of valid reasoning, by the aid of symbols, are:
1st, That a fixed interpretation be assigned to the symbols employed in the expression of the data; and that the laws of the combination of those symbols be correctly determined from that interpretation.
2nd, That the forms processes of solution or demonstration be conducted throughout in obedience to all the laws determined as above, without regard to the question of the interpretability of the particular results obtained.
3rd, That the final result be interpretable in form, and that it be actually interpreted in accordance with that system of interpretation which has been employed in the expression of the data. [2, pp. 67–68]

Whatever may be the importance of this connection between mathematics and logic and of its reflection on the automatism of the calculus, it remains true that when Boole is searching for a logical instrument modelled on the algebraic writing, he assumes the validity of the logical laws of deduction. Moreover, he remains faithful to the Aristotelian analysis of the proposition, with its subject and predicate, despite the fact that its algebraic writing leads him accept an extensive interpretation of the copula.

7.6 Philosophical Consequences of this Ontological Conception of Operations

Peacock attempted to provide an ontological basis for the instrumentality of an algebra based on operations that he regarded as combinations which constituted the processes of the very faculties of the mind. The problems thus posed proved a fertile field for subsequent researchers and led to the production of new objects directly obtained by this consecutive extension of the operating prospects involved. All the research carried out by this specific philological or symbolical school is profoundly marked by this fundamental separation between syntactic and semantic properties of the algebraic calculus, and assumes a hierarchical system in which, as it is legitimated by the operations of the mind, the syntactic logical frame predetermines all possible interpretations. With an ontological approach which stresses the operations in their mechanical aspect, it is essential to bear in mind that, to ensure the profoundly human character of their instrumentality, they should be attributed to the constitution of the mind itself, which, since it is of divine origin, was essentially conceived for obtaining knowledge of a created world. The last chapter of Boole's *Investigation on the Laws of Thought* makes this point very clear. He is essentially involved in the debate concerning the instrumentality of mathematics and, from his point of view, mathematics are necessarily included in a larger system of knowledge of the world, which of course, is not limited to secular knowledge:

Perhaps the most obviously legitimate bearing of such speculations would be upon the question of the place of Mathematics in the system of human knowledge, and the nature and office of mathematical studies, as a means of intellectual discipline. No one who has attended to the course of recent discussions can think this question an unimportant one. ...
The laws of thought, in all those operations of which language is the expression or the instrument, are of the same kind as are the laws of the acknowledged processes in Mathematics. ...
But upon the very ground that human thought, traced to its ultimate elements, reveals itself in mathematical forms, we have a presumption that the mathematical sciences occupy, by the constitution of our nature, a fundamental place in human knowledge, and that no system of mental culture can be complete or fundamental, which altogether neglects them.

But the very same class of considerations shows with equal force the error of those who regard the study of Mathematics, and of their applications, as a sufficient basis either of knowledge or of discipline. ... As truly ... as the cultivation of the mathematical or deductive faculty is a part of intellectual discipline, so truly is it only a part. ... Much of this error, as actually existent among us, seems due to the special and isolated character of scientific teaching – which character it, in its turn, tends to foster. ... It is impossible, however, not to contemplate the particular evil in question as part of a larger system, and connect it with the too prevalent view of knowledge as a merely secular thing, and with the undue predominance ... of those motives ... which are founded upon a regard to its secular advantages. [2, pp. 422–24]

We are not to regard Truth as the mere creature of the human intellect. The great results of Science, and the primal truths of religion and of morals, have an existence quite independent of our faculties and of our recognition. We are no more the authors of the one class than we are of the other. It is given to us to discover Truth – we are permitted to comprehend it; but its sole origin is the real connecting link between Science and Religion. It has seemed to be necessary to state this principle clearly and fully, because the distinction of our knowledge into Divine and Human has prejudiced many minds with the belief that there is a mutual hostility between the two – a belief as injurious as it is irrational. [MacHale, 1985, 43][8]

The question here posed concerns the place of truth, and the responsibility of the mathematician in the production of new objects. For the network of these British algebraists, and clearly for Boole himself, truth is always to be referred to the created world. The instrumentality of operations, if it is revealed, or "suggested", by an experience which never asks to be considered, is quite simply the sign itself, a testimony of the existence of structures even more profound, whose manifestations act as signals to the mathematician who is attempting to discover them. Such a mathematician cannot be perceived as a creator. He can only exert his freedom in terms of his own experience. He can only invent some means that will lead him to knowledge more efficiently, means that are constituent parts of reason and language.

7.7 How Frege Claims the Existence of an Objective and Significant Logic

Two decades later, Frege's approach to meaning precisely and explicitly sets itself apart from that of Boole[9]. What he objects to in Boole's conception

[8] This address was given by Boole in 1847 at a meeting of the Lincoln Early Closing Association, held to celebrate the "abridgement of the hours of business in all trades, with a view to the physical, mental and moral improvement of those engaged there". The limit was ten hours a day.

[9] His recently published *Nachgelassene Schriften* contains a thorough comparison between the logical calculus of Boole and his *Begriffsschrift*. His paper, "Sur le

is the fact that it operates on symbols which are deprived of any meaning. On the contrary, his own aim is to associate in the most intimate manner logical forms and their contents. His criticism takes in all the authors, such as Schröder for instance, who developed the algebra of logic after Boole since, as he says, their symbolism is unable to express the contents of thought:

I did not want to represent an abstract logic in formulas, but to express a content through written signs in a more precise and surveyable fashion ... Boole's formula-language symbolic logic represents only the formal part of language, and that only incompletely [5, p. 65]

When, in 1879, Frege published his *Begriffsschrift, eine der arithmetischen nachgebildete Formelsprache des reinen Denkens*[10], known as the *Ideography*, the seven volumes of the *Mathematische Schriften* of Leibniz had already appeared. Indeed, the sub-title of the *Ideography* is a direct reference to Leibniz and Kant[11]. Referring to Leibniz, Frege takes care to make clear that he is not working on a *calculus ratiocinator*, but on a *lingua characteristica* [16, p. 71]. In opposition to Kant, he first refutes any resort to intuition, even a pure intuition, and sets out to show that logic is not sterile, in the sense in which Kant regarded as sterile the analytical propositions of formal logic. It must be underlined that Frege is convincing on both these specific points: he will define a whole number as something other than an intuitive object, and will prove that *Ideography* can build new contents.

Frege wished to free himself from the widespread but naive conception of a whole number universally accepted as natural by the philosophical and scientific community, and to elude the evidence of the "1+1 = 2" of the arithmetical textbooks. He wanted to break with the confusion between a rational development of arithmetic and the techniques of arithmetical calculus. On a more profound level, he was extremely suspicious of any resort to psychologism, underlining the shortcomings of mathematicians and philosophers, of John Stuart Mill or of Husserl[12] as much as those of Hankel and Weierstrass. They all recycle arguments directly derived from empiricism, and do so without explicitly identifying the process involved. His cautious approach to the question is linked to his specific logical definition of a cardinal number, in his *Grundlagen der Arithmetic* (1884). Frege thus shifts the boundary Kant had drawn between analytical and synthetic propositions and, by defining succession in logical terms, he calls into question the Kantian precept according to which time belongs to pure intuition.

but de l'Idéographie", which was published in a local review of Iena in 1882–83, compares the directive principles of the two systems [16, p. 11]

[10] Ideography, a formula language, modelled upon that of arithmetic, for pure thought.

[11] He summarizes the Leibnizian project of a logical calculus, and explicitly refers to Kant's attempt to describe the powers of pure thought systematically.

[12] Husserl's *Philosophie der Arithmetik* investigates anew some of the arguments which derive from British Empiricism, probably from Mill.

Thus his *Ideography* sets out to motivate directly an arithmetic – and subsequently a mathematics – based on logic. By so doing, it reverses the previously existing relation between mathematics and logic. Presented in a very original symbolism, which will not survive to its author because of its bi-dimensional aspect, its approach to logic is completely transformed. It replaces an analysis of the proposition in terms of grammatical elements such as subject–copula–predicate with another analysis founded on the notion of the function of several variables. This new analysis is based on the state of development of this notion of function in the second part of the Nineteenth-Century, when it was no longer perceived as merely a relation between numerical variables, as it was when Boole and Babbage wrote[13], but as a correspondence of whatever kind between two sequences of objects. For the first time, Frege distinguishes the unasserted propositional content and its assertion. More essentially, he gives an account of logic as a deductive system, founded on six axioms whose independence will be called into question only later, when explicit rules will be formulated to discipline those signs which represent judgements. He introduces negation and implication as primitive operators, expresses the other logical connectors in the light of them both, and goes on to introduce a universal quantifier seen as a primitive notion from which the existential quantifier itself derives.

More essentially for our point of view, Frege undertook a semantic analysis of sentences, and obtains it by applying to any unit of common language, and in the first instance to arithmetical equality, the distinction between meaning and reference. To this end, he uses the notion of function, which he transfers to what he names a "conceptual term", and the criterion of substitutability in a sentence, in order to distinguish between a sign, its meaning and its reference or denotation. A conceptual term, such as " ... is the man who conquered the Gauls", is an unsaturated expression which has no reference. In this way, the reference for a proper noun is nothing else than the object which it designates, as Frege regards as an object everything which in the expression does not contain any empty place. The concept is then considered as a function whose arguments are "true" and "false", and becomes the reference of a conceptual term, which is by no means the object falling under this concept. So, Frege distinguished between the concept and its extension, between the representative power of a concept and the subsumption of which it constitutes the foundation. As for the sentence, its meaning is a thought, and its reference is its truth value. This concern for a systematic reference constitutes for Frege the ultimate goal of the scientific approach.

It seems to be demanded by scientific rigour that we should have provisos against an expression's possibly coming to have no reference; we must see to it that we never perform calculations with empty signs in the belief that we

[13] Both Babbage and Boole placed this notion of function at the centre of their investigations, seeing it as it was seen when they worked, as a relation between numerical variables.

are dealing with objects. People have in the past carried out invalid procedures with divergent infinite series. ...
This involves the requirement as regards concepts, that, for any argument, they shall have a truth-value as their value; that it shall be determinate, for any object, whether it falls under the concept or not. ...
The requirement of the sheer delimitation of concepts thus carries along with it this requirement for functions in general that they must have a value for every argument [17, pp. 32–33].

However, Frege does not really resolve the question of meaning, because his criterion of substitutability is not founded on synonymy, but only on extensionality [8, pp. 32–59].

In all this work, Frege is first of all profoundly adverse to any psychological conception of thought, against which he writes with some bitterness, particularly in his 1894 account for the *Philosophie der Arithmetik* of E. Husserl (1859–1938). The distinction that Frege sets up between meaning and reference seems there to be specifically intended to work on objective thought, not only a collective, but a universal thought, which, it is his intention, should be separated from subjective thought, which can only be personal:

The reference and sense of a sign are to be distinguished from the associate idea. If the reference of a sign is an object perceivable by the senses, my idea of it is an internal image, arising from memories of sense impressions which I have had and acts, both internal and external, which I have performed. Such an idea is often saturated with feeling; the clarity of its separate parts varies and oscillates. The same sense is not always connected, even in the same man, with the same idea. The idea is subjective: one man's idea is not that of another. There result, as a matter of course, a variety of differences in the ideas associated with the same sense ...

[Several men] are not prevented from grasping the same sense; but they cannot have the same idea ... If two persons picture the same thing, each still has his own idea. ... An exact comparison is not possible, because we cannot have both ideas together in the same consciousness [17, pp. 59–60].

For Frege, thought is not produced by an individual subject. It constitutes an independent object that any particular subject can discover. It seems that all his work is undertaken in order to escape from the notion of mental image as representation, which characterized the psychological approach as a whole, here translated as "idea", and distinguished from "thought" and from "sense" of Frege's own vocabulary. Frege refuses to consider logic as a part of psychology. He attempts to make the existence of objects and concepts independent from the existence of human mind. For Frege, the notion of mental image leads to confusion:

Thus we have a blurring of the distinction between image and concept, between imagination and thought. Everything is transformed into something subjective. But just because the boundary between the subjective and the objective is obliterated, what is subjective acquires in its turn the appearance of objectivity. People speak, e.g., of such and such a mental image, as if it could be in public view, detached from the imagining mind. And yet a man never has somebody else's mental image, but only his own; and nobody even knows how far his image (say) of red agrees with somebody else's; for the peculiar character of the image I connect with the word "red" is something that I cannot convey. In order to be able to compare one man's mental images with another's, we should have to have united them into one and the same state of consciousness, and to be sure that they had not altered in the process of transference. It is quite otherwise for thoughts; one and the same thought can be grasped by many men. The constituents of the thought, and a fortiori things themselves, must be distinguished from the images that accompany in some images that each man forms of things [17, p. 317].

Whatsoever we may think of the distinctions between the syntactic character of Boole's approach and the semantic character of Frege's work, neither the one, nor the other, was prepared to renounce the ontological conception of thought as a universal instrument of reasoning.

7.8 Where is the Issue of Meaning Located in Contemporary Approaches?

What changes in the passage from the Nineteenth to the Twentieth Century is precisely the conception of the relations between syntax and semantics, as well as of their respective epistemological statute. This transformation is essentially at work with Gödel and the failure of the reductionist trend, in particular with the constitution of the model theory, and the acceptance on his part of the possibility of non-classical logics.

Although he was in close contact with the Vienna Circle, whose project was clearly inscribed in the logicist trend, Kurt Gödel (1906–78), in 1931, formulated his famous theorem of incompleteness, by which he eliminates the existence of any prospect of confounding exact language and universal language. He demonstrates that some propositions can be established as true without being demonstrable inside an axiomatic formal system. A meaning has to be conferred on such propositions in order to establish these results of incompleteness. With his work, truth and demonstrability become definitively dissociated concepts: the first belongs to the semantic part of logic, whereas the second belongs to the syntactic part.

At the same time, and independently of Gödel's approach, Alfred Tarski (1901–83) – a member of the great Polish Logical School who, in 1939 when it was wrecked by Nazism and the Second World War, moved to the United States – analyses the paradoxes of set theory from the self-referential character of language, stipulating that "a language which contains its own semantics must inevitably be inconsistent" [Tarski, 1974, 134]. Universalism only characterizes the ordinary everyday language, but formalized languages not at all: indeed, they are obliged to abandon it. Tarski introduces a clean break between syntax and semantics, classifying explicitely the truth of a proposition amongst its semantic properties. He carries forward an investigation of new semantic concepts such as the definisability of sets, which until then had been stigmatised by the logical paradoxes of set theory.

His approach is greatly influenced by the logical algebraic trend developed by C.S. Peirce (1839–1914), and by E. Schröder (1841–1902), both of whom followed in the footsteps of Boole, but Tarski was able to take advantage on the twentieth-century mathematical developments involving algebraic structures and to go beyond the mimetism of their algebraic writing. He is able to analyse in greater depth what he defines as the elementary equivalence between theories, generalizing thereby the concept of an isomorphism between mathematical structures. His constant concern is to distinguish, in all algebraic concepts and methods, those which do not involve notions from set theory. Thus, Tarski constantly contrives to effect a twofold investigation of problems, which leads either from the axioms, or from the interpretations or models, of deductive theory. He does not produce the concept of model, which had been used by several authors since the *Grundlagen der Geometrie* of Hilbert in 1899, but first gives a technical definition of it, when, in 1935 he speaks of the realization of a formal language and of satisfaction of formulas in this language [Tarski, 1974, II, 149–59]. In 1954 he will clearly define what he means by "model theory":

Within the last years a new branch of meta-mathematics has been developing. It is called the theory of models *and can be regarded as a part of the semantics of formalized theories. The problems studied in the theory of models concern mutual relations between sentences of formalized theories and mathematical systems in which these sentences hold. Every set Σ of sentences determines uniquely a class K of mathematical systems; in fact, the class of all those mathematical systems in which every sentence of Σ holds. Σ is sometimes referred to as a* postulate system *for K; mathematical systems which belong to K are called models of K.* [38, III, p. 30]

He goes on to carry out a syntactic analysis, similar to that of Russell and Alfred North Whitehead (1861–1947), and a semantic analysis, drawn from algebra, which leads him to identify what is a model and what is a structure, all the more readily indeed because the theory of most algebraic structures can be written in an elementary language, i.e., in a logical language of the first order.

But, in considering the theory of models as "a part of the semantics of formalized theories" and, as such, as an investigation between their properties and the mathematical structures in which these properties are satisfied, Tarski makes manifest a radical change from the preceding points of view, because he essentially eliminates the hierarchy between syntax and semantics, or between logic and mathematics which had in different ways always been reformulated, from Peacock to the Hilbertian meta-mathematics, whose mathematics were still dominated by the theory of demonstration. What is now considered is that the syntactic part, which defines a dumb language, is never built *in abstracto*, but in order to make explicit the logical structure of a particular theory. Hence, syntactic choices, as well as semantic ones, are possible. The ontological question, which was so essential for Boole, disappears and gives way to a new technical instrumental treatment, submitted to a requirement of applicability and efficiency, and assumed as such. This reversal of prospect involves several essential transformations, especially the withdrawal of the universalistic assumption of truth, its subordination to other concepts or theoretical processes, such as, in logic, the concepts of consequence of a theory, of satisfaction of a formula in a model, and perhaps above all, a constant concern with constructivity. For instance, in defining the completeness of an elementary theory as the fact that all its models verify the same first order statements, Tarski shows that any complete theory makes possible a principle of transfer and that this enables us to obtain new theorems automatically from one model to another model with the same elementary theory. Thus, recourse to a simple analogy can be replaced by a constructive and demonstrative principle which permits either the transfer of properties, or the building of new models. Such a trend will be developed in detail by Abraham Robinson (1918–74), the founder of the non-standard analysis.

Conclusion

From the Nineteenth to the Twentieth Century, the progressive abandonment of any idea of an ontological foundation profoundly modifies relations between the realm of symbolical laws, and the realm of its interpretations. New links are recognized, and assumed, between the inventive process and the demonstrative formulation, with a more systematic investigation of the relations between logical syntax and logical and mathematical semantics. In the field of relations between logical and mathematical semantics, the Twentieth Century no longer discusses, as did Peacock, either chronological precedence, or any "suggestion" process for discovering a precedence literally "revealed" by the identification with algebraic formulas: the only, but essential, question is now one of the constitutive interactions which lead to the building new concepts. Thus, the theory of models breaks with the requirement of legitimating the certainty of contents by formalism alone, by mathematical theories with their sole syntax, inventive activity by demonstrative activity. For the working mathematician and logician both make sense.

Since truth became a semantic property, the question of certainty can no longer be monopolized, or even taken over, by the only syntactic part of logic, and the question of meaning relegated towards its interpretative part. I see in this a major contribution of the evolving relations between logic and mathematics during the Twentieth Century, one which perhaps is not sufficiently appreciated. And I have chosen to develop this point in particular, because it seems to me that such a characteristic aspect is likely to find the most important echo for the practitioners or theoreticians of musical logics: when a large part of their work is concerned with new automatic processes, the issue of meaning cannot be excluded from any theoretical consideration of the creative act.

References

1. Babbage, Ch.: *The Works of Ch. Babbage.* 11 vols. Ed. by Campbell-Kelly, M. London: Pickering 1989
2. Boole, G. (1854): *An Investigation of the Laws of Thought, on which are founded the mathematical Theories of Logic and Probabilities.* London: Walton & Maberly 1992. *Les lois de la pensée*, trad. fr. S.B. Diagne. Paris: Vrin
3. Cantor, G. (1932). *Gesammelte Abhandlungen mathematischen und philosophischen Inhalts.* Berlin: Springer. Ed. by E. Zermelo.. Réimp. Hildeshein: Olms 1962
4. Cavaillès, J.: *Philosophie mathématique* Paris: Hermann 1962
5. Coffa, J.A.: *The Semantic Tradition from Kant to Carnap.* Cambridge: C.U.P. 1995
6. Corsi, P.: The heritage of Dugald Stewart: Oxford philosophy and the method of political economy. Nuncius. **iii**, 89–144 (1987)
7. L. Courant: Opuscules et fragments inédits de Lb. Rééd. Publ. ENS. 1983, p. 97 et 143
8. De Rouilhan, P.: *Les paradoxes de la représentation.* Paris: Minuit 1988
9. Descartes, R. (1637): *Discours de la Méthode. Œuvres et Lettres.* Gallimard 1953
10. Duchesneau, F.: *L'empirisme de Locke.* La Haye: M. Nijhoff 1973
11. Durand(-Richard), M.J.: Genèse de l'Algèbre Symbolique en Angleterre: une Influence Possible de John Locke. Revue d'Histoire des Sciences, **43**(2–3), 129–80 (1990)
12. Durand-Richard, M.J., Charles Babbage (1791–1871): De l'Ecole algébrique anglaise à la 'machine analytique'. Mathématiques, Informatique et Sciences humaines **118**, 5–31; Erratum, **120**, 79–82 (1992)
13. Durand-Richard, M.J.: L'Ecole Algébrique Anglaise: les conditions conceptuelles et institutionnelles d'un calcul symbolique comme fondement de la connaissance. In: Goldstein, C., Gray, J., Ritter, J. (eds.): *L'Europe Mathematique – Mythes, histoires, identité. Mathematical Europe – Myth, History, Identity.*, pp. 445–77. Paris: Eds. M.S.H 1996
14. Frege, G. (1879): Begriffsschrift, a formula language, modeled upon that of arithmetic, for pure thought. In: Heijenoort, J. van: *From Frege to Gödel: A Source Book in Mathematical Logic.* Harvard Univ. Press 1967
15. Frege, G.: *Nachgelassene Schriften.* Hamburg: Meiner Verlag 1969. Ed. Hermes, H., Kambartel, F., Kaulbach, F.: "Booles rechnende Logic und Begriffsschrift (1880–81)".
16. Frege, G.: *Ecrits logiques et philosophiques*, trad. Cl. Imbert. Paris: Points Seuil 1994

17. Geach, P., Black, M. (1952): *Translations from the Philosophical Writings of G. Frege*, Oxford, B. Blackwell: "Function and Concept", 21–41; "On Sense and Reference"; "Illustrative extracts from Frege's review of Husserl's 'Philosophie der Arithmetik'" (C.E.M. Pfeffer, Leipzig, 1891), Zeitschrift für Philosophie und phil. Kritik **13**, 313–332 (1894)
18. Gödel, K. (1931): On formally undecidable propositions of *Principia Mathematica* and related Systems I. In: Heijenoort, J. van: *From Frege to Gödel: A Source Book in Mathematical Logic.* Harvard Univ. Press 1967
19. Gregory, D.F.: On the Real Nature of Symbolical Algebra. Transactions of the Royal Society of Edinburgh **14**, 208–16 (1840)
20. Gregory, D.F. (1839): On the solution of linear differential equations with constant coefficients. Cambridge Mathematical Journal **1**, 2nd ed., 25–36 (1846)
21. Hamilton, W. 1833 (1852): (Anonyme), Recent Publications on Logical Science. Edinburgh Review **58**(115), 194–238 (avr. 1833). Republié dans *Discussions on Philosophy and Literature, Education and University Reform, Chiefly from the Edinburgh Review*, pp. 116–174. London 1852
22. Hamilton, W.R. (1834): On conjugate functions, or algebraic couples, as tending to illustrate generally the doctrine of imaginary quantities, and as confirming the results of Mr. Graves respecting the existence of two independent integers in the complete expression of an imaginary logarithm. Transactions of the Royal Irish Academy **17**, 203–422 (1834). Halberstam, H., Ingramn, R.E. (eds.): *The Mathematical Papers of Sir William Rowan Hamilton*, vol. 3: *Algebra*, pp. 3–96. Cambridge: Cambr. Univ. Press 1967
23. Heijenoort, J. van: *From Frege to Gödel: A Source Book in Mathematical Logic.* Harvard Univ. Press 1967
24. Hilbert, D. (1926): Über das Unendliche. Mathematische Annalen **95**, 161–90 (1926). Trad. frçse (in): Largeault, J.: *Logique Mathématique, Textes*, pp. 215–245. Paris: A. Colin 1972
25. Hobbes, Th., 1904 (1651): *Leviathan, or the Matter, Forme and Power of a Commonwealth, ecclesiasticall and civill.* Ed. by Waller, A.R., Cambridge, CUP
26. Kline, M.: *Mathematical Thought from Ancient to Modern Times.* New York 1972
27. Locke, J.: *Essay on Human Understanding.* 2nd ed. London 1690
28. Lovelace, A.A. (1843): Sketch of the Analytical Engine invented by Charles Babbage. In: Babbage, *Works*, 3, 89–170
29. Mac Farlane, A.: George Peacock (1791–1858). *Lectures on ten British Mathematicians.* New York 1916
30. Menabrea, L.F. (1842): Notions sur la machine analytique de M. Charles Babbage. *Bibliothèque Universelle de Genève*, t. xli, n° 82, octobre 1842, (in) Babbage, Ch. (1989) *The Works of Ch. Babbage*, Ed. Campbell-Kelly, London, Pickering, vol. 3
31. Mosconi, J.: Charles Babbage: vers une théorie du calcul mécanique. Revue d'Histoire des Sciences **XXXV**(1), 69–107 (1983)
32. Newton, I.: *The Mathematical Works of I. Newton*, assembled by Whiteside, D.T. Vol. 2. New York, London: Johnson reprint corporation 1964. B.N.: 4 R 11517(3)
33. Peacock, G. (1830): *Treatise of Algebra.* 2 vols. Cambridge: Reed 1842–45

34. Peacock, G.: A Report on the recent progress and actual state of certain branches of analysis. *Proceedings of the British Association for the Advancement of Science*, pp. 185–351. London 1833
35. Peacock, G.: *Report on the recent progress and actual state of certain branches of analysis.* Cambridge 1834
36. Sinaceur, H.: *Corps et Modèles.* Paris: Vrin 1994
37. Stewart, D.: *Elements of the philosophy of the human mind*, Ld, 3 vols. 1792–1814–1827
38. Tarski, A.: *Collected Papers.* Ed. by Givant, S.R., McKenzie, R.N.. 4 vols. Birkhauser 1986
39. Tarski, A. (1935): art. XVI: "Sur le concept de conséquence logique". *Logique, Sémantique, Métamathématique*, t. 2, pp. 141–52. Paris: Armand Colin 1974
40. Tarski, A. (1935–36). art. XV: "La construction d'une sémantique scientifique". *Logique, Sémantique, Métamathématique*, t. 2, pp. 131–39. Paris: Armand Colin 1974
41. Tarski, A. (1954): Contributions to the theory of models. I. *Indagationes Mathematicæ*, vol. 16, (*Koninklijkle Nederlandse Akademie van Wetenschappen, Proc.*, Series A, Mathematical Sciences, vol. 57), pp. 572–81. *Collected Papers*, vol. 3, 517–25, p. 517.
42. Tarski, A.: *Logique, sémantique, métamathématique*, 1923–44. Trad. ss dir. Granger, G.G. Paris: Armand Colin 1972–1974
43. Whately, R.: *Elements of Logic.* London 1826. Also published in *Encyclopædia Metropolitana.* London 1849

8 Musical Analysis Using Mathematical Proceedings in the XXth Century

Laurent Fichet

In the twentieth-century, musical theories using mathematical proceedings are different from those of the Renaissance or the Baroque period. Musicians or philosophers such as Mersenne, Descartes, Rameau, D'Alembert, Euler were essentially concerned with the problem of finding out the numerical relationship hidden in music or in sound. Their basic assumption was the direct link between the use of ratio 3/2 or a limited number of simple intervals in the ground bass and a possible rational explanation of music.

Other coincidences between numbers and music were soon explored, and, as in the nineteenth-century, this kind of research began to seem rather useless. As Helmholtz pointed out, "*in the middle of last century, showing that a thing was natural seemed enough to prove its beauty and its necessity*" [1]. Although some aspects of musical practice seemed to conform to mathematical or physical principles, which could justify a certain way of composing, this would not at all allow a precise approach to composition.

A further step consisted of using mathematics as a tool for a better analysis of music. This was an essentially artistic task, if compared, for example, with Leibnitz' aim which was to find out the unknown motives of soul (*Musica est exercitium arithmeticae occultum nescientis se numerare animi*).

Not everybody in the twentieth-century adopted this change of nature in the use of mathematics. Many composers were attempty to find other coincidences between music and numbers in order to apply them in a compositional context. We might mention Ligeti's Second Quartet for the use of number nineteen as well in the pitch domain as in the organisation of bars. The best known case in undoubtedly that of Xenakis, who applied for some years statistical formulas such as Poisson law or Markov processes in his compositions. In all these cases, the esoterical flavour takes us some centuries back. We will thus confine ourselves to methods proposed for analysis, where mathematics appears to be a tool rather than an objective. We briefly present some historically relevant musical propositions by discussing, in particular, Allen Forte's "*Set Theory*" through an extract from the Second Sonata by P. Boulez.

HINDEMITH

In spite of their lack of rigour, Hindemith's theories [2] represent an exceptional opportunity to obtain interesting information on music through a mathematical approach.

Hindemith's reasoning is based on acoustic study as he starts from differential sounds. He then proposes to classify chords according to their degree of consonance, by means of some mathematically expressible processes. Here are some chords he classified as depending on a progressively increasing then decreasing dissonance.

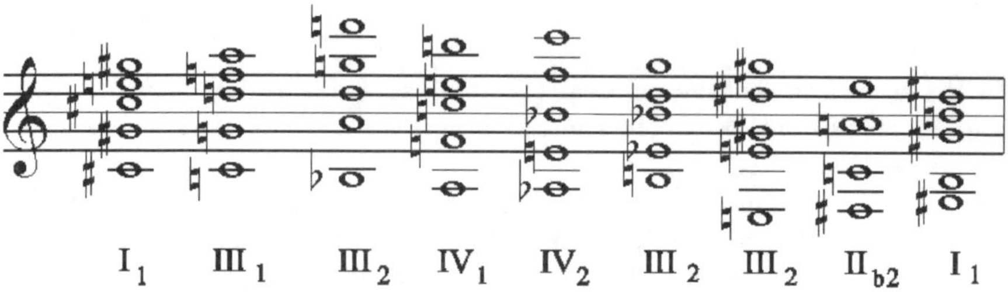

Fig. 8.1.

His classification is questionable and there is no proof that chord progressions are more interesting when they follow his principles, but Hindemith's theories are an interesting approach.

SCHILLINGER and ANSERMET

In the case of Schillinger or Ansermet, the principles of the theories themselves are erroneous. Schillinger's theories [3] seem so incongruous nowadays, that there is little need to go into details.

In particular, he asserted that he could improve old compositions by applying to them a very shallow geometrical analysis. This is how he improves J.S. Bach's Invention n° 8:

Fig. 8.2.

Ansermet uses mathematical notions that are more elaborate than simple symmetry [4], but he seems to have problems in mastering them.

For example, he may set down amazing equations such as:
Log 1 = Log 2 (that is Log 2 = 0).

He gives the value 1 to the fifth *Do-Sol*, and the value 2 to the second *Do-Re* (*Do-Re* = *Do-Sol*+ *Sol-Re*), and −1 to the diminished fifth *Do-Fa#*. But according to these principles, the same result would unfortunately hold for any chord including a fifth between its extreme notes.

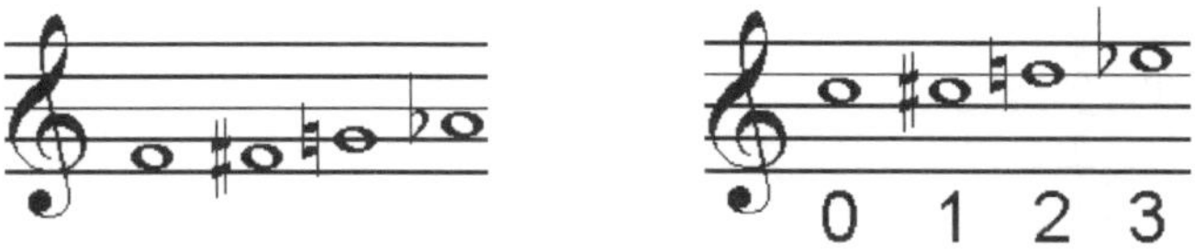

Fig. 8.3.

As a matter of fact, Ansermet and Schillinger belong to those musicians who used mathematics to give the impression that their theories were scientific. They could not lead to interesting results since this was not their objective. Ansermet clearly confessed [5] that through his researches he intended to deal a fatal blow to dodecaphonic, as well as to electroacoustic, music.

THE SET THEORY BY FORTE

The numerous applications of Information theory, inspired by C.E. Shannon's researches [6], should also be mentioned, but we prefer to concentrate on A. Forte's Set Theory. This theory seems to be overestimated in the Anglo-Saxon countries that consider it, by quoting Ian Bent, one of the most important contribution to development of musical analysis [7]. On the contrary, we mean to show that Forte's theory can offer very few interesting elements compared with a more intuitive analysis which uses mathematics in a less rigid way.

Forte's Pitch Class Set Theory [8] is conceived to analyse atonal compositions from a mathematical viewpoint. Once the score has been segmented into sets from 3 to 9 notes, his theory allows first to classify those sets in a rather limited number of units, then to find relationships between those Pitch Class Sets.

Such analysis is supposed to clarify the structure of pieces that are not easily susceptible to traditional analysis. We will see what a mathematical analysis of this kind can offer in the case of the second movement of the Second Sonata by P. Boulez. Then, we will check if the logic of that theory allows adequate understanding of the basic points of the musical logic in that piece.

One of the weakest part of this theory has probably to do with the arbitrariness of the segmentation process, where "*the term primary segment will be used to designate a configuration that is isolated as a unit by conventional means, such as rhythmically distinct melodic figure*" [9]. Fortunately, in our movement, the choice is fairly obvious, the first bars giving well-drawn units of four notes. The dodecaphonic aspect of these first bars makes the problem of cutting even easier.

Bars 1 to 3:

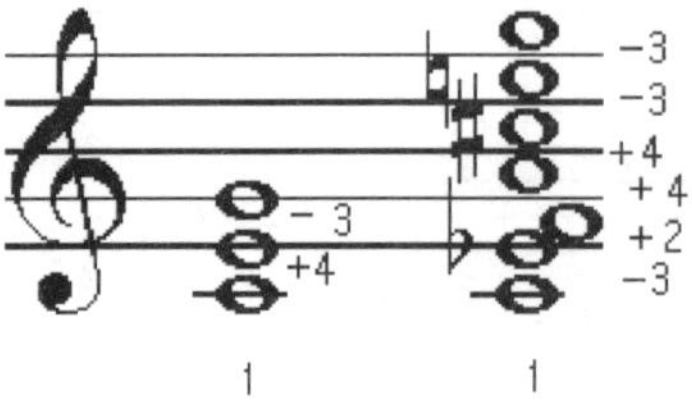

Fig. 8.4.

How are we to compare these different units? Forte suggests to reduce every unit to its simplest expression by placing the notes on a possibly tightly restricted ambitus and then by taking Do as the starting note. With the first unit:

Fig. 8.5.

Integer numbers from 0 to 11 are used in order to replace the notes of the chromatic scale. The Set $0, 1, 2, 3$ is so called "4–1" as it comes first in the classification of units of four notes by Forte. In this case, we notice that the four notes of the third bar can also be reduced to the four notes $0, 1, 2, 3$ so that they belong to the same Pitch Class Set 4–1. The four notes of the second bar $(0, 1, 2, 7)$ would be indicated with "4–6". This is what happens for the first nine bars:

For the first six bars, this method seems to give satisfactory results, for it enables us to reduce the score to combinations of only two different units of notes: 4–1 and 4–6. But, as soon as some notes no longer correspond to an obvious unit, shall we count a new Set of five notes? The seventh bar embodies ambiguities as well. Subsequently, the system no longer works so well: the question of the formation of units becomes in fact rather tricky.

Fig. 8.6.

It is no use going further now to notice that, even when the score goes along with it, Forte's theory involves overlappings, but can not lead us to draw valid inferences concerning the structure of this piece.

We have added a "root" to every set by choosing each time the lowest of the four notes reduced to the most restricted ambitus. It is obvious that, from one identical unit (according to Forte) to the other, something like a cycle (sixths) appears.

Bars: 1 2 3 4 5 6 7
Set: 4–1 *4–6* 4–1 *4–6* 4–1 4–1 *4–6*
Root: Fa ***Re*** Si b ***Sol*** Re # Sol# ***Do***

The restricted units undoubtedly led us to notice the similarity between some groups of four notes, but the process by translating everything to 0 finally makes us neglect an important indication: all the Sets 4–1 have probably a similar structure, but, according to P. Boulez, the exact pitch of the composing notes plays its part in the composition. The question is not to state that there would be a "tonal" approach in this kind of sequences of sixths. We can merely remark that such an interval is really pertinent in this context, since is is to be found clearly expressed in the first notes of the two others movements.

So, we can regret that the logic of analysis leads to neglect the composer's logic. Nevertheless, we used the principle of formation of units, but we must point out the fact that these units can be easily defined without the help of this theory. At times, the score will make the analyst form such units instinctively, but not at all to use them as Forte did.

We now concentrate on bars 49 and 50 in the Second movement:

Fig. 8.7.

According to Forte's theory, units C and D have no relationship at all. Even Forte's "vectors" which give the interval content (212100 for C, and 122010 for D) show no similarity. But, by considering no more than the three intervals which appear successively, we get a minor second, a minor third, and a minor second for C, and a reverse order for D: minor third, minor second and minor third. The unit D appears to be a plain derivation of C, which is not unrelated to Set 4–1, as it appeared in bar 9.

In addition, if we do not lock ourselves into a rigid code of analysis like Forte's theory, from the very first bars, we will detect a symmetry (concerning pitches as well as rhythm ...) between bar 1 and bar 3, with bar 2 as a centre. We will then see that such symmetry is to be found in the entire movement (bar 103 corresponds to bar 1, 102 to 2, 101 to 3 ...), with a centre round about bar 53, and that small units of 4 notes like C other D are finally themselves small palindromes.

It is undoubtedly unkind to reproach an analysis for not detecting components which are irrelevant to its concerns. On the other side, ignoring the essential relationship between well defined units is all the more unfortunate, since that theory is complex, hard to understand for musicians, and should be able to prove its efficiency precisely in the case of a work like Boulez' Second Sonata.

Forte's method corresponds rather well to what Yizhak Sadaï calls "hermetic analysis": "*it is reliant on a method, a protocol that fixes a series of predetermined procedure. Therefore, that kind of analysis is still ineffective in regard to an important number of phenomenous which stand outside its operational frame*" [10].

We may also rightly suspect that most analysis inspired by mathematical processes tends to miss essentials. The more rigorous they are, the more their field of action will be limited. In the case of Forte's Set Theory, they will prove incapable of detecting important components, even when they are closely allied to that field of action.

(Translated by Miss Pandellé)

References

1. Helmholtz, H.: *Théorie physiologique de la musique*, p. 300. Paris: Masson 1868
2. Hindemith, P.: *Unterweisung im Tonsatz.* 260 p. Mainz: Schott 1940
3. Schillinger, J.: *The Schillinger system of musical composition.* New York: Carl Fischer 1941
4. Ansermet, E.: *Les fondements de la musique dans la conscience humaine.* 1119 p. Neuchâtel: La Baconnière 1961. New Ed.: Paris: Laffont 1989
5. Ansermet, E.: *Ecrits sur la musique*, p. 92 et 107. Neuchâtel: La Baconnière 1971
6. Fichet, L.: *Les théories scientifiques de la musique aux XIXe et XXe siècles.* Paris: Vrin 1996
7. Bent, I.: *L'analyse musicale.* 1998 in France. Ed. MacMillan 1987
8. Forte, A.: *The structure of atonal music.* 224 p. New Haven: Yale University Press 1977
9. Ibidem, p. 83
10. Sadaï, Y.: De l'analyse pour l'analyse et du sens de l'intuition. *Musurgia* **65** (Nov. 1995)

9 Universal Prediction Applied to Stylistic Music Generation

Shlomo Dubnov and Gérard Assayag

Abstract. Capturing a style of a particular piece or a composer is not an easy task. Several attempts to use machine learning methods to create models of style have appeared in the literature. These models do not provide an intentional description of some musical theory but rather use statistical techniques to capture regularities that are typical of certain music experience. A standard procedure in this approach is to assume a particular model for the data sequence (such as Markov model). A major difficulty is that a choice of an appropriate model is not evident for music. In this paper, we present a universal prediction algorithm that can be applied to an arbitrary sequence regardless of its model. Operations such as improvisation or assistance to composition can be realised on the resulting representation.

9.1 Introduction

Machine learning is the process of deriving a set of rules from data examples. Being able to construct a music theory from examples is a great challenge, both intellectually, and as a means for a whole range of new exciting applications. Such models can be used for analysis and prediction, and, to a certain extent, they can generate acceptable original works that imitate the style of their masters, recreating a certain aspect of music experience that was present in the original data set. The process of composition is a highly structured mental process. Although it is very complex and hard to formalise, it is not completely random. The task of this research is to try to capture some of the regularity apparent in the composition process by applying information theoretic tools to this problem.

Mind-Reading Machines

In early 50's at Bell Labs David Hagelbarger has built a simple 8 state machine, whose purpose was to play the "penny matching" game. The simple machine tried to match the future choices of a human player over a long sequence of random "head" or "tail" choices. Mind-reading was done by looking at similar patterns in opponent's past sequence that would help predict the next guess. The achieved rate of success was greater that 50%, since human choices could not be completely random and analysing patterns of previous choices could help foretell the future.

Inspired by Hagelbarger's success, Shannon has built a different machine with improved performance. An account of Shannon's philosophy on mind-reading machines can be found in [1].

It is important to note that if the model of the data sequence was known ahead of time, an optimum prediction could be achieved. The difficulty with most real situations is that the probability model for the data is unknown. Therefore one must use a predictor that woks well no matter what data model is. This idea is called "universal prediction".

Music Generation and Style Replication

Generative theory of music can be constructed by explicitly coding music rules in some logic or formal grammar [15–17]. This approach is sometimes called "expert system" or "knowledge engineering approach". A contrasting approach is the statistical learning or empirical induction approach. Several researchers have used probabilistic methods, notably Markov models, to model music [12–14]. Pinkerton used a small corpus of diatonic major- key nursery rhyme to learn a Markov model, which he later used to generate nursery rhymes. Because he used a small alphabet (seven symbols of the diatonic scale and a tied note symbol), he was able to use a high- order (long context) Markov model up to order eight. Conklin and Witten 1995 ([12]) used trigrams[1] to generate chorale melodies from parameters on a corpus of Bach chorale melodies. A more recent Markov model experiment was done by [14]. Like Conklin and Witten, they worked with chorale melodies, and like Pinkerton, they experimented with orders up to eight. Their corpus was of 37 hymn tunes (giving perhaps 5000 note transitions). To capture similarities between pieces in different keys (but the same mode), all pieces were into C. The experiment showed that at very low orders (e.g., unigram), generated strings do not recognisably resemble strings in the corpus, while at very high orders, strings from the corpus are just replicated.

An interesting "compromise" between the two approaches is found in more recent works of [11]. Cope uses grammatical generation system combined with what he calls "signatures", melodic micro-gestures common to individual composers. By identifying and reusing such signatures, Cope is able to reproduce the style of past composers in reportedly impressive ways.

Predictive Theories in Music

Following the work of Meyer [2] it is commonly admitted that musical perception is guided by expectations based on the recent past context. Predictive theories are often related to specific stochastic models which estimate the probability for musical elements to appear in a given musical context, such

[1] An n-gram is a sequence of symbols of length n. The first $n-1$ of these are the context.

as Markov chains mentioned above. If one is dealing with a data sequence whose probabilistic model is known, then one can optimally predict the next samples in the sequence. If one does not know the model, there are two possible solutions. One is to estimate the model first and use it for prediction. The second approach is to use a predictor that works well for every model or at least works as good as any other predictor from a limited class of prediction methods.

In music applications the model is unknown. Considering the context or the past samples for prediction, one of the main problems is that the length of musical context (size of memory) is highly variable, ranging from short figurations to longer motifs. Taking a large fixed context makes the parameters difficult to estimate and the computational cost grows exponentially with the size of the context. In order to cope with this problem one must design a predictor that can deal with arbitrary observation sequence and is competitive to a rather large class of predictors, such as Finite State Machine Predictors and Markov Predictors. Philosophically, we take an agnostic approach: do the best we can relative to a restricted class of strategies.

Finite State Prediction

In order to describe the theory of prediction for a completely arbitrary data model we need to define the concept of finite-state predictor. Let us define a set S and two functions $f : S \times A \to A$ and $g : S \times A \to S$, such that the predictions $\hat{x}_i$ for a sequence $x_1, x_2, \dots, x_n$ are generated by the following mechanism:

$$\hat{x}_i = f(s_i)$$
$$s_i = g(s_{i-1}, x_{i-1})$$

The initial state s_0 is given as well. In the finite state (FS) predictor the predicted value depends only on the current state s_i according to the prediction function f. For each new observation the machine moves to a new state according to the transition rule g. The error between a sequence of predictions and the actual data is defined by

$$d_n(x_1^n, \hat{x}_1^n) = n^{-1} \sum_{i=1}^{n} \delta(x_i, \hat{x}_i)$$

where $\delta(x, \hat{x})$ is the error count, i.e a Hamming distance function that equals 0 if $x = \hat{x}$ and 1 otherwise. The minimal fraction of errors for an S-state predictor is called "S-state-state predictability" and is denoted by $\pi_S(x_1^n)$. If we want to consider the performance of FS predictor for increasing S, the length of the sequence must be increased. Growing n first and S second, FS predictability is defined as

$$\pi(x) = \lim_{s \to \infty} \limsup_{n \to \infty} \pi_s(x_1^n) \ .$$

FS predictors are examined in detail in [10]. They consider the problem of constructing a universal predictor that performs as well as any finite state predictor. By definition, $\pi(x)$ depends on the particular sequence x. The surprising result is that a sequential predictor can be found that does not depend on x and yet achieves asymptotically FS predictability $\pi(x)$. Similarly, when the class of FS predictors is further confined to Markov predictors[2] then the corresponding prediction performance measure is called Markov predictability. It is further shown by Feder et al. (1992) [10] that the finite-state predictability and the Markov predictability are always equivalent, which means that it is sufficient to confine attention to markov predictors in order to achieve the finite-state predictability. For a treatment of nonparametric universal prediction theory the reader is invited to consult also additional references [8,9].

In our work we present a dictionary-based prediction method, which parses an existing musical text into a lexicon of phrases/patterns, called *motifs*, and provides an inference method for choosing the next musical object following a current past *context*. The parsing scheme must satisfy two conflicting constraints. On the one hand, one wants to maximally increase the dictionary to achieve better prediction, but on the other hand, enough evidence must be gathered before introducing a new phrase, so that a reliable estimate of the conditional probability is obtained. The secret of dictionary-based prediction (and compression) methods is that they cleverly sample the data so that most of the information is reliably represented by few selected phrases. This could be contrasted to Markov models that build large probability tables for the next symbol at every context entry. Although it might seem that the two methods operate in a different manner, it is helpful to understand that basically they employ similar statistical principles.

Predictability and Compression

The preceding discussion might seem needlessly complicated to someone current in compression and coding methods. It is widely known that prediction serves as the basis for modern data compression and it seems just natural that an opposite analogy would exist, i.e. a good compression method would be also useful for a good predictor. A standard measure for compression quality is coding redundance or how close the entropy of the coded sequence approaches the entropy of the data source. Intuitive link between predictability and entropy is easy to establish. Entropy (also sometimes called “uncertainty”) measures the minimal number of bits needed to describe a random event. For a completely random, i.i.d binary sequence, one must transmit all bits in order to describe the sequence. If the probability for ones is greater then for zeros (or vice versa), one can devise a scheme where long sequences of ones

[2] Markov predictor of order k is FS predictor with 2^k states where $s_i = (x_{i-k}, \dots, x_{i-1})$.

are assigned to short codewords, thus saving on the total number of bit, i.e. achieving on the average less then one bit per symbol. The entropy function $H(p)$ for a sequence with probability p to see "1" is given by

$$H(p) = -\{p \log p + (1-p)\log(1-p)\} \ .$$

Predictability on the other hand measures the minimum fraction of errors that can be made by some prediction machine over long data sequences. For instance, optimal single state predictor employs counts $N_n(0)$ and $N_n(1)$ of zeros and ones occurring along the sequence x_1^n. It predicts "0" if $N_n(0) > N_n(1)$ and "1" otherwise. The predictability of this scheme is

$$\pi\left(x_1^n\right) = \frac{1}{n}\min\left\{N_n(0), N_n(1)\right\} \ ,$$

where $N_n(x)$, $x \in \{0,1\}$ is the joint count of ones and zeros occurring along the sequence x_1^n. Comparing the behaviour of prediction to entropy is best demonstrated in the following graph:

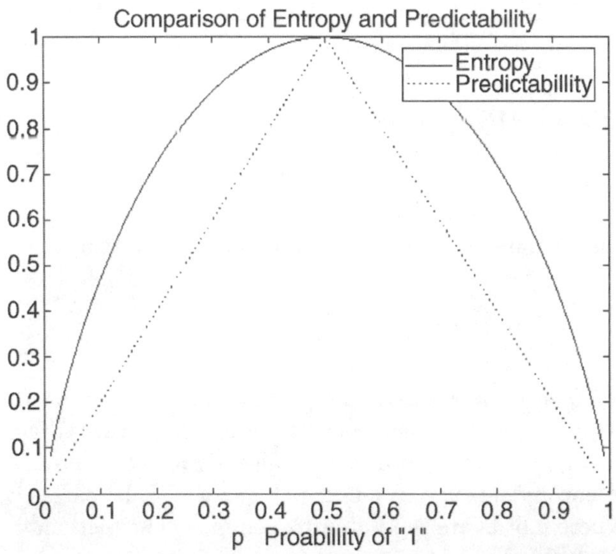

Fig. 9.1. H and π drawn as a function of the probability p

The predictability is related to the error probability in guessing the outcome of a variable, while the compressibility is related to its entropy. It can be further shown that a lower limit to predictability exist in terms of the entropy. For the binary case discussed above, it can be shown that $\rho/2 \geq \pi \geq h^{-1}(\rho)$, where ρ is the compressibility, π is the predictability and $h(.)$ is the binary entropy function. While the two quantities are not functionally dependent, it is evident that they do coincide on the extreme points.

Predictability and Complexity

We will terminate this long introductory section by a brief discussion of relations between predictability and some other complexity measures. As we stressed in the beginning, one of the great advantages of the universal method is its applicability to arbitrary sequences, including deterministic sequences. The complexity of sequences that are not governed by a probabilistic model (sometimes called "individual" sequences) can be considered in terms of the Solomonoff–Kolmogorov–Chaitin complexity. This measure defines complexity of a sequence as the length of a shortest program for a universal Turing machine that outputs the sequence. In the same spirit we have a complexity definition by Lempel Ziv who considered the shortest code needed to reproduce an individual sequence by an FS encoder. Their well-known Lempel–Ziv algorithm ([6]) has been shown to achieve finite-state compressibility for every sequence. The details of the LZ incremental parsing algorithm, that will serve as the basis for our prediction method, will be discussed below. Feder et al. (1992) [10] prove that in a similar manner to the compression property of the incremental parsing method, a predictor which uses the conditional probabilities induced by the LZ scheme attains Markovian predictability and this FS predictability for any individual sequence.

9.2 Dictionary-Based Prediction

As we have explained above, we use dictionary based methods for assessing the probability of the next sample given its context. In the following sections we will describe in detail the parsing algorithm and its application to stylistic music generation.

Incremental Parsing

We chose to use an incremental parsing (IP) algorithm suggested by [6]. IP builds a dictionary of distinct motifs by sequentially adding every new phrase that differs by a single next character from the longest match that already exists in the dictionary. For instance, given a text $\{a\,b\,a\,b\,a\,a \ldots\}$, IP parses it into $\{a, b, ab, aa, \ldots\}$ where motifs are separated by commas. The dictionary may be represented as a tree.

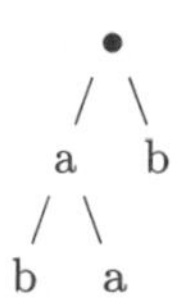

Fig. 9.2. IP Tree

Probability Assignment

Assigning conditional probability $\hat{p}^{\mathrm{LZ}}(x_{n+1}|x_1^n)$ of a symbol x_{n+1} given x_1^n as context is done according to the code lengths of the Lempel Ziv compression scheme. Let $c(n)$ be the number of motifs in the parsing of an input n-sequence. Then, $\log(c(n))$ bits are needed to describe each prefix (a motif without its last character), and 1 bit to describe the last character (in case of a binary alphabet). For example, the code for the above sequence is $(00, a), (00, b), (01, b), (01, a)$ where the first entry of each pair gives the index of the prefix and the second entry gives the next character. Ziv and Lempel have shown that the average code length $c(n)\log(c(n))/n$ converges asymptotically to the entropy of the sequence with increasing n. This proves that the coding is optimal. Since for optimal coding the code length is 1/probability, and since all code lengths are equal, we may say that, at least in the long limit, the IP motifs have equal probability.

Thus, taking equal weight for nodes in the tree representation, $\hat{p}^{\mathrm{LZ}}(x_{n+1}|x_1^n)$ will be deduced as a ratio between the cardinality of the subtrees (number of subnodes) following the node x_1^n. As the number of subnodes is also the node's share of the probability space (because one codeword is allocated to each node), we see that the amount of code space allocated to a node is proportional to the number of times it occurred.

In our example, the probability on the arc from the root node to $\{a\}$ is $3/4$, root to $\{b\}$ is $1/4$, probability from node $\{a\}$ to $\{aa\}$ is $1/2$ and from $\{a\}$ to $\{ab\}$ is $1/2$.

Seen in the bin representation, the probabilities are simply the relative portion of counts of characters $N_c(x)$, $x \in \{a, b\}$ appearing in bin with label c, giving

$$\frac{N_c(x)}{N_c(a) + N_c(b)} .$$

Sometimes a corrected count is preferred, considering the probability for a next symbol x to enter a current bin, giving

$$\frac{N_c(x) + 1}{N_c(a) + N_c(b) + 2} .$$

This is equivalent, in the tree representation, to adding the count of a current node to cardinality of the subtrees in every direction. For large counts, the two probabilities are very close.

Growing the Context in IP vs. Markov Models

An interesting relation between Lempel–Ziv and Markov models was discovered by [7] when considering the length of the context used for prediction. In IP every prediction is done in the context of earlier prediction, thus resulting in a "sawtooth" behavior of the context length. For every new phrase the first character has no context, the second has context of length one, and so on. In contrast, the Markov algorithm makes predictions using a totally flat context line determined by the order of the model. Thus, while a Markov algorithm makes all of its prediction based on 3- or 4-character contexts, the IP algorithm will make some of the predictions from lower depth, but very quickly it will exceed the Markov constant depth and use a better context. To compensate for its poor performance in the first characters, IP grows a big tree that has the effect of increasing the average length of the phrase so that beginnings of the phrase occur less often. As the length of the input increases to infinity, so does the average length, with the startling effect that at infinity it converges to the entropy of the source. In practice though, the average phrase length does not rise fast enough to provide for reliable short-time predictions. On the other hand, it behaves surprisingly well for long sequences. Our experiments show that this IP scheme, along with the appropriate linear representation of music, provides with patterns and inferences that successfully match musical expectation.

Another important feature of the dictionary-based methods is that they are "universal". If the model of the data sequence was known ahead of time, an optimum prediction could be achieved at all times. The difficulty with most real situations is that the probability model for the data is unknown. Therefore one must use a predictor that works well no matter what the data model is. This idea is called "universal prediction" and it is contrasted to Markov predictors that assume a given order of the data model. Universal prediction algorithms make minimal assumptions on the underlying stochastic sources of musical sequences. Thus, they can be used in a great variety of musical and stylistic situations. Our IP based predictor is one such example of universal predictor. This differs also from knowledge-based systems, where specific knowledge about a particular style has to be first understood and implemented [5].

9.3 The Incremental Parsing (IP) Algorithm

The *IPMotif* function computes an associative dictionary (the *motif dictionary*) containing motifs discovered over a text.

```
Parameter text, a list of objects
dict = new dictionary
motif = ()
While text is not empty
  motif = motif ! pop (text)
  If motif belongs to dict
    Then value(dict,motif)++
    Else add motif to dict with value 1
      motif = ()
return dict
```

`dict` is a set of pairs (key, value) where the keys are motifs and values are integer counters. `text` and `motif` are ordered lists of untyped objects (we don't restrict to characters). `value(dict,motif)` retrieves the value associated with motif in dict. $W!k$ notates the list obtained by right-appending object k to list W. `Pop(var)` returns the leftmost element from the list pointed to by `var` and advances `var` by one position to the right.

The text is processed linearly from left to right, object after object, without any backtracking or look-ahead. At any current time, the variable `motif` contains the current motif W being discovered and the variable `text` contains the remaining text, beginning just after W. Now a new object k is popped from the text and appended to the right of `motif`, which value changes to $W!k$. If $W!k$ is not already in the dictionary, it is added to it and `motif` is reset to an empty list (), thus being prepared to receive the next motif. The LZ78 compression algorithm would, at that time, output a codeword for W, depending on W's index in the dictionary, along with the object k. Compression would occur because W, which must have been previously encountered, is now output as a simple code. But since we are not concerned with compression, we do nothing more. If $W!k$ is already in the dictionary, we increment the counter associated with it and iterate. By doing this, we compute for each motif $W!k$ the frequency at which object k follows motif W in the text. It is an IP property that, if motif W is in the dictionary, then all its left prefixes are there. So, if for instance motifs *ABC*, *ABCD*, *ABCE*, *ABCDE*, are discovered at different places, the frequency of C following AB will be equal to 4. Another way to look at it is to consider that, for each motif W in the dictionary, for which there exists other motifs $W!k_i$ in the dictionary, we will easily get the (empirical) conditional probability distribution $P(k_i|W)$ (probability of occurrence of k_i knowing that W has just occurred).

In order to achieve this, we have to transform the motif dictionary into another one, called a *continuation dictionary*, where each key will be a motif W from the previous dictionary, and the corresponding value will be a list of couples $(\ldots(k, P(k|W))\ldots)$ for each possible k in the object alphabet, representing in effect the empirical distribution of objects following W.

The *IPContinuation* function computes a continuation dictionary from a motif dictionary.

```
Parameter dict1, a dictionary
dict2 = new dictionary.
For each pair (W!k, counter) in dict1
   If W belongs to dict2
   Then value(dict2,W) =
         value(dict2,W) !(k counter)
   Else add W to dict2
          with value ( (k counter) )
Normalize (dict2)
Return dict2
```

The function `Normalize` turns the counters in every element of dict2 into probabilities.

Exemple
Text = $(abababcabdabcdabce)$
Motif dictionary = $\{((a)6)((b)1)((ab)5)((abc)3)((abd)1)((abcd)1)((abce)1)\}$
Continuation dictionary = $\{((a)((b\,1.0)))((ab)((c\,0.75)(d\,0.25))((abc)$
$((d\,0.5)(e\,0.5))\}$

As can be seen in the previous example, a single pass IP analysis on a short text is not sufficient to detect a significant amount of motifs. There is no information on continuations for motif b or motif ba. Due to the asymptotic nature of IP, these motifs will eventually appear when analyzing long texts. Another way to increase redundancy and to detect more motifs is to parse several times the same text using the same motif dictionary, rotating each time the text to the left by one position.

The *IPGenerate* function generates a new text from a continuation dictionary. Suppose we have already generated a text $(a_0a_1 \ldots a_{n-1})$. There is a parameter p which is an upper limit on the size of the past we want to consider in order to choose the next object.

1. Current text is $(a_0a_1 \ldots a_{n-1})$
   context = $(a_{n-p} \ldots a_{n-1})$.
2. Check if context is a motif in the continuation dictionary.
3. If found, its associated value gives the probability distribution for the continuation. Make a choice with regard to this distribution and append the chosen object k to right of text.
   text = text!k. Iterate in 1.
4. If context is not found in dictionary, shorten it by popping its leftmost object.
   context = $(a_{n-p+1} \ldots a_{n-1})$. If motif becomes () generate a *failure* otherwise iterate in 2.
5. Upon failure either stop or append a random object to text, then iterate in 1.

9.4 Resolving the Polyphonic Problem

The IPGenerate algorithm works on any linear stream of objects. It was successfully tested on linear streams of midi pitches from solo pieces or isolated voices of polyphonic pieces. In order to be able to process polyphony, thus fully capturing rythmical, countrapuntal and harmonic gestures, we had to find a way to linearize multivoice midi data in a way that would musically make sense and take advantage of the IP scheme. The best results were achieved by using a variant of the superposition languages defined by Chemillier & Timis [4].

To understand this, take the 2-voice example shown below.

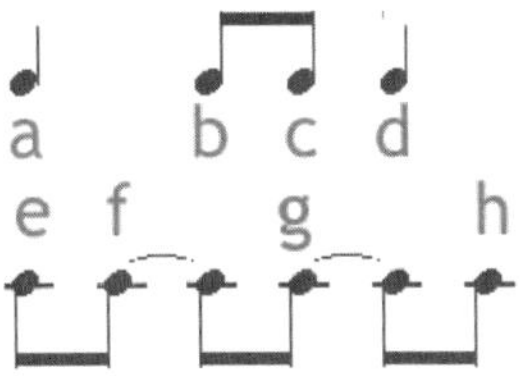

Fig. 9.3.

Only the rhythm is notated. Pitch, as well as other relevant information are coded with letters a through h. If we slice time with respect to the common time unit (the gcd of the durations, i.e. the eighth note) we may code the sequence using 2 parallel words:

$a\ \boldsymbol{a}\ b\ c\ d\ \mathbf{d}$
$e\ f\ \boldsymbol{f}\ g\ \boldsymbol{g}\ h$

where the letter **x** in bold means the continuation of the previous (contiguous) letter x (which is either a beginning symbol or itself a continuation). In order to linearize, we go from the normal alphabet, augmented by continuation symbols, $S = \{a, b, c, \ldots, \mathbf{a}, \mathbf{b}, \mathbf{c}, \ldots\}$ to the cross-alphabet $S \times S$. Now the sequence is: $(a, e)(\mathbf{a}, f)(b, \mathbf{f})(c, g)(d, \mathbf{g})(\mathbf{d}, h)$.

In order to cope with any arbitrary time structure and to optimize the parsing, we use the following variant.

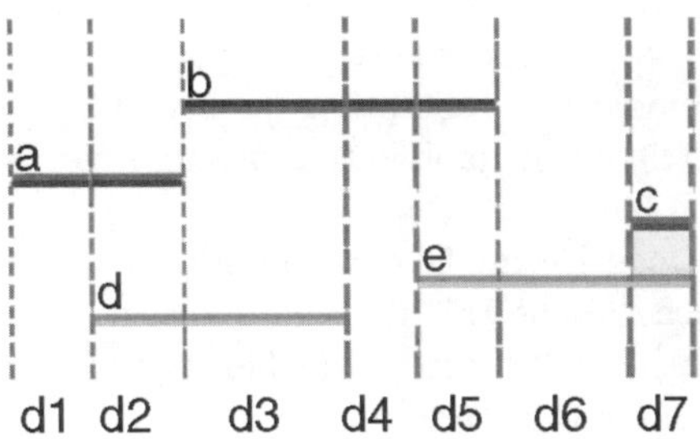

Fig. 9.4.

Time is sliced at each event boundary occuring in any voice. A set of durations $D = \{d_1, \ldots d_7\}$ is thus built. Using the cross alphabet $S \times S \times D$ we build the linear triplet sequence: $(a, -, d_1)(\mathbf{a}, d, d_2)(b, \mathbf{d}, d_3)(\mathbf{b}, -, d_4)(\mathbf{b}, e, d_5)$ $(-, \mathbf{e}, d_6)(c, \mathbf{e}, d_7)$, where – denotes the empty symbol (musical rest).

These triplets can easily be packed into 3 bytes numbers if we code only the pitches along with the durations. In order to optimize the duration alphabet, we quantize the original durations into a reasonable set of discrete rhythmic values. The idea is then easily generalized to n-voice polyphony.

9.5 Experiments

Once a multi-voice midi file is transformed into a linear text based on the cross alphabet, it is presented to the IPMotif/IPcontinuation algorithm. The resulting continuation dictionary can then be randomly walked by IPGenerate to build variants of the original music.

The cross-alphabet representation used has proven to fit decisively into the IP framework. In particular, the continuation symbols encode the fact that certain notes, in certain contexts, have a certain probability of being sustained while other notes are playing on other voices. The result is that countrapuntal gestures, as well as harmonic patterns, tend to be generated in a realistic way with regard to the original. Another characteristic of IP is that if not only one text but a set of different texts are analyzed using the same motif dictionary, the generation will "interpolate" in a space constituted by this set. This interpolation is not a geometrical one, but rather goes randomly from one model to another when there exists a common pattern of any length and a continuation from the second model is chosen instead the first one.

IPGenerate has been tested, in normal and interpolation mode, over the set of 2-voices Bach Inventions, normalized for tonality and tempo. While the lack of overall harmonic control do not favors consistant harmonic progression in the resulting simulations, these should be seen as "infinite" streams where interesting subsequences, show original and convincing counterpoint and harmonic patterns.

On the Bach material, we have established empirically that 0 rotation[3] of the original text would lead to a poor, unusable, continuation dictionary; 3–4 rotations are optimal, in that whole phrases from the original may be generated; more rotations do not improve the generation quality. This is certainly due to the way phrases are built from combination of small motifs in this style of music.

In the Jazz domain, a new piece by Jean-Réemy Guedon, miniX, has been created recently at Ircam by the French "Orchestre National de Jazz" with the assistance of Frederic Voisin. In this 20 mn piece, about half of the solo parts were IPGenerated and transcribed on the score.

These experiments were carried-out using OpenMusic, a Lisp-based visual language for music composition [ASS99]. Some results are available at: http://www.ircam.fr/equipes/repmus.

References

1. Shannon, C. (1953): A Mind Reading Machine, Bell Laboratories memorandum. Reprinted in the Collected Papers of Claude Elwood Shannon, IEE Press, pp. 688–689, 1993

[3] that is, repeating the analysis after a rotation of the text by one symbol

2. Meyer, L. (1961): Emotion and Meaning in Music, University of Chicago Press, Chicago
3. Assayag, Agon, Laurson, Rueda: Computer Assisted Composition at Ircam: PatchWork & OpenMusic. Computer Music Journal, to come, (1999)
4. Chemillier, M.: Structure et méthode algébriques en informatique musicale. Doctorat, LITP 90-4, Paris VI, 1990
5. Cope, D.: *Experiments in Musical intelligence.* Madison, WI: A-R Editions, 1996
6. Ziv, J., Lempel, A.: Compression of individual sequences via variable rate coding. IEEE Trans. Inf. The. **24**(5), pp. 530–536 (1978)
7. Williams, R.N.: *Adaptive Data Compression.* Norwell, Massachusetts: Kluwer Academic Publishers 1991
8. Blackwell, D.: Controlled Random Walk. In: *Proceedings of the 1954 International Congress of mathematics*, Vol. III, pp. 336–338, Amsterdam, Holland
9. Hannan, J.F.: Approximation to Bayes Risk in Repeated Plays. In: *Contributions to the theory of Games*, Vol. 39, pp. 97–139. Princeton 1957
10. Feder, M., Merhav, N., Gutman, M.: Universal Prediction of individual sequences. IEEE transactions on Information Theory **38**, 1258–1270 (1992)
11. Cope, D.: *Computers and musical style.* Oxford University Press 1991
12. Conklin, D., Witten, I.: Multiple viewpoint systems for music prediction. Interface **24**, 51–73 (1995)
13. Pinkerton, R.: Information theory and melody. Scientific American **194**, 76–86 (1956)
14. Brooks Jr., F., Hopkins Jr., A., Neumann, P., Wright, W.: An experiment in musical composition. In: Schwanauer, Levitt (eds.): *Machine Models of Music*, pp. 23–40. MIT Press 1993
15. Cope, D.: An expert system from computer-assisted composition. Computer Music Journal **11**(4), 30–46 (1987)
16. Ebicoglu, K.: An Expert System for Harmonization of Chorals in the style of J.S. Bach. PhD Dissertation, Department of Computer Science, SUNY at Buffalo, 1986
17. Lidov, D., Gambura, J.: A melody writing algorithm using a formal language model. Computer Studies in Humanities **4**(3-4), 134–148 (1973)

10 Ethnomusicology, Ethnomathematics. The Logic Underlying Orally Transmitted Artistic Practices

Marc Chemillier

Ethnomathematics is a new domain that has arisen during the last two decades, at the crossroad between history of mathematics and mathematics education. This domain consists in the study of mathematical ideas shared by orally transmitted cultures. Such ideas are related to number, logic and spatial configurations [9,11]. My purpose is to show how ethnomusicology could turn musical materials in this direction. Music will be considered here as a mean of organizing time through patterns of sound events. Thus we shall focus on musical forms and structures, rather than on other aspects of music (such as social aspects for instance). We will ask whether particular forms of traditional music share specific properties, namely combinatorial properties, that could be of some interest from an ethnomathematical point of view.

The study of mathematical ideas of non-literate peoples goes against persisting notions in the mathematics literature, which are strongly influenced by the late nineteenth-century theory of classical evolution. According to this theory, cultures can be ordered on an intellectual scale from primitive peoples to Western culture [43]. These ideas have been quite influential in mathematics literature and continue to be cited [30]. Another idea developed by [33,34] introduced a distinction between the Western mode of thought, and a "prelogical" mode of thought (or "unscientific" as Evans-Pritchard said) characterizing traditional peoples. A rich debate has arisen from this controversial theory, involving anthropologists, cognitive psychologists, and philosophers, on the nature of "rationality", with special attention paid to the witchcraft problem of the Zande peoples from Sudan [23,46,28,29,26], with echoes in [27,31] and others. Ethnomathematics has grown in the wake of this epistemological debate, gathering mathematicians from different parts of the world including the southern hemisphere. A research program has been sketched, and an International Study Group was founded [2].

The efforts made by ethnomathematicians in order to correct erroneous theories on the ability of human thought to think abstractly or logically rely greatly on the works of former ethnologists who have recorded information involving mathematical ideas while doing field work at the end of the nineteenth or during the twentieth century. Not being especially engaged with mathematics in their own culture, these ethnologists did not extract the whole mathematical content of their recorded material. Thus a great amount of work remains in the study of this field material from a mathematical point of view. In the case of music, recorded material will consist in various forms of written transcriptions of the music, as we shall see in the examples discussed in this paper.

Talking about "mathematics" in the context of orally transmitted societies requires some preliminary remarks, since mathematics are sometimes considered a Western invention [42]. There is an implicit assumption underlying this approach asserting that the practices we are studying share something in common with Western mathematics. In fact, both are linked through phenomena we have called "mathematical ideas". The concept of number, for instance, is a mathematical idea which seems to be universal [18,47]. But the scope of mathematical ideas is not clearly delineated. Furthermore, the way these ideas are expressed and their context in human thought vary from culture to culture. As Daniel Andler pointed out during his lecture at the Diderot Forum, there may exist a gap between the formal properties of traditional objects (such as geometric drawings, for instance), which are discovered by ethnomathematicians and expressed in their own mathematical language, and the cognitive processes of peoples who produced these objects. This is particularly true since the studies are generally based on recorded materials collected during field works made in the past, without interacting with the native peoples. This leads to the question whether the formal properties discovered in these field materials reflect a conscious activity of the mind. The answer to this question determines the cognitive level of our ethnomathematical descriptions. Even in Western mathematics, one can distinguish different cognitive levels, as pointed out in [21] following the ideas of [32]. One of these levels is the formalized text written by mathematicians in professional publications. But the mathematician's activity involves many other levels, including simple "reveries" in which mathematical ideas are put together in an involuntary way.

Ethnomathematical studies attempt to order mathematical practices of nonliterate peoples on the scale of different cognitive levels. Since the work of Piaget [36], new methods have been developed in psychology for cross-cultural studies [17], as applied in the analysis of the strategies of players of a well-known African game called "awele" [38]. The development of cognitive anthropology is the result of this growing interest for the cognitive aspects of ethnological studies [5,41]. In the examples discussed in this paper, we shall try to indicate to what extent the formal properties we are studying are explicit in the mind of the native peoples, but as we shall see, this is not always possible without interacting with them.

My presentation is divided into three parts. The first part deals with sand drawing from the Vanuatu, a non-musical example presented here in order to show what kind of ideas is studied in ethnomathematical writings. Decorative arts have been the first activity of nonliterate peoples which became the subject of mathematical investigations, from the point of view of their symmetry properties, involving the classification developed by crystallographers and adapted by [37] to the two dimensional case [40,45]. In the case of sand drawings, the mathematical analysis involves formal languages, as it is described for the *kolam* of India in [39], but also graph theory. The beautiful Vanuatu figures we shall present in this paper have been studied by various mathematicians who have been interested in the properties of their tracing paths [10,11,24].

Following this preliminary example taken from visual art, I will present two musical examples, in order to show how musical practices could bring new insights in the development of ethnomathematics. The first musical example concerns harp music from Nzakara people of Central African Republic. I have been working on this repertory for ten years in collaboration with Éric de Dampierre [20]. As we shall see, the short harp formulas played by traditional musicians share interesting combinatorial properties.

The last example deals with polyrhythmic music from the Aka Pygmies, which has become famous since the works of ethnomusicologist Simha Arom [6]. He has discovered interesting properties of asymmetric rhythms underlying this music, and we shall focus on the combinatorics of these rhythmic patterns.

10.1 Sand Drawings from the Vanuatu

10.1.1 The Guardian of the Land of Dead

We first turn to a country where there is a rich tradition of tracing figures in the sand, the Republic of Vanuatu, called the New Hebrides before its independence in 1980. This chain of some eighty islands is located in the South Pacific, 200 kilometres northern-east of New Caledonia. On some of these islands (mainly Malekula, Ambrym and Pentecost), the tradition of tracing figures in the sand has produced many interesting and sophisticated figures.

The technique used to draw these figures simply consists in tracing on the sand with the finger. Often a framework of a few horizontal and vertical lines is drawn before tracing the figure itself. This practice is still in use, as one can see by looking at recent pictures of traditional sand drawings in the catalogue of the exhibition devoted to arts from the Vanuatu at the Musée des Arts d'Afrique et d'Océanie in Paris in 1997 [44], or in a little book by Jean-Pierre Cabane [13]. This part of my paper took originally the form of a concert-conference at the Musée, which was initiated by composer Tom Johnson who wrote a piece of music for contrabass saxophone following the tracing of the tortoise (which is reproduced in Fig. 10.3 below) [16].

The tracing paths of these figures satisfy a rule, which is strongly related to graph theory. Figures "are to be drawn with a single continuous line, the finger never stopping or being lifted from the ground, and no part covered twice" [11, p. 45]. Moreover when the drawing ends at the point from which it began, it is called *suon*. This rule for tracing figures corresponds to what is called *Eulerian path* in graph theory, that is a continuous path that covers each edge of a graph once and only once. Finding a Eulerian path in a given graph is not trivial, and sometimes not even possible in situations such as the famous seven bridges of Königsberg, as Euler proved it in 1736.

Euler's statement provides a good example for illustrating the different cognitive levels found in mathematical activity. The necessary condition of the statement relies on a simple idea: finding a Eulerian path in a graph requires that for each vertex the number of adjacent edges is even, except two of them, since a vertex being reached by one edge must be left by another. Thus adjacent edges can be grouped by pairs, except those emanating from the beginning and ending points. The sufficient condition is a more formal result, asserting that whenever all numbers of edges emanating from the same vertex are even, except for two vertices, then a Eulerian path can be found. The proof develops the previous simple idea in the form of a more technical induction argument on the number of vertices.

An important point concerning this tracing rule is that most of the drawings are related to myths, some of them emphasizing the mathematical property of the figures expressed by the tracing rule. Often tracing figures in the sand is achieved while somebody is telling a story associated with the figure. Thus myths appear to be comments of related figures. The tracing rule itself is sometimes part of the myth, as it is the case for a specific figure related to the passage to the Land of the Dead. In the myth, this figure is said to be traced on the sand by the guardian of the Land of the Dead. The guardian challenges those who try to enter. When the ghost of a newly dead person approaches, the guardian traces half the figure as shown in Fig. 10.1 [13, p. 18]. The ghost has then to complete the bottom part of the figure with a continuous path. If he fails, he is eaten by the guardian. As one can see, the challenge consists precisely in finding a Eulerian path in a graph, thus being strongly related to the mathematical property of the figure.

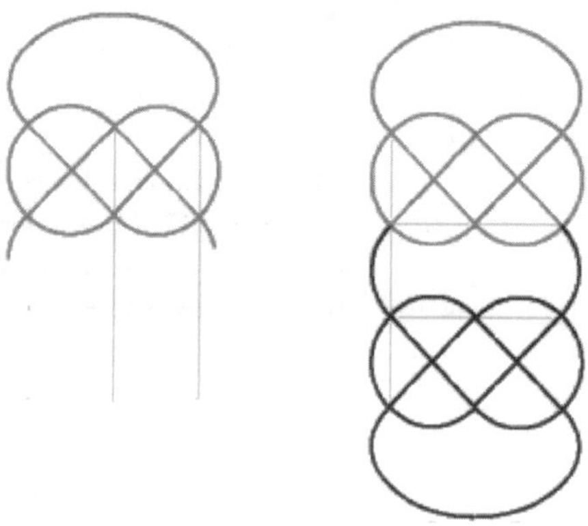

Fig. 10.1. Challenge to access the Land of the Dead

Other figures are associated with stories that refer directly to the tracing of the figure. One of them is called "Rat eats breadfruit, half remains": "First a figure described as a breadfruit is drawn completely and properly. Then the retracing of some edges is described as a rat eating through the breadfruit. Using the retraced lines as a boundary, everything below it is erased as having been consumed" [11, p. 47]. The result is shown in Fig. 10.2 [13, p. 53]. In this case the retracing of some edges is part of the story. This clearly demonstrates the fact that backtracking is considered to be improper.

Fig. 10.2. Rat eats breadfruit, half remains

Our knowledge of the tracing of these figures is greatly indebted to the works of a young British ethnologist, Bernard Deacon, who spent two years in the Vanuatu islands in the early thirties, at the age of 22. He made decisive works concerning kinship, but he also recorded about one hundred sand drawings. What is important for our ethnomathematical studies is that Deacon not only reproduced the figures themselves in his field notes. He also had the intuition to record the exact tracing path of these figures. He did so by adding numbers to the edges of the figures, so that the sequence of numbered edges corresponds exactly to the tracing path. Unfortunately, Deacon died of blackwater fever as he was awaiting transportation home. His works are known thanks to the field notes that were found in his luggage. His annotated figures of sand drawings were published in 1934 [22].

The drawing of the tortoise is one of the most well-known and beautiful drawings of the Vanuatu tradition. If we analyse its tracing path as recorded by Deacon, we can find some regularities in it. The tracing is made of subgraphs with constraints similar to the whole graph itself. In addition to these intermediate stages, one finds connecting paths that combine more elementary shapes.

The tracing paths recorded by Deacon provide field materials of great interest for ethnomathematical investigations. In the next section, we shall study in detail a simple but quite ingenious geometric construction that can

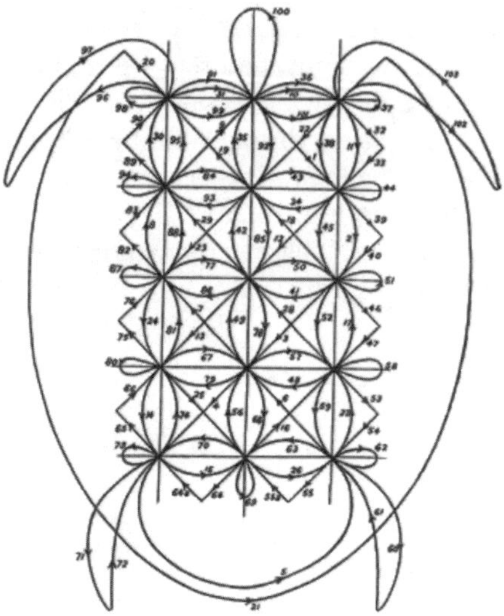

Fig. 10.3. The tortoise as recorded by Deacon

be deduced from these sand drawings. But this requires that we first travel to another part of the world, to Angola, where there is also a rich tradition of sand drawings.

10.1.2 The Logic of the Long Line

This tradition of sand drawings from the Tshokwe of Angola has also been studied from an ethnomathematical point of view by Paulus Gerdes, a mathematician from Mozambic, who wrote a book on the subject describing a lot of figures and studying their properties. The tracing of theses drawings obeys a rule similar to the one we have encountered in the Vanuatu tradition. Figures have to be drawn with a continuous path, the finger being kept in contact with the sand. In the Tshokwe tradition, this rule corresponds to what Gerdes calls the *monolinearity* property. It slightly differs from what mathematicians call a Eulerian path in that lines can cross one another, but are not allowed to touch each other without crossing [24, p. 20].

The drawing reproduced in Fig. 10.5 is similar to others one can find in Angola. Its shape looks like a lattice. The figure has 9 rows and 7 columns of points (with additional secondary columns and rows). Drawings similar to this one can be found in Angola, but they do not have the same number of rows and columns. The reason why is that when one tries to trace this specific figure with a continuous path satisfying the monolinearity property, the path joins its starting point before it can complete the figure.

What is interesting from an ethnomathematical point of view is that for particular lattices of points with numbers of rows and columns which do not

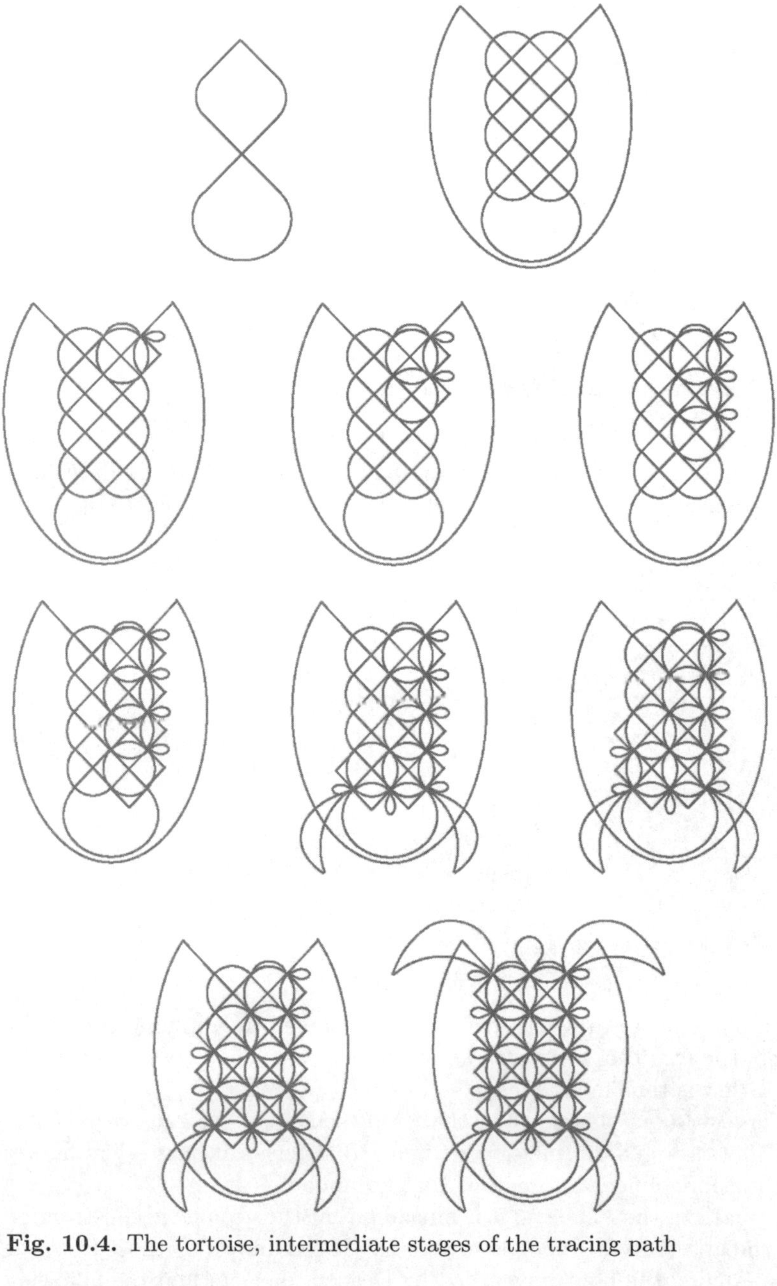

Fig. 10.4. The tortoise, intermediate stages of the tracing path

permit monolinear figures, the Tshokwe have discovered a geometric construction which transforms the figure into one which is monolinear although the numbers of rows and columns remain the same. This construction can be

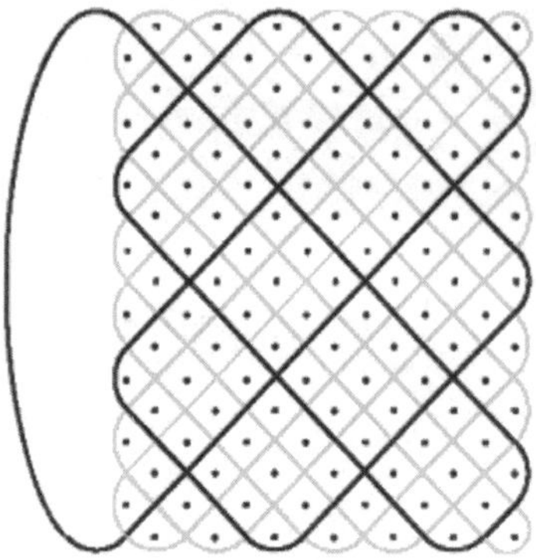

Fig. 10.5. A non-monolinear figure (lattice 9×7)

described as follows: a column is chosen, and each pair of crossing lines along this column is replaced by two quarter circles, as shown in Fig. 10.6.

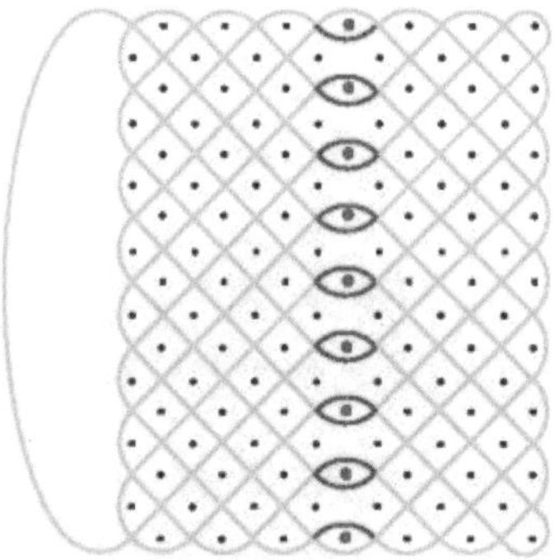

Fig. 10.6. A geometric algorithm

When one tries to draw the new figure, obtained by applying this transformation, the resulting path covers the entire figure. The figure produced by this algorithm is thus monolinear.

In Gerdes' book, this construction is presented as an "algorithm" [24, p. 205], that is a general method designed to transform lattice-like figures that are not monolinear into figures that are monolinear.

Going back to the Vanuatu, it is amazing that the geometric construction used by the artists from Angola seems to have been known by those from the Vanuatu. Among the figures recorded by Deacon, one can find the following one (Fig. 10.8), which is the same [22]. This one is entitled "Bird in the nest", as it represents a bird sitting on her eggs. As indicated by Deacon, the eggs correspond precisely to the ovals introduced by the algorithm. On the left

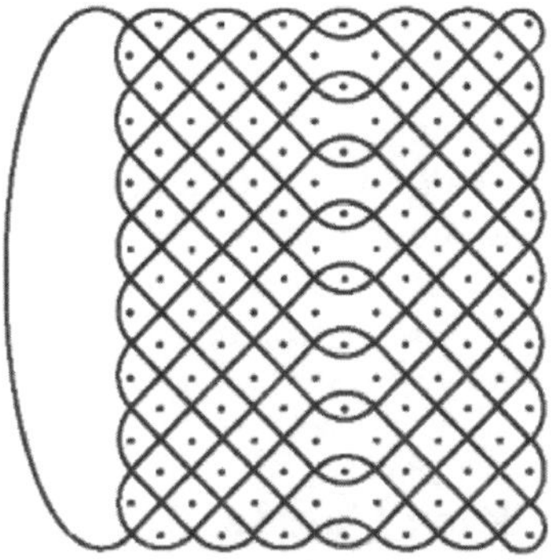

Fig. 10.7. The resulting monolinear figure

side is the (large) head of the bird. On the right side are the feathers of the tail, added as ornaments.

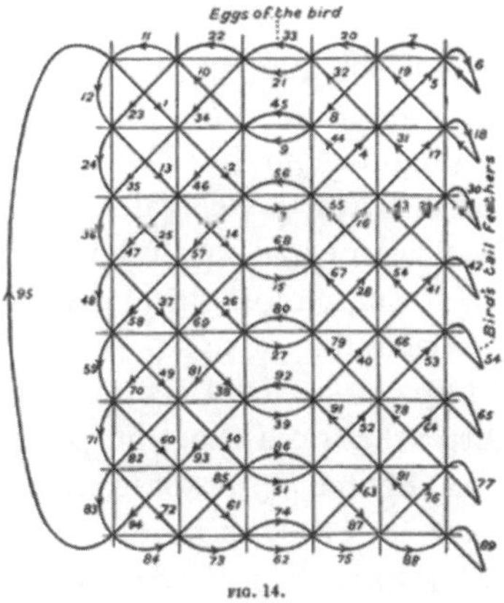

FIG. 14.

Fig. 10.8. Bird in the nest as recorded by Deacon (lattice 8×5)

As one can see in the examples of sand drawings discussed earlier, there are some clear indications establishing that the graph property of the figures is deliberately produced by the tracing path. It was explicitly mentioned in the figure entitled "Rat eats breadfruit, half remains" in Vanuatu, and this is also the case in Angola, as reported by Gerdes. Concerning the geometric construction that has been described in this section, one can verify that many drawings from Angola are based on the same principle, with different numbers of rows and columns, and that from a mathematical point of view, the construction has the property to transform non-monolinear figures into monolinear ones. Gerdes presents this construction as being a "very powerful algorithm" explicitly used by the native artists, but we do not know exactly what the artists have in mind when they draw these figures. To what extent are they conscious of the property of the construction? The level of cognition involved in this practice is not yet clear.

In the following two parts of my paper, we shall try to adapt the same approach to music. As we shall see, some amazing properties can be found in the musical examples which will be discussed, but the problem of identifying the correct cognitive level involved in the related mathematical constructions is much more difficult to resolve.

10.2 The Harp of the Former Nzakara Courts

10.2.1 The Art of Poet-Harpists

We now turn to Central Africa, where the Nzakara people live, in order to study some aspects of their harp music. Nzakara and Zande harps are well-known, because of the beautiful sculpted heads which adorn their necks. The Musée de la musique in Paris presented in 1999 an exhibition devoted to these instruments. While having fallen into neglect, these traditional instruments were still played upon request in 1993 by some old harp players, as one can hear by listening to the record entitled *Music from the former Bandia courts* published in the collection CNRS/Musée de l'homme founded by Gilbert Rouget [49]. The Nzakara-Zande territory is spread between Central African Republic, Congo (ex-Zaire) and Sudan. The Nzakara peoples are related to Zande, whose belief in witchcraft has triggered a strong philosophical discussion on "rationality", as we have recalled at the beginning of this paper.

The Nzakara harpists' repertoire is divided into categories. Each piece of poetry sung with the accompaniment of the harp relies on a short harp formula played repetitively as an *ostinato*. There are different types of formulas designated by terms such as *ngbakia, limanza, gitangi.* These terms also refer to traditional dances, the harp formulas being adapted from rhythms and musical elements borrowed from the dance repertoire played on the portable xylophone or the drum.

The formulas belonging to the categories *limanza* and *ngbakia* are adapted from the portable xylophone repertoire. In fact, one knows this repertoire of xylophones only through formulas that have been adapted to the harp, since the portable xylophone has now completely disappeared from the Nzakara region. We know the portable xylophone was an important instrument in the former Nzakara kingdom, as it was played by King Bangassou himself, but we have no recordings of its repertoire. The formula reproduced in Fig. 10.9 belongs to the category *limanza.* An example based on this formula can be heard on the CD *La parole du fleuve*, track 5 [49]. The piece is played by Maliba in 1969, and was recorded by Éric de Dampierre. We shall present some amazing properties of this apparently simple formula, studied in detail in [14,15].

The harp formula shown in Fig. 10.9 is interesting when one looks closely at its structure. The strings are plucked by pairs, one with another. The formula can thus be considered the superimposition of two melodic lines, one played with the left and one played with the right. Something remarkable appears when one compares the upper line with the lower line shifted six notes ahead: they are nearly identical. Strange enough, this is what is called a "canon" in Western classical music.

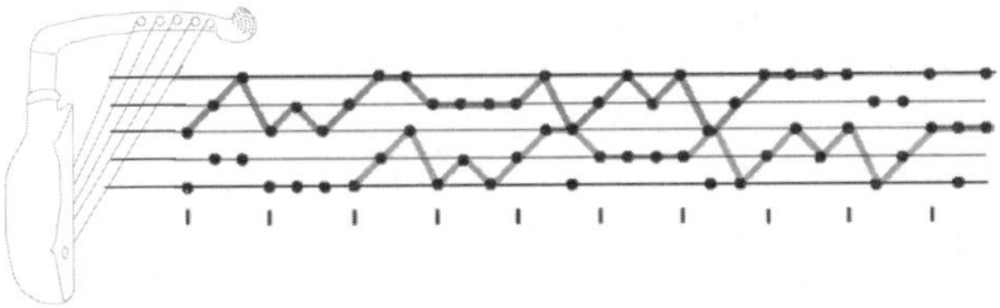

Fig. 10.9. A *limanza* canon: the two lines are nearly identical

Now let us go further into the analysis of this harp formula by describing an algorithm that can explain its construction. If one selects pairs of strings plucked simultaneously, taking the first one, then the sixth one, then the twelfth one and so on, up to five pairs denoted by numbers 0, 1, 2, 3, 4 (as shown in Fig. 10.10), the resulting sequence takes the form of a kind of ascending staircase.

Fig. 10.10. Five pairs of strings plucked simultaneously

A property of the formula appears when one carries on this process by taking into account all the pairs occurring in the sequence. The formula being repeated as a loop, one can consider it an infinite periodic word over an alphabet, which is the finite cyclic group $\mathbf{Z}_5$. To such an infinite word, denoted by u, taking values into a finite cyclic group $\mathbf{Z}_p$, one can associate the *difference word* denoted by Du such that for every integer n, one has $Du(n) = u(n+1) - u(n)$. There are some interesting relations between the periodicity of the word u and the periodicity of the word Du, depending on the structure of the finite group $\mathbf{Z}_p$ [1]. In the case of Nzakara harp canons, the infinite words may be called *redundant*, since the period of Du is strictly less than the period of u. In fact, the word represented in Fig. 10.10 has a period of 30, whereas its associated difference word has a period of 6 (with values 2 1 2 1 4 1). This leads to an algorithm for generating harp canons. It suffices to consider a given word (for instance 0 2 3 0 1 0 in the formula represented in Fig. 10.10), and to translate it several times by adding the same value to its elements (for instance 1, which gives 1 3 4 1 2 1, then 2 4 0 2 3 2, and so on), until we reach the initial word again.

The interesting fact about this construction is that it is not limited to the harp formula shown in Fig. 10.10. It is a general construction that can be observed in different harp formulas of the Nzakara repertoire. Figure 10.11 shows another example, which belongs to the category called *ngbakia* (another category of music played formerly on the xylophone). This harp formula is built on the same principles. Here the initial sequence has only two pairs corresponding to 0 1, and the translation value is 3, which gives the resulting periodic word 0 1 3 4 1 2 4 0 2 3. We thus have more than one pattern illustrating the same construction, and the whole repertoire contains six such harp formulas. Hence, translating a given sequence by adding the same value to its elements appears to be a general algorithm applied in different contexts.

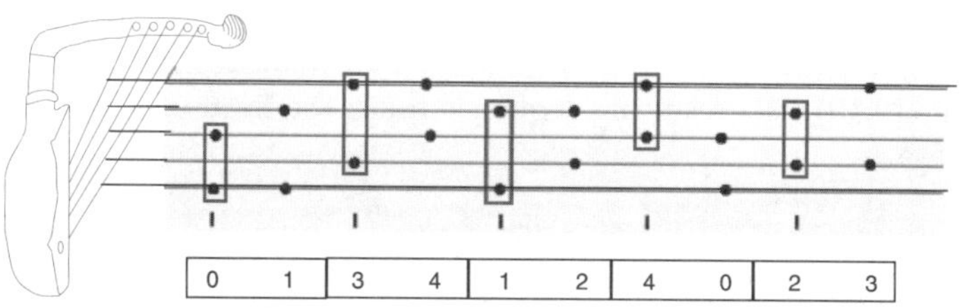

Fig. 10.11. A *ngbakia* harp formula

There is a surprising additional uniqueness property if one translates a word having only two pairs (as it is the case for the *ngbakia* formula shown in Fig. 10.11). Let us enumerate the different possible combinations, adding a few constraints to eliminate "degenerating" sequences. First, we demand that the resulting combination contain no repetition of a pair. Secondly, we demand that it not be factorised into two equal subsequences (these two conditions are motivated by the fact that harp formulas of the repertoire never fail to satisfy them). Having these conditions in mind, we enumerate all possible combinations obtained by translating different words containing two pairs. Since one can fix the initial pair to 0, there are only five possible values of the second pair, as shown in the table below.

The first combination is excluded, because it contains the repetition of a pair 0 0. The second is exactly the harp formula used by Nzakara harpists and shown in Fig. 10.11. The third is excluded, because it is a cyclic permutation of the previous one (as one can see by deleting its three initial elements 0 2 3). Since harp formulas are repeated endlessly, there is no way to

repetition	0 0 3 3 1 1 4 4 2 2
Nzakara solution	0 1 3 4 1 2 4 0 2 3
cyclic permutation of the solution	0 2 3 0 1 3 4 1 2 4
repetition	0 3 3 1 1 4 4 2 2 0
factorisation into two equal subsequences	0 4 3 2 1 0 4 3 2 1

distinguish two different cyclic permutations of the same formula by defining an initial point. Whereas there exist musical situations where the initial point of a sequence is a criteria for distinguishing different cyclic permutations (for instance, in ancient Greek theory of rhythm, "long-short" and "short-long" were considered distinct rhythms), no such reason exists in the repertoire of Nzakara harpists, so the second and third combinations are considered the same harp formula. The fourth combination is excluded, because it contains the repetition of a pair 3 3. The fifth combination is excluded because of its factorisation into two equal subsequences 0 4 3 2 1. The conclusion is then the uniqueness of the Nzakara solution.

10.2.2 The Plant-of-the-Twins

There is no way to connect the quite remarkable combinatorial properties of this harp formula to modes of thought of native people, since as we have already said, traditional harp music has fallen into neglect, and there are not enough Nzakara harpists still in activity to give us information about this question. All one can say is that according to the works of Éric de Dampierre, the musical structure of canon that has been discovered among those special harp formulas (in fact, there are six harp formulas which are canon in the same way as the previous example) could be related to some geometric considerations that Nzakara peoples make about a special plant used in the ritual for twins.

The plant called *bisibili* is represented in Fig. 10.12. As Éric de Dampierre has shown in his book *Penser au singulier*, this plant was of great importance in the old Nzakara tradition. It was the plant-of-the-twins, which means that when a pair of twins was born, a branch of it was planted in front of their house. It is interesting to point out that this plant was used in the ritual for twins for a purely "geometric" reason. What the Nzakara are interested in is the relative spatial positions of the two lines of leaves of the plant. These lines are not on the same plane. Furthermore, one of them is shifted so that the leaves are not attached to the stalk at the same points [19, p. 14].

There is a noticeable similarity between the shape of this plant and the musical structure of canon. The way one line of leaves is shifted along the stalk of the plant-of-the-twins could be related to the way the lower melodic line of harp formulas is shifted to produce a canon. Whereas we have no information that could establish this connection, we can assert that the ritual for twins reveals the existence of a clear interest for mathematical

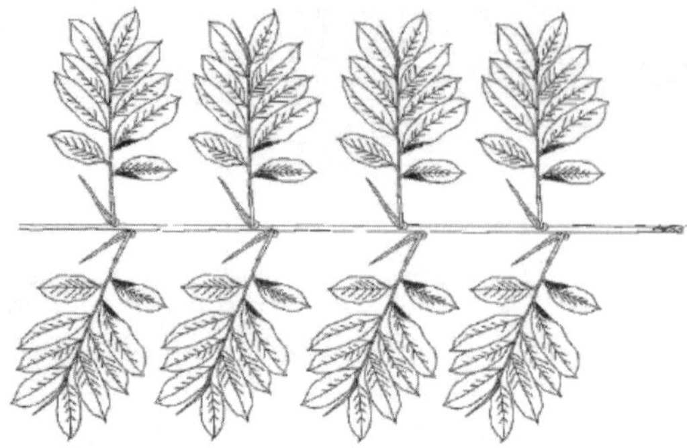

Fig. 10.12. The plant-of-the-twins (drawing by J.-M. Chavy)

ideas in the Nzakara tradition, involving geometric transformations such as glide-reflections. But the link between music and geometry has never been confirmed by native peoples. In a recent work on the same subject [12], Klaus-Peter Brenner has rejected an interpretation of the harp formulas as "canons". For various musico-cognitive considerations, he has developed a "permuting cell-sequencing structure", which is related to an algorithm based on the iterated translation of a given sequence.

10.3 African Asymmetric Rhythms (after the works of Simha Arom)

10.3.1 Asymmetric Rhythm of the Aka Pygmies

Our last section will be devoted to examples taken from the CD-ROM entitled *Pygmées Aka. Peuple et musique* realized by Simha Arom and his team [50]. These examples are four-part vocal polyphonies accompanied by a rhythmic combination the graphical representation of which is shown in Fig. 10.13. The corresponding piece of music, entitled *mbenzele*, can be heard in the section of the CD-ROM devoted to music analysis. We shall focus on a rhythmic pattern which is hidden in the rhythmic combination, but which can be played separately on the CD-ROM. This rhythmic pattern is marked with black squares in the graphical notation, Fig. 10.13. It is played with pairs of clashed metal blades called *diketo* [6, p. 438].

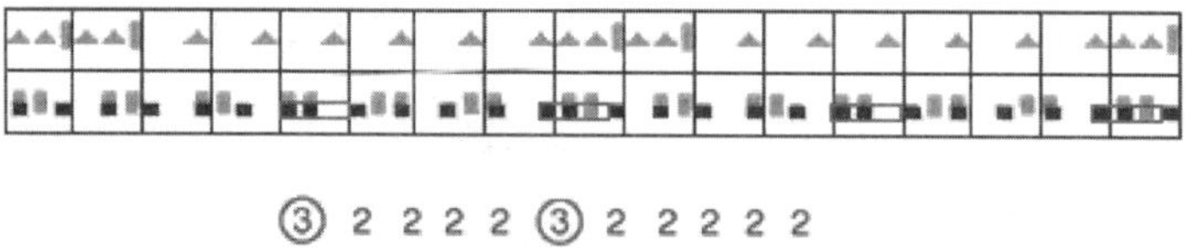

Fig. 10.13. Aka Pygmies polyrhythm from the piece *mbenzele*

There is a kind of limp in this sequence, due to a shorter rhythmic duration inserted in the sequence of equal durations. This additional rhythmic value breaks the regularity of the sequence. Its duration is about half the duration of the one behind it. Bringing them together results in a duration equal to three units, marked with small rectangles in Fig. 10.13, whereas other durations are equal to two. Hence, the sequence as a whole can be viewed as groups of two-unit elements separated by a three-unit element.

This rhythmic pattern shares an interesting property of asymmetry. The three-unit elements of the sequence are not spaced in a strictly regular manner. The groups of two-unit elements, which separate them, contain four elements on one side, and five elements on the other side. Thus the pattern denoted as 3 2 2 2 2 3 2 2 2 2 2 contains two groups of unequal length. There exist other interesting asymmetric patterns of this type in traditional African music. They have been studied by Simha Arom who pointed out a specific property called the "rhythmic oddity" property ("imparité rythmique" in French) that will be presented in the next section. The following table shows all the patterns of this type that can be found in Arom's book [6], taken from various repertoires played in Central African Republic (Aka Pygmies p. 439 and 839, Zande repertoire of the *kponingbo* p. 470, Gbaya repertoire of the sanza, and Ngbaka repertoire of the harp p. 435 and 474).

3 3 2	Zande
3 2 3 2 2	Aka, Gbaya, Nzakara
3 2 2 3 2 2 2	Gbaya, Ngbaka
3 2 2 2 3 2 2 2 2	*not in use*
3 2 2 2 2 3 2 2 2 2 2	Aka

As one can see in the table above, one of the patterns is excluded (thus being written in italics in the table). It corresponds to value $k = 3$, where k and $k + 1$ denote the numbers of two-unit elements on both sides of the sequence, the only possible values for integer k being 0, 1, 2, and 4. The reason why $k = 3$ is not accepted is that patterns of the asymmetric type are always based on some regular pulse. One can note that this makes a great difference between them and other asymmetric rhythmic patterns called *aksak* used in Central Europe. African asymmetric rhythms are divided into two unequal parts, but are associated with a regular pulse, which can be grouped into two parts of equal length.

The numbers of beats in each pattern form a regular *geometric* progression 2, 4, 8, with no missing value. The corresponding numbers of units contained in each sequence depends on the division of the beat, which can be binary or ternary. The values for theses numbers are 8, 12, 16 and 24 (corresponding respectively to 2×4, 4×3, 4×4 and 8×3), and more generally, each number of units takes one of the possible two forms 2^a and $2^a \times 3$. The missing value

Fig. 10.14. Asymmetric African patterns and their corresponding pulse

in the preceding table is excluded, since it corresponds to a total number of units equal to 20.

10.3.2 The Rhythmic Oddity Property

In this last section, we shall be interested in some combinatorial properties of asymmetric patterns of the "rhythmic oddity" type. The rhythmic oddity property can be stated by placing the two- and three-unit elements of the sequence on a circle (thus expressing the fact that the pattern is played as a loop). For the Aka Pygmies sequence, denoted as 3 2 2 2 2 3 2 2 2 2 2, we get a three-unit element on top, a three-unit element on bottom, and groups of two-unit elements which separate them on both sides (four on the right side, and five on the left side). The property asserts that *if one attempts to break the circle into two parts, it is not possible to have two equal parts.* Whatever the chosen breaking point, there is always one unit lacking on one side.

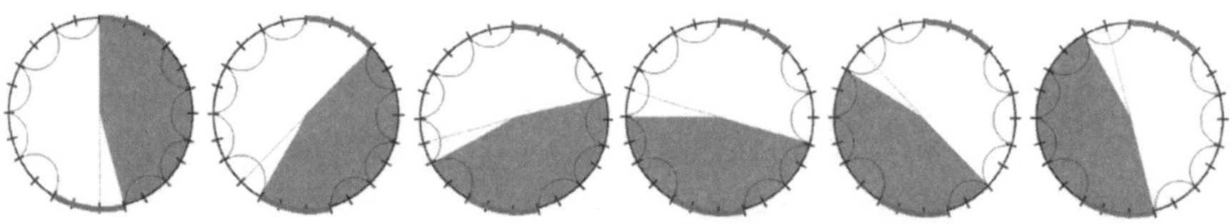

Fig. 10.15. Rhythmic oddity property: no breaking point giving two equal parts

This is what Simha Arom called the *rhythmic oddity* property. The asymmetry of the pattern is to some extent intrinsic, in the sense that there exists no breaking point giving two equal parts. Every division of the pattern gives two unequal parts, "half minus one" on the one side, and "half plus one" on the other side [7,8]. Note that the oddity property requires that the circle is divided into an even number of units, so that it is possible to find patterns of the "half minus one / half plus one" type. Denoting by n the total number of units (n is even), a pattern can be viewed as an infinite n-periodic word on a two-letter alphabet, one letter corresponding to an attack and the other meaning "no attack". The rhythmic oddity property can then be stated as the fact that if there is an attack for integer i, then there must be no attack for integer $i+n/2$, a statement which means that no breaking point occurres in the sequence.

Now we turn to a combinatorial problem: is it possible to generate other patterns similar to the one in Fig. 10.15 by placing elements of three or two units on a circle, so that the resulting figure cannot be broken into two equal parts? Figure 10.16 shows the 24-unit circle, with two three-unit elements placed side by side. The resulting pattern does not satisfy the rhythmic oddity property, since it can be broken into two equal parts.

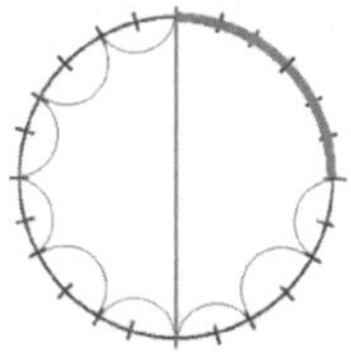

Fig. 10.16. Breakable pattern with three-unit elements placed side by side

It is easy to verify that there are only two possible ways to place the two three-unit elements on the circle, as shown in Fig. 10.17. But in fact, these two solutions are the same pattern, since they only differ by a cyclic permutation. Hence, there exists only one pattern with two elements of three units, which satisfies the oddity property, and this pattern is the one that was found in the Aka Pygmies repertoire.

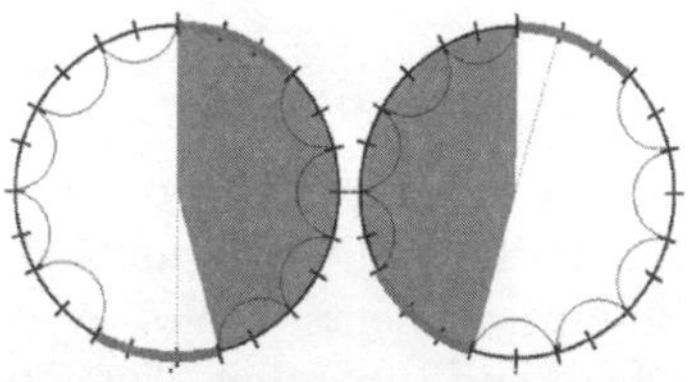

Fig. 10.17. Two ways of placing two elements of three units

If we look for patterns with more than two elements of three units, we must notice that the number of those elements must be even (if it is not, the total number of units cannot be even). Moreover, it can be proved that the number of two-unit elements must be odd. For instance, Fig. 10.18 shows a pattern where the numbers of elements of two and three units are equal to six and four, respectively. We have placed these elements on the circle in such a manner that the resulting pattern looks asymmetric. But although this pattern seems to satisfy the oddity property, there exists one point on the circle where it is possible to break it into two equal parts. This is a consequence of a simple argument related to the "pigeon hole principle" (the basic idea of Ramsey Theory asserting that n balls cannot be distributed into p boxes so that all boxes contain less than n/p balls) [25]. This argument implies that when the number of three-unit elements on one side of the circle is less than half the total number, then it must be greater than half this number on the other side. The initial breaking line, corresponding to $i = 0$, contains one element of three units on the right side, and three elements on the other side. When one rotates this line from point to point around the circle, the number of elements of three units contained in each side of the circle can be increased by one, decreased by one, or left unchanged. In the part of the circle which is initially on the right side, the number of elements of three units has grown from one to three as the breaking point is displaced halfway around the circle. Therefore, there exists an intermediate position where this number takes the value of two. This position is reached for $i = 4$, and at this point, we have two elements of three units on both sides of the circle. The number of remaining elements of length two being even, the circle can thus be broken into two parts of equal length. Finally, when the number of two- and three-unit elements are both even, the pattern cannot satisfy the oddity property. Since the number of three-unit elements must be even, as noticed above, it implies that the number of two-unit elements is odd.

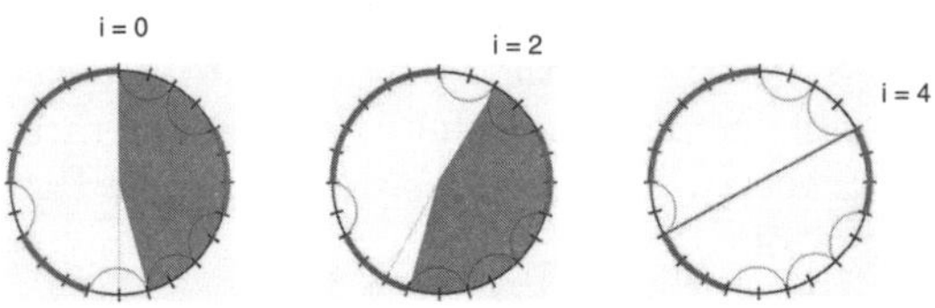

Fig. 10.18. Breakable pattern with four elements of three units

On the 24-unit circle, on can place only two, four or six elements of three units (recall that this number must be even). But the value four must be rejected, since in this case, the number of two-unit elements is also even. Indeed, denoting by n_2 and n_3 the numbers of two- and three-unit elements respectively, one has $2n_2 + 3n_3 = 24$, so that if $n_3 = 4$, then $n_2 = 6$ is even. The remaining values are then two and six. In the case of two elements of three units, we have found that there is only one solution, whatever being the total number of units on the circle, and the corresponding patterns are represented in Fig. 10.19.

In the case of six elements of three units (the total number of units on the circle being equal to 24), a computation has proved that there are only two patterns represented in Fig. 10.20.

One of them is split into two forms, a direct one and its mirror image:

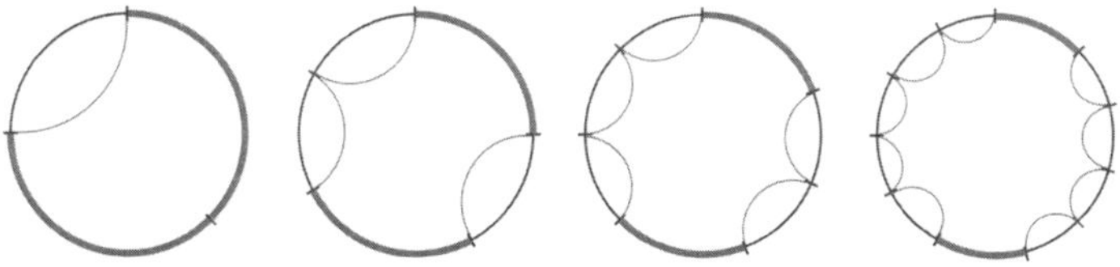

Fig. 10.19. All patterns with two elements of three units

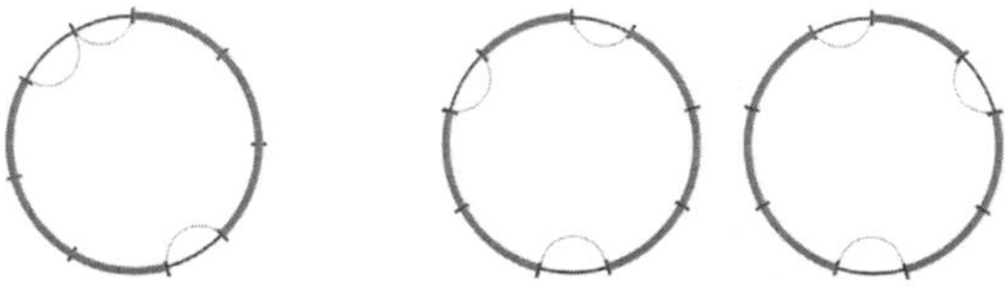

Fig. 10.20. All patterns with six elements of three units

pattern #1 = 3 3 3 2 3 3 3 2 2

pattern #2 = 3 3 3 2 3 3 2 3 2
mirror = 2 3 2 3 3 2 3 3 3

This abstract enumeration (done by computer) is obviously disconnected from the way native peoples think. But it is interesting to go back to their music with our generative result in mind to see what patterns are used in the repertoires of the Central African Republic. As we have already seen in the table of the previous section, all possible patterns involving two elements of length three are used. But what is much more surprising is that one of the two patterns involving six elements of length three (pattern #2) is also used. It contains groups of one, two and three elements of three units, separated by isolated elements of two units. This pattern is called *mokongo*, and it is used in the repertoire of the Aka Pygmies [6, p. 436].

The just explained *mokongo* is played in superimposition with the pattern we have already presented in the previous section (Fig. 10.13). The combination of the two is part of the piece *zoboko*, played during divination for the hunters (an example of which can be heard on the CD-ROM *Pygmées Aka*).

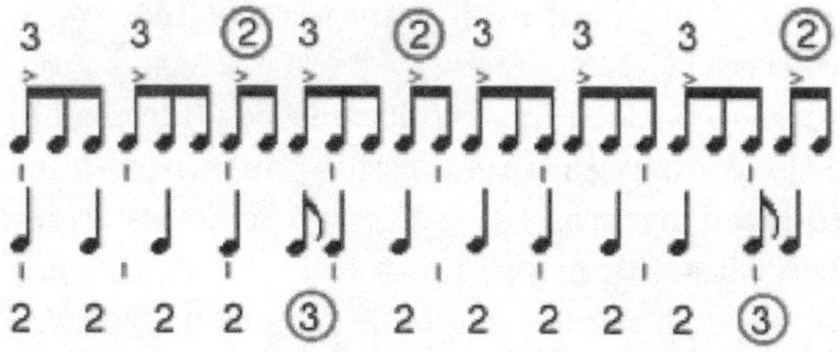

Fig. 10.21. Unbreakable pattern with six elements of three units

This example leads to a question related to perception. The rhythmic patterns shown in Fig. 10.21 are played in a fast tempo. The oddity property asserts that a pattern cannot be broken into two parts of equal length. But when one tries to break this pattern into two nearly equal parts, the length of those parts differ from half the total length of the pattern by just a single unit, which is a very short duration. It is impressive that such a short difference in duration is perceived so clearly that the oddity principle is universally observed in Pygmy music.

10.4 Conclusion

In both repertoires that we have studied, the formal structures are far too complex to be easily perceptible when one listens to the corresponding musical sequences. It is not realistic to assume that musical sequences of this form can be distinguished from others just by listening to them. Thus we have to admit that there might be a distance between musical forms produced by the human mind, and more elementary musical forms that human ear can identify. This statement is quite obvious in the case of Western classical music. It relies on the well-known fact that the score can be viewed from two different points of view, the composer's and the listener's (these points of view corresponding respectively to the "poïetique" and "esthesique" levels introduced in the French semiology of music, the score being associated to the "neutral" level) [35]. The composer's point of view must be distinguished from the listener's, because the richness of a musical masterpiece relies greatly on the fact that those points of view are not the same, which allows the listener to adopt many different points of view, each giving new qualities to the music. In the case of orally transmitted music traditions, this statement is less expected because no written score can be clearly identified with the neutral level, upon which the listener's multiple points of view are based. In some sense, Western classical music is closer to the Vanuatu sand drawing case. The figures traced in the sand could be to some extent associated with the neutral level. The fact that one of these figures can be traced using a continuous path is a property that is hardly recognizable when one looks at the figure. In this case, the formal property of the figure (the existence of a continuous path) is disconnected from qualities directly perceptible, and it remains a more speculative property.

As a last remark, I would like to go back to the cognitive issue addressed at the beginning of this paper. All the properties stated in this paper are related to visual or auditory forms produced by orally transmitted societies. None of them directly concerns the way these forms are produced, except to some extent the Eulerian property of sand drawings from the Vanuatu, which is mentioned explicitly in the mythology of the native peoples. From a cognitive point of view [3,4], our description focuses on the objects produced, and not on the cognitive processes producing them. When we describe what we call an "algorithm" producing an object a (this object being a lattice-like figure obtained by the geometric transformation we have studied in Angola, or a harp formula based on the translation of sequences in the Nzakara case), we construct a sequence of intermediate objects $a_0, a_1, \ldots$ with $a_n = a$. But this sequence of a_i can be viewed as an algorithm only if one can prove that the intermediate objects a_i have conscious existence in the native peoples' mind. Otherwise, we have only demonstrated a formal property of the given material.

In order to bring more cognitive insight to our investigations, we would need more information about the way people think about these objects and their related properties. Without such information, we only have a probabilistic understanding of the number of patterns satisfying some specific properties. Most of the properties we have studied were stated in combinatorial terms, by enumerating sets of combinations of elements, and by selecting only those satisfying specific properties. We have demonstrated that as soon as a formal property is discovered, the particular combinations satisfying this property were practiced quite generally. Among the countless number of possible combinations, those satisfying the given property that were used were more numerous than what would result from a random choice of these combinations (it was the case for the harp formulas built upon the translation of sequences, or the asymmetric rhythmic patterns satisfying the oddity property). There clearly exists a cognitive process underlying native people's choice, which favours this particular property, by selecting the corresponding patterns, whatever this selection process may be.

Funding for this study was provided by a grant from the French Ministry of Culture (1998–2000). I thank Tom Johnson who gave me the opportunity to study the arts of the Vanuatu during the exhibition at the Musée des Arts d'Afrique et d'Océanie in 1997. I thank all members of the Laboratoire d'ethnomusicologie (Musée de l'Homme) for their advices. I also thank Simha Arom for his help and inspiration, and Susanne Fürniss who provided me with Aka Pygmies musical examples from Simha Arom's archives. A Web page including animated figures illustrating this paper is available at the following URL: http://www.info.unicaen.fr/~marc/publi/diderot/index.en.html. I am greatly indebted to Tom Johnson for his comments on this text.

References

1. M. Andreatta, D.T. Vuza: Tata Mountains Math. Publications **22** (to appear, 2001)
2. U. D'Ambrosio: ISGEm Newsletter, **4**, 1 (1988) pp. 5–8 (http://web.nmsu.edu/~pscott/isgem41.htm)
3. D. Andler (ed.): *Introduction aux sciences cognitives*, coll. Cerisy, Paris, Folio, 1992
4. D. Andler: 'Logique, raisonnement et psychologie', J. Dubucs, F. Lepage (eds.), *Méthodes logiques pour les sciences cognitives* (Hermès, Paris 1995) pp. 25–75

5. R. D'Andrade: *The Development of Cognitive Anthropology* (Cambridge University Press, 1995)
6. S. Arom: *Polyphonies et polyrythmies d'Afrique Centrale. Structure et méthodologie* (Selaf, Paris 1985) (English translation: *African polyphony and polyrhythm*, Cambridge, 1991)
7. S. Arom: 'Symétrie et ruptures de symétrie dans la musique de tradition orale: le cas de l'Afrique Centrale', *Quadrivium. Musiques et sciences* (Institut de pédagogie musicale et chorégraphique, Paris, 1992) pp. 209–215
8. S. Arom, J. Khalfa: Revue de musicologie, **84**, 1 (1998) pp. 5–17
9. M. & R. Ascher: 'Ethnomathematics', *History of Science*, xxiv (1986) pp. 125–144
10. M. Ascher: 'Graphs in cultures: a study in ethnomathematics', *Historia Mathematica*, xv (1988) pp. 201–227
11. M. Ascher: *Ethnomathematics. A multicultural view of mathematical ideas* (Chapman & Hall, New-York, 1991) (French translation by K. Chemla, S. Pahaut, *Mathématiques d'ailleurs*, Seuil, Paris, 1998)
12. K.P. Brenner: *Die kombinatorisch strukturierten Harfen- und Xylophonpattern der Nzakana (Zentralafrikanische Republik) als klingende Geometrie – eine Alternative zu Marc Chemilliers Kanonhypothese* (Holos-Verlag, Bonn, to appear)
13. J.-P. Cabane: *Ululan, les sables de la mémoire* (Grains de sable, Nouméa, 1997)
14. M. Chemillier: 'La musique de la harpe', E. de Dampierre (ed.), *Une esthétique perdue* (Presses de l'Ecole Normale Supérieure, Paris, 1995) pp. 99–208
15. M. Chemillier: 'Mathématiques et musiques de tradition orale', H. Genevois, Y. Orlarey (éds.), *Musique & Mathématiques* (Aléas-Grame, Lyon, 1997) pp. 133–143
16. M. Chemillier: *La logique de la longue ligne Vanuatu, conférence au Musée des Arts d'Afrique et d'Océanie*, 30 octobre 1997 (http://www.info.unicaen.fr/~marc/publi/vanuatu/ephemere.html)
17. M. Cole, J. Gay, J.A. Glick, D.W. Sharp: *Cultural Context of Learning and Thinking* (Basic Books, NY, 1971)
18. T. Crump: *Anthropology of numbers* (1990) (French translation by P. Lusson, *Anthropologie des nombres* (Seuil, Paris, 1995))
19. É. de Dampierre: *Penser au singulier* (Société d'ethnologie, Nanterre, 1984)
20. E. de Dampierre (ed.): *Une esthétique perdue* (Presses de l'Ecole Normale Supérieure, Paris, 1995)
21. P.J. Davis, R. Hersh: *The mathematical experience* (Birkhäuser, Boston, 1982) (French translation by L. Chambadal, *L'univers mathématique* (Gauthier-Villars, Paris, 1985))
22. B. Deacon: J. of the Royal Anthro. Institute, **64** (1934) pp. 129–175
23. E.E. Evans-Pritchard: Bull. of the Faculty of Arts, University of Egypt, Cairo, **ii**, 1 (1934) pp. 1–36
24. P. Gerdes: *Une tradition géométrique en Afrique. Les dessins sur le sable* (L'Harmattan, Paris, 1995)
25. R. Graham, B. Rotschild, J. Spencer: *Ramsey Theory* (Wiley, 1980)
26. J.L. Grootaers: *Witchcraft substance and "Zande logic"*, unpublished
27. J. Habermas: *Théorie de l'agir communicationnel*, tome 1: *Rationalité de l'agir et rationalisation de la société* (Fayard, Paris, 1987) (original edition 1981)
28. M. Hollis, S. Lukes (eds.): *Rationality and relativism* (Basil Blackwell, Oxford, 1982)

29. R. Horton (ed.): *La pensée métisse* (IUED, Genève, 1990)
30. G. Ifrah: *Histoire universelle des chiffres* (Seghers, Paris, 1981) (English translation *From One to Zero: A Universal History of Numbers* (Viking, New York, 1985)
31. P.N. Johnson-Laird, R.M.J. Byrne: *Deduction*, Essays in Cognitive Psychology (LEA, 1991)
32. I. Lakatos: *Proofs and Refutations* (Cambridge University Press, 1976) (French translation by N. Balacheff, J.-M. Laborde, *Preuves et réfutations* (Hermann, Paris, 1984))
33. L. Lévy-Brühl: Les *fonctions mentales dans les sociétés inférieures* (Paris, 1910) (English translation *How Natives Think* (George Allen and Unwin, London, 1926))
34. L. Lévy-Brühl: *La mentalité primitive* (Paris, 1922)
35. J.-J. Nattiez: *Fondements d'une sémiologie de la musique* (Paris, coll. 10/18, 1975) (English translation *Music and Discourse: Towards a Semiology of Music* (Princeton University Press, 1990))
36. J. Piaget, B. Inhelder: *La naissance de la pensée logique de l'enfance à l'adolescence* (1958) (English translation *The Growth of Logical Thinking from Childhood to Adolescence* (Routhledge and Kegan Paul, London, 1958))
37. G. Pòlya: Z. f. Kristal., lx (1924) pp. 278–282
38. J. Retschitzki: *Stratégies des joueurs d'awélé* (L'Harmattan, Paris, 1990)
39. A. Rosenfeld, R. Siromoney: Languages of design **1** (1993) pp. 229–245
40. A. Speiser: *Die Theorie der Gruppen von endlicher Ordnung*, 2e ed. (Springer, Berlin, 1927)
41. D. Sperber: *La contagion des idées* (Odile Jacob, 1996)
42. A. Szabo: *Les débuts des mathématiques grecques* (Vrin, Paris, 1977) (original edition Budapest, 1969)
43. E.B. Tylor: *Primitive Culture*, 2 vol. (London, 1871)
44. *Vanuatu. Océanie. Arts des îles de cendre et de corail*, catalogue de l'exposition, (Musée des Arts d'Afrique et d'Océanie, Paris, 1997)
45. H. Weyl: *Symétrie et mathématique moderne* (Flammarion, Paris, 1964)
46. B.R. Wilson (ed.): *Rationality* (Basil Blackwell, Oxford, 1970)
47. C. Zaslavsky: *Africa counts. Number and Pattern in African Culture* (Prindle, Weber & Schmidt, Boston, 1973) (French translation by V. Henderson (Ed. du Choix, Paris, 1995))

Audio CDs and CD-ROMs

48. *Central African Républic. Music of the former Bandia courts*, recordings, texts and photographs by M. Chemillier & E. de Dampierre, CNRS/Musée de l'Homme, Le Chant du Monde, CNR 2741009, Paris, 1996
49. *La parole du fleuve. Harpes d'Afrique Centrale*, Cité de la musique, CM001, Paris, 1999
50. *Pygmées Aka. Peuple et musique*, CD-ROM realized by S. Arom, S. Bahuchet, A. Epelboin, S. Fürniss, H. Guillaume, J. Thomas, Montparnasse Multimédia, Paris, 1998

11 Expressing Coherence of Musical Perception in Formal Logic

Marc Leman

Abstract. Formal logic can be used for expressing certain aspects of musical coherence. In this paper, a framework is developed which aims at linking expressions in the formal language to an underlying interpretation in terms of musical images and image transformations. Such an interpretation characterizes truth within a framework of spatio-temporal representations and perception-based musical information processing. The framework provides a way for defining a semantics for the coherence of musical perception.

11.1 Introduction

Cognitive musicology aims at understanding the nature of musical information processing and imagination in composing, listening and performing. Of particular relevance is the choice of a proper description system.

Early cognitivist approaches of the seventies took formal (propositional and predicate) logic as the formalism for musical information processing [7,8]. Later on, and based on linguistic inspired ideas and models (see e.g. [14] for an overview), mathematical logics, extended formal logical systems (temporal, modal, non-monotonic, ...), and derived forms such as grammars, semantic networks, and Petri nets have been used as descriptive tools for musical conceptualization and subsequent reasoning [2]. What these approaches aim to achieve is the description of musical objects in connection to human information processing. In modeling perception and music analysis, assumptions often fit with the ontology of classical music theory (which is much *note* or *score* oriented).

Meanwhile, alternatives have been formulated on *ecological* and *naturalistic* grounds [9]. They ground human musical information processing on dynamic interactions with a musical sound environment, and model the physical and physiological processes using tools from the natural sciences. The so-called *functional equivalence models* give an account of the causal physical/physiological processes underlying musical behavior. (Examples of such equivalence models are dynamic systems, neural networks, cellular automata, and physical models.)

This paper is concerned with the question whether it is possible to connect the logical approach to ecological/naturalistic models, in particular, whether it is possible to express the coherence of musical perception in a formal logic, by adopting a semantics grounded in functional equivalence models. Such a linking could provide possible new ways of dealing with relationships between musical concepts in computational systems.

A basic motivation for developing a logic of musical coherence pertains to its possible role in grasping in a concise way the inherent compelling forces of human musical perception in computational musical information processing and associated musical content extraction and data mining [1]. For example, may we deduce from the occurrence of a clarinet sound that its timbre is nasal? Given a sequence of chords, may we conclude that the addition of a new chord will create tension release? A proper logic of musical coherence may furthermore contribute to creative applications in composition, performance, and interactive computer music systems [4,3,5,15].

In what follows, we first discuss the role of formal logics in expressing musical coherence. Then we introduce a framework for describing musical coherence in terms of musical images, spatio-temporal memories, and transformation principles. A final section explores the idea of interpreting the truth of formal expressions of musical coherence in terms of a mapping onto spatio-temporal representations and transformations.

11.2 Formal Logical Accounts of Musical Coherence

An early representative example is the musical logic, developed by Jos Kunst (1978), intended to represent the *dynamics* of a musical conceptualization. We take it as a representative example of a logical approach. The logic takes modalities, such as *necessary* ($\Box$), and *possible* ($\Diamond$) as operations on musical propositions ($p, q, \dots$) in the temporal domain. The propositions stand for any kind of conceptualization of heard music. $\Box p$, for example, would mean that 'p is always with respect to past and present', and $\Diamond p$ would mean 'p is at some time in the past or present'. The modal (temporal) logic allows for the description of a changing conceptualization. For example, when some new sonoric object is encountered while 'listening' to music, the conceptualization of the music may change, and what formerly was conceptualized as $\Box p$ may change into $\Diamond\Box p$. The sudden change in conceptualization is part of the formal description. The so-called bi-valence function describes processes of conceptual *un-learning* and *learning* situation. For example, when a concept p is attributed to a sonoric object at a starting time t_0, and this conceptualization is maintained until t_1, then $\Box p$ will be instantiated. However, at a certain moment, owing to a new sonoric situation and related conceptualization, the expression $\Box p$ may no longer hold true due to $-p$, hence at a next time step, the expression $-\Box p \wedge \Diamond\Box p$ is derived. It means that $\Box p$ is not longer valid and that at some time in the past we had $\Box p$. The *learning* part, then, introduces a new concept, say q, and the derivation of a new law $\Box q$. The conceptualization and subsequent unlearning of p is conceived of in terms of a time line which is distinct from the time line in which q is conceptualized.

The (non-monotonic) bivalence function $BivF(p, q)$ provides a means to cross both time lines. It expresses the change of conceptualization as an unlearning and learning process.

The example illustrates some strong as well as weak points of a formal mathematical logical approach. One of the strong points is that the propositions may relate to any kind of musical concept, from general concepts such as *I hear a Sonata composed by Pierre Boulez*, to *Clarinet sound!*, or *Piece in C major*, up to *I like this rough percussion section*, to simply *Pitch C*. Musical content and relationships between content can thus be described at a *meta-level*. *Clarinet sound* is furthermore a very concise representation of an object in the external world whose properties as physical object and auditory image representation may be harder to define. The conceptualization and its associated reasoning based on general laws thus provides a very concise means of representation and reasoning.

The logical approach is appealing for the representation of musical imagery, a mental act in which a sonoric object or process is represented as a concept in a person's mind. Many composers acknowledge the use of metaphorical descriptions and general conceptualization before they may come to a more concrete auditory imagery of the imagined object of process. Hence, also for composition, the formal logical approach is relevant.

Yet any abstract formal description of music has a problematic character as well. The expression *Clarinet sound*, formally expressed as p, has nothing in common with an actual sounding clarinet, nor with an imagined sounding clarinet. p merely points to a content, which is furthermore not represented in the language. The classical formal logic furthermore does not specify a criterion for testing the truth of p, it merely specifies a semantics through a mapping on truth-values leaving open the way in which p is actually evaluated. In a similar way, the logical constants ($\vee, \wedge, \supset$, etc...) do not connect content but to symbols (= pointers to contents), and their mapping on truth-values is purely formally defined, in terms of the truth-values of the symbols.

11.3 Reasoning about Musical Coherence

Reasoning about musical coherence, in computational terms, involves knowledge about the meaning of the coherence. For example, if we say that "a clarinet sound has a nasal timbre", then we don't simply say something about the clarinet sound, and about the nasal timbre, but also about the intrinsic connection that a clarinet sound *perceptually* implies a nasal timbre. More over, if "clarinet sound" is given as a concept, we should be able to deduce from it "nasal timbre", provided that we can rely on some knowledge which says that "clarinet sound *perceptually* implies nasal timbre". Such a deduction may be realized through non-explicit, dispositional knowledge that is provided by a perceptual model which can test the truth of the connection between "clarinet sound" and "nasal timbre". Provided that such a connection is found, then the deduction from "clarinet sound" to "nasal timbre" is accepted to be a valid one.

A possible way to proceed is to define the semantics of musical perceptual coherence in terms of interpretations within a mathematical model (a functional equivalence model of physical and physiological causality) that handles content and transformations between content. The truth of a statement expressing musical coherence then depends on its realization within that model.

11.4 Characterizing Coherence

Musical perceptual coherence is constrained by the *embedded physical environment*. It is the environment in which musical information processing takes place. It comprises the world of sounds and the constraints imposed by cultural conditions which allow us to characterize certain sounds (or sound sequences) as music.

The type of constraints that a logic of musical perceptual coherence should be able to grasp are those that define meaning when dealing with music perception and cognition. Meaning, however, is grounded in the way in which acoustical signals are dealt with through human sensory, perceptive and cognitive information processing principles and associated memories. A logic of musical coherence, therefore, has to deal with information processing. A particular example of coherence relates to aspects of tension creation and tension relaxation. Tension effects are typically based on the comparison of sound images that listeners build up in a memory. The coherence of tension effects then follows from the image transformation principles which model causal neuronal information processing.

It seems straightforward to define coherence as a logical concept for underlying causal interactions. In particular, transformation of sound into auditory images may be considered as a type of coherent transformation because *causal* possible transformation. As far as images are concerned, transformation of images into other images rely on auditory principles whose properties are grounded on causal principles (flow of neuronal *activation*). These transformation are called *coherent* when speaking about them.

11.5 Representations of Musical Content

In defining a semantics of musical perceptual coherence it is necessary to specify how musical content can be formally dealt with. Our strategy is indeed that an expression involving musical coherence is interpreted through a formal description of its content. This is achieved through the notion of *musical image*. Musical images are conceived of as spatio-temporal representational entities whose dynamics can be described in terms of image processing functions within different memory pools. In our approach, these entities provide a link between sound and its conceptualization. Below are listed some main features and observations concerning musical images.

- Definition of Images
  Images are conceived of as the result of feature extraction processes that ultimately rely upon sound waves. Images, therefore, reflect properties of the sound environment in a representational form carried by neurons. The activation of neurons is registered in terms of numerical values that account for the image information. The neural patterns that carry images may be described in different ways, occupying a defined space and time. Neural patterns occur as spike-trains, but in our modeling approach, a firing *rate-code* can be a useful encoding format for images. It allows the representation of images as vectors (see below). The rate-code refers to the amount of neuronal firing per time interval. Information can furthermore be time-coded as in the auditory periphery and then transformed into a space-encoded format. Transformation from time-code to space-code may be an important aspect of the musical logic.

- Image Processing
  It is possible to describe image processing in the auditory system independently from the actual physical (neuronal) carriers in terms of signal processing such as filtering, integration, and correlation. The signal processing mathematics accounts for physiological plausible, coherent, and often causally justified image transformations, including an initial sound to image transformation. Higher-brain processing dynamics can be described by other functional equivalent models, such as cellular automata, self-organizing networks, or just by image processing principles.

- Flux of Musical Sensations
  Low-level musical features, such as loudness, pitch, rhythm, and timbre are no fixed conceptualizations of sonoric objects. This is a problem for any formal logical approach to music. Instead, these musical features are to be understood as emerging primary features of underlying auditory processing. In the present model, musical images are based on the processing of signals in the time-domain, taking particular features of neuronal signal transmission in the auditory periphery such as the rate-code as a starting point. In this concept, the neural synchronization to temporal micro-structures in the neuronal patterns provides a foundation for dealing with the flux of musical sensations and hence of images as different as loudness, sensory dissonance, roughness, pitch, content-dependent pitch, density, volume, timbre, etc ...

- Different Time-Scales and Memories
  Images and associated processing provide a domain for logical reasoning. But it should be noted, however, that images can be related to different time-scales of signal processing. Images at the periphery of the auditory

system are assumed to change faster than at the center, where we may conceive of images in terms of memory structures formed at long term. We may also note the use of different memory systems in which images are operating. Memories are containers of images. The echoic memory systems accumulates sensory information and uses it for local contextual information processing [10], while the long-term memory may store invariant information into a sort of statistical memory system [11]. Episodic memory may hold salient episodes of musical information.

- Factualization
  Different time scales and memories suggests that a account of musical images should incorporate the notion of factualization, or the idea that images may have different degrees of abstraction from a concrete sound. Ultimately, images may connect to entities that can operate as abstract symbols in a formal logical reasoning process.

The many different types and appearances of spatio-temporal representational entities, and their particular emergent character suggest that a semantics of musical coherence has to rely on an integrated framework of image transformation operators.

To summarize thus far: A spatio-temporal representaitonal system for musical images provides a formal model for dealing with the sub-symbolic description of musical content. Its specification as a functional equivalence model for causal processing guarantees coherence. In establishing a connection between the symbolic level and the sub-symbolic level we have to deal with the following objects:

- extra-logical and logical symbols
- images and image transformations
- vectors and operations

The symbols are objects at the formal logical level. The extra-logical symbols point to sounds, images, or image-transformation principles, while logical symbols specify logical connections between symbols. Images are objects that describe a state-of-the-art at the neural processing level. They reflect properties of sound objects that occur both in space and in time. A next level is the level at which images and image transformation are implemented. This level is vector-oriented. The vectors representing images are the objects of numerical operations that have a functional equivalence with causal processing.

11.6 Implementation

A spatio-temporal representational system for musical images has been implemented in a Matlab-based environment currently under development, which provides an analysis tool for musical content using auditory-based musical signal processing. The toolbox comprises a description at three different levels: functional-logical, signal processing, and implementation.

- The *functional-logical* level provides a description of image transformation concepts in terms of a functional formal logic. It allows a concise description which resembles the way in which the Matlab modules are to be used while programming in Matlab. Thus far, however, reasoning has not been implemented.
- The *signal processing* level provides a description of the Matlab functions in terms of the mathematics of signal processing where a trade-off exists between psychoacoustical accurateness and processing efficiency, hence the notion of functional equivalence model which assumes that different degrees of functional equivalence may exist, e.g. models of the inner ear can be based on filters, or on physical modeling of resonance.
- The implementation level concerns the way in which the function is actually implemented in Matlab. Signal processing functions of the Matlab signal processing toolbox have been used where possible.

11.7 Image-transformations

Having set the general framework, we now give a few examples of images and image-transformation principles and their expression in a functional logical language.

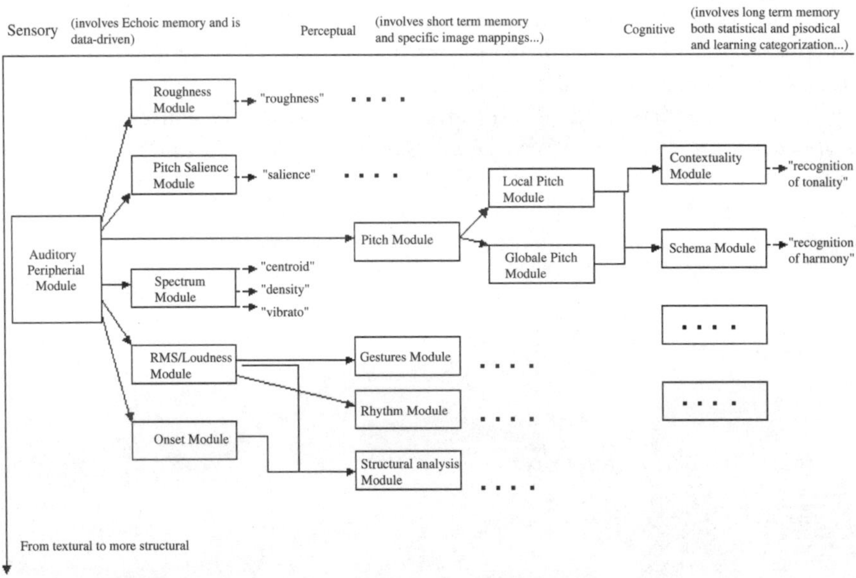

Fig. 11.1. Chart of image transformation modules, organized along the distinction between sensory, perceptual, and cognitive information processing (horizontal axis), and from textural to structural (vertical axis). The chart is not exhaustive

Figure 11.1 gives an overview of modules as conceived from the viewpoint of a distinction between sensory, perceptual, and cognitive processing. This chart, showing some processing modules as organized from sensory to cognitive (horizontal axis) and from textural to more structural (vertical axis), is not exhaustive. The modules are depicted as boxes, the associated indices as text, such as "roughness", "centroid", etc ... Important for the current purpose is that each module, which stands for a signal/image function or operator, has one or more associated image types from which indices are derived. In what follows, an example is given of a signal to image operator in the pitch domain. In the next section, its possible use within a formal logical reasoning system is discussed.

11.7.1 The "Primary" Auditory Images

The auditory peripherial module (APM) is a basic module in that it transforms sounds into a basic format on which all other images are based. The APM can be described as a function which transforms a musical sound signal $s(t)$ into a set of patterns $e_n(t)$ (with $n = 1 \dots 40$) that encode the responses of an array of auditory neuronal fibers spread along the basilar membrane (see (11.1)). The auditory filters, sometimes also called auditory *channels* thus perform an analysis of the sound in frequency sub-bands according to a critical band scale. They encode the signals as a spike-rate in time and frequency (time rate-code, and space rate-code). The rate-code amounts to the probability of neuronal firing during a short interval of time (in our implementation the interval is 0.4 ms). Space and time refer to the viewpoint: the *space code* considers a pattern in terms of an assembly of neurons, while the *time code* concerns the pattern of a single neuronal entity (such as a neuron or nerve fiber). The frequency range covered by the space code in terms of the center frequencies of the auditory filters is from 140 Hz to 8877 Hz but the highest frequency encoded in rate code is about 1250 Hz.

$$\mathrm{APM} : s(t) \longrightarrow \begin{bmatrix} e_1(t) \\ e_2(t) \\ \dots \\ e_{40}(t) \end{bmatrix} = \tilde{e}(t)$$

or, alternatively, we can write

$$\mathrm{APM} : s(t) \longrightarrow \tilde{e}(t) \tag{11.1}$$

or

$$\tilde{e}(t) = \mathrm{APM}\,(s(t)) \tag{11.2}$$

The whole pattern, denoted $\tilde{e}(t)$, is called the *primary image* of $s(t)$. It should be considered the auditory counterpart of D. Marr's (1982) *primal sketch* in the domain of visual perception. The tilde character denotes a vector and each vector-component corresponds to one auditory filter. Hence the tile character points to space code. A *running* vector means that the values of the vector-components change over time, where (t) denotes time. The pattern $\tilde{e}(t)$ thus represents the activation of a group of neurons at the level of the auditory nerve over subsequent time steps. All further information processing is based on these images, hence the connotation *primary image.*

The APM operator is effectively implemented in terms of signal processing functions that aim at simulating an auditory system at the periphery. We don't go deeper into the description of this module at the signal processing level, in order to have a more detailed focus at a next level, which relates to pitch.

11.7.2 Pitch Processing

Pitch processing can be captured by a formal logical description starting from the primary images. Starting from a neural pattern $e_{cf}(t)$, where cf represents an auditory channel, we may describe the particular transformation as:

$$e_{cf}(t) \longrightarrow \tilde{p}_{cf}(t)$$

where $\tilde{p}_{cf}(t)$ is the result of a periodicity analysis on a section of that pattern. The detected periods in that section are registered along a spatial array of periods. Hence, the transformation involves a transition from images in the temporal domain (time-code) to images in the spatial domain (space-code). At each time step, one thus obtains a periodicity analysis of the primary pattern $e_{cf}(t)$ that fits the window.

A periodicity analysis is performed in all $e_{cf}(t)$ patterns, so that we can write the periodicity analysis (PA) as:

$$\text{PA}: \begin{bmatrix} e_1(t) \\ e_2(t) \\ \dots \\ e_{40}(t) \end{bmatrix} \longrightarrow \begin{bmatrix} \tilde{p}_1(t) \\ \tilde{p}_2(t) \\ \dots \\ \tilde{p}_{40}(t) \end{bmatrix}$$

Secondly, a *coincidence mechanism* (CM) sums up the $\tilde{p}_{cf}(t)$ patterns and stores the result in a *summary auto-correlation* pattern called $\tilde{p}(t)$.

$$\text{CM}: \begin{bmatrix} \tilde{p}_1(t) \\ \tilde{p}_2(t) \\ \dots \\ \tilde{p}_{40}(t) \end{bmatrix} \longrightarrow \tilde{p}(t)$$

The resulting pattern $\tilde{p}$ at time t is also called the *completion image* at time t because the pattern will focus on common periodicity and in a sense complete a spectrum in function of its common sub-frequencies. The completion image is related to the notion of *virtual pitch pattern.* It gives an account of the common periodicity along the auditory neurons. Given an incomplete harmonic tone complex, its analysis in terms of a common periodicity of the auditory resolved harmonics will point to the fundamental of the harmonic complex, in a way which resembles a model proposed by Van Noorden (1982). The completion images are of interest in musical analysis, hence the function may be summarized as:

$$\mathrm{PM} : \tilde{e}(t) \longrightarrow \tilde{p}(t)$$

From the point of view of logical reasoning, care should be taken that the primary images $e(t)$ have a different dimensionality and meaning than the completion images $\tilde{p}(t)$. The space-code (as well as time-code) has a different nature and meaning.

Similar to the auditory peripherial module, a next level of detail allows the description in terms of a coherent signal processing mathematics. We do not enter into the details of signal processing here, suffice to say that the notion of periodicity can be implemented in different ways.

11.8 Perceptually Constrained Logical Reasoning

A semantics of formal logical reasoning can be set up which relates certain formal expressions to image transformations, hence providing a model theory in terms of mappings to a domain of image transformations or interpretations that characterize truth.

In that sense, the truth of a particular formal expression can be given by an interpretation in which coherence can be checked. Reasoning about musical concepts in the formal logical sense thus relies on a model which characterizes the derivations in terms of perceptually constrained transformation principles. These principles guarantee logical coherence in terms of causal transformation principles. The implementation of this idea in terms of a computational approach requires a linking between logical inference and the mathematics of signal processing.

In the logical language, the expression $p \longrightarrow q$ is read as "p coherent-implies q". It defines an operation on concepts whose truth-value is not grounded on the truth of p and q, but on an instantiation of the relation as a whole. The expression is validated (or falsified) by creating an interpretation of the expression in the domain of image transformation operators. This can be expressed by saying that in order to check the validity of $p \longrightarrow q$ we must check the validity of $\exists p', q', \Phi : p' \longrightarrow_{\Phi} q'$, where p' and q' are image instantiations (or interpretations) of p and q and where Φ refers to $a(n)$ (chain of) image transformation operations such as APM, PM, etc ... The interpretation thus maps the expression $p \longrightarrow q$ onto a statement which has a truth value.

For example, the expression "clarinet sound coherent-implies nasal timbre" is valid if nasal timbre can be obtained from an auditory-based signal analysis of the clarinet timbre. It involves a set of distinguished image transformation principles that start with the auditory peripheral module and subsequent analysis of the formant positions in derived spectral images. The truth of $p \longrightarrow q$, obviously, is contingent, which means that there are possible instantiations where it is false.

Inference rules can be based on the coherence-implication, provided that on all instantiations, it cannot be false. In dealing with such an approach, we may also involve the classical logical operators (such as conjunction ($\wedge$), disjunction ($\vee$), negation ($\sim$), and implication ($\supset$), and their truth semantics in the classical way, and specify new rules of consequence for the coherent-implication operator. Such rules can be established if the expression allows any interpretation of the coherent-implication.

For example, we may add the following deduction rule:

$$p, p \longrightarrow q \vdash q$$

which says that if p is true, and $p \longrightarrow q$ is true, then we can conclude that q is true. For example, if we have a 75 Hz amplitude modulated sound, and a 75 Hz amplitude modulated sound coherent-implies high roughness, then we may conclude that we have high roughness. $p \longrightarrow q$ can be seen as a disposition. The idea is indeed that q may be derived from p, provided that $p \longrightarrow q$ holds.

In a similar way, we may say, for example, that if a 75 Hz amplitude modulated sound coherent-implies high roughness, and we don't have high roughness, then we may conclude that we don't have a 75 Hz amplitude modulated sound. Hence, the following rule could be added to the inference system:

$$p \longrightarrow q, \quad \sim q \vdash \sim p\,.$$

In a similar way, we may add $p \longrightarrow \sim q, q \vdash \sim p$.

If a particular sound excerpt coherent-implies to be not in the key of C, and we are in the key of C, then we can conclude that we don't have this particular sound excerpt.

On the other hand, the following derivation can not be accepted:

$$\sim (p \longrightarrow q) \vdash q \longrightarrow p\,.$$

For example, if it is not true that a particular performance of a musical piece coherent-implies the expression of aggression, then we cannot derive that the expression of aggression coherent-implies that particular performance of a musical piece.

However, if $p \longrightarrow q$ is true then it follows that p implies q:

$$p \longrightarrow q \vdash p \supset q\,.$$

In other words, from the coherence between p and q we may conclude that if p is true, q cannot be false. The reverse is not accepted.

From p to deduce that $q \longrightarrow p$ is not acceptable either, because $q \longrightarrow p$ is checked as a whole, not solely on the true-value of p. The occurrence of a concept in the logics does not imply that this concept is coherent-implied.

Although the coherent implication is rather restricted, it can be combined as in the following general inference rules of propositional nature:

$$p \longrightarrow (q \wedge r) \vdash (p \longrightarrow q) \wedge (p \longrightarrow r)$$
$$(p \vee q) \longrightarrow r \vdash (p \longrightarrow r) \vee (q \longrightarrow r)$$
$$p \longrightarrow q, p \longrightarrow r \vdash p \longrightarrow (q \wedge r)$$
$$p \longrightarrow q, r \longrightarrow q, \vdash (p \vee r) \longrightarrow r$$
$$p \longrightarrow q, q \longrightarrow r, \vdash p \longrightarrow r$$

The propositional system may furthermore be extended with temporal operators: X (= next time), G (= always), F (= at some time), U (= until), B (= before). Extensions towards modal and temporal logic may involve the following rules:

$$F(p \longrightarrow q) \vdash F(p) \supset F(q)\,.$$

If at some time p coherent-implies q then we may conclude that if p some time, then q some time.

$$G(p \longrightarrow q) \vdash G(p) \supset G(q)$$

If always p coherent-implies q, then we may conclude that if always p, then always q.

$$(p \longrightarrow q)Br \vdash (pBr) \supset (qBr)\,.$$

If p coherent-implies q happens before r, we may conclude that if p happens before r, then q happens before r.

$$pB(q \longrightarrow r) \vdash (pBq) \supset (pBr)\,.$$

If p happens before q coherent-implies r, then p happens before q implies that p happens before r.

$$(p \longrightarrow q)Ur \vdash (pUr) \supset (qUr)\,.$$

If p coherent-implies q happens until r, then p happens until r implies that q happens until r.

$$pU(q \longrightarrow r) \vdash (pUq) \supset (pUr)\,.$$

If p happens until q coherent-implies r, then p happens until q implies that p happens until r.

The above examples show that the coherent-implication operator can be incorporated in a logical inference system. Its semantics can be defined in terms of a mathematical theory of perceptually constrained signal processing and inference rules can be specified which allows reasoning in a conceptual domain.

11.9 Discussion

The concern for the development of a concrete framework has forced us, thus far, to focus exclusively on *perceptually* coherence. Similar ideas, inspired by the idea that musical perception and musical action are strongly interconnected, could indeed be developed in view of *motoric* coherence (see Godøy, 1997 who argues along these lines). Yet, in the current framework, such an approach should involve equivalence models of bodily movement (e.g. physical models of movement) which we don't have at our disposal at this moment. In addition, we are well aware that the present account leaves open a number of problems and difficulties. Of particular interest also is the connection with modal logic and temporal logic. The coherence-implication discussed in the present paper seems to have a lot of properties in common with the implication in a modal logics called S4. Connections with temporal logics can be clarified in view of Marsden (2000).

11.10 Conclusion

Expressing musical coherence in a formal logic can be based on the representational entities of an underlying coherence system of spatio-representational entities. The framework consists in a link between lower-level perception-driven spatio-temporal representational structures on the one hand, and high-level conceptualizations on the other hand (See Fig. 11.2). A major advantage of this approach is that it ensures coherence between conceptualization in terms of inherent compelling forces (relating to sound and sound processing), rather than imposed meta-level relationships (established by an interpreter providing meaning to symbols).

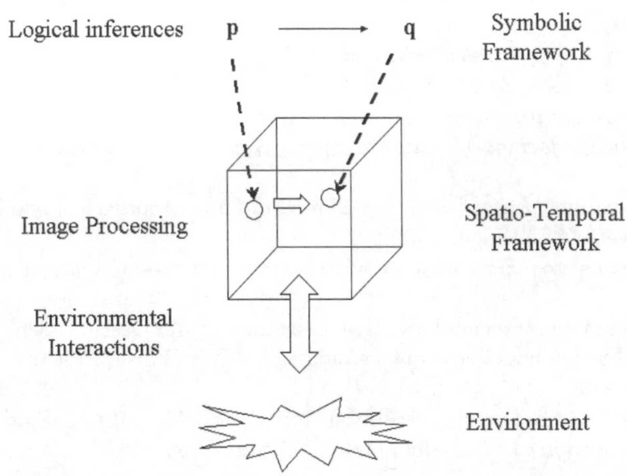

Fig. 11.2. Connection between concept, images and sonoric objects

Acknowledgments

Thanks to Francesco Carreras and Jean Paul Vanbendegem for reading a first draft.

References

1. Aigrain, P. (ed.): *Musical Content Feature Extraction.* Swets & Zeitlinger 1999. (Special issue of *Journal of New Music Research*)
2. Balaban, M., Ebcioglu, K., Laske, O.: *Understanding Music with AI: Perspectives on Music Cognition.* Cambridge, MA: MIT Press 1992
3. Camurri, A.: Music Content Processing and Multimedia: Case Studies and Emerging Applications of Intelligent Interactive Systems. Journal of New Music Research **28**(4), 351–363 (1999)
4. Camurri, A., Ferrentino, P.: Interactive Environments for Music and Multimedia. Multimedia Systems **7**, 32–47 (1999)
5. Camurri, A., Leman, M.: AI-Based Music Signal Applications. In: Roads, C., De Poli, G., Pope, S. (eds.): *Musical Signal Processing.* Lisse, The Netherlands: Swets & Zeitlinger 1997
6. Godøy, R.: Knowledge in Music Theory by Shapes of Musical Objects and Sound-Producing Actions. In: Leman, M. (ed.): *Music, Gestalt, and Computing: Studies in Cognitive and Systematic Musicology.* New York: Springer 1997
7. Kunst, J.: *Making sense in music: an enquiry into the formal pragmatics of art.* Ghent: Communication and Cognition 1978
8. Laske, O.: *Music, memory and thought.* Ann Arbor, MI: University of Microfilms International 1977
9. Leman, M. (ed.): *Music, Gestalt, and Computing: Studies in Cognitive and Systematic Musicology.* New York: Springer 1997
10. Leman, M.: An Auditory Model of the Role of Short-Term Memory in Probe-Tone Ratings. Music Perception **17**(4), 481–509 (2000)
11. Leman, M., Carreras, F.: Schema and Gestalt: Testing the Hypothesis of Psychoneural Isomorphism by Computer Simulation In: Leman, M. (ed.): *Music, Gestalt, and Computing: Studies in Cognitive and Systematic Musicology,* pp. 144–168. Heidelberg, New York: Springer 1997
12. Marr, D.: *Vision: a computational investigation into the human representation and processing of visual information.* San Francisco: W.H. Freeman and Company (1982)
13. Marsden, A.: *Representing Musical Time: A Temporal-Logic Approach.* Lisse, The Netherlands: Swets & Zeitlinger 2000
14. Monelle, R.: *Linguistics and Semiotics in Music.* Chur: Harwood Academic Publishers 1992
15. Paradiso, J.: The Brain Opera Technology: New Instruments and Gestural Sensors for Musical Interaction and Performance. Journal of New Music Research **28**(2), pp. 130–149 (1999)
16. van Noorden, L.: Two channel pitch perception. In: Clynes, M.: *Music, mind and brain: the neuropsychology of music.* London: Plenum Press 1982

12 The Topos Geometry of Musical Logic

Guerino Mazzola

Abstract. The logic of musical composition, representation, analysis, and performance share important basic structures which can be described by Grothendieck's functorial algebraic geometry and Lawvere's topos theory of logic. We give an account of these theoretical connections, discuss and illustrate their formalization and implementation on music software. Three issues are particularly interesting in this context: First, the crucial insight of Grothendieck that "a point is a morphism" carries over to music: Basically, musical entities are transformations rather than constants. Second, it turns out that musical concepts share a strongly circular character, meaning that spaces for music objects are often defined in a self-referential way. Third, the topos-theoretic geometrization of musical logic implies a progressively geometric flavour of all rational interactions with music, in particular when implemented on graphical interfaces of computer environments.

Das ist wohl schon die Mathematik des "Neuen Zeitalters".
Alexander Grothendieck [12] on "Geometrie der Töne" [20]

Introduction

When methods of modern algebra and functorial algebraic geometry were first applied to music theory in the early eighties [18,19], the central concern was not logic but geometry, i.e., the investigation of categories of local and global compositions which formalize the relevant objects and relations for harmony (cadence and modulation), counterpoint (Fux rules), melody, rhythm, large musical forms (in particular the classical sonata theory), including their classification, and the paradigmatic semiotics of musical structures as described by Ruwet [37] and Nattiez [31].

This approach included satisfactory theorems which model modulation, counterpoint, and string quartet theory in coincidence with the classical knowledge and tradition, which yield classification for some interesting global structures [20,23], and which have been operationalized in music composition software [21] and corresponding CDs [4–6,22]. It was however incomplete and too narrow in its concept framework for many musical problems.

Firstly, the Yoneda point of view was not properly developed. This defect became virulent after Noll's reconstruction of Riemann harmony [32]. Secondly, the development of the music platform RUBATO® for analysis and performance [24,27] enforced a critical review of music data models for universal purposes from score representation to performance [25] and the definition of an extended concept framework whose elements were described in [25,40] and implemented in RUBATO®'s PrediBase DBMS. Thirdly, the differenciation between mathematical fiction and musical facticity had to be explicated and led to the concept of textual and paratextual predicates [28,29]. At this point, logical and geometric perspectives were forced to unite. This approach is centered around topos-theoretic construction of musical predicates by means of logical and geometric operations and also targets at the design of universal composition tools. Presently, several research groups (e.g. TU Berlin, IRCAM Paris, Univ. Osnabrück, UNAM Mexico, ETH Zürich, Univ. Zürich) are collaborating in the theoretical and software design of these extensions.

In what follows, I shall first expose the concept framework of denotators and forms which mimic the Aristotelian concept of substance and form or – in more mathematical flavor – point and space. Follows the description of the central category of (functorial) local compositions, a special type of denotators, and its extension to 'musical manifolds', the (functorial) global compositions. The next section is devoted to a simple but motivating example: the unification of fundamental structures for Fux counterpoint and Riemann harmony. The concluding section focuses on the seamless transition between musical objects and truth objects in the language of denotators.

The present exposition is based on the forthcoming book *The Topos of Music* [30].

12.1 Galois Theory of Concepts

Forms and denotators are based on the category **Mod** of (left) modules over associative, rings[1] with identity. The morphisms of this category are the diaffine morphisms. This means that if M, N are modules over rings R, S, respectively, a diaffine morphism $f : M \to N$ is the composition $f = e^n \circ f_0$ of a dilinear morphism f_0 with respect to a ring homomorphism $r : R \to S$ and a translation e^n on the codomain N. The morphism set from M to N is denoted by $M@N$. The category of presheaves over **Mod** is denoted by $\mathbf{Mod}^{@}$; in particular, the representable presheaf of a module M is denoted by $@M$. More generally, for any presheaf F in $\mathbf{Mod}^{@}$, its value at module M will be denoted by $M@F$. In the context of $\mathbf{Mod}^{@}$, we shall call a module an *address*, a terminology which stresses the Yoneda philosophy, stating that the isomorphism class of a module is determined by the system $@M$ of all the 'perspectives' it takes when 'observed' from all possible addresses. Recall [17] that $\mathbf{Mod}^{@}$ is a topos whose subobject classifier Ω evaluates to $M@\Omega = \{S \mid S = \text{sieve in } M\}$. Its exponential Ω^F for a presheaf F evaluates

[1] The empty module (!) is included in this category to guarantee universal constructions.

to $M@\Omega^F = \{S|\ S = \text{subfunctor of } @M \times F\}$, and for a representable $F = @N$, we have $M@\Omega^{@N} \overset{\sim}{\to} (M \times N)@\Omega$, the set of sieves in $M \times N$. For a subfunctor $S \subset @M \times F$, an address B, and a morphism $f : B \to M$, we write $f@S = \{(f,s)|\ (f,s) \in B@S\}$, i.e., $B@S = \coprod_{f \in B@M} f@S$.

To construct the formal setup of denotators and forms, we consider the set $Mono\mathbf{Mod}^@$ of monomorphisms in $\mathbf{Mod}^@$. We further consider the set $Types = \{\mathbf{Simple}, \mathbf{Syn}, \mathbf{Limit}, \mathbf{Colimit}, \mathbf{Power}\}$ of form types. We then need the free monoid $Names = \langle UNICODE \rangle$ over the *UNICODE* alpabet[2]. We next need the set *Dia*(*Names*) of all diagram schemes with vertices in *Names*. More precisely, a diagram scheme over *Names* is a finite directed multigraph whose vertices are elements of *Names*, and whose arrows $i : A \to B$ are triples (i, A, B), with $i = 1, \ldots$ natural numbers to identify arrows for given vertices.

Next, consider the set $Dia(Names/\mathbf{Mod}^@)$ of diagrams on *Dia*(*Names*) with values in $\mathbf{Mod}^@$. Such a diagram is a map

$$dia : D \to \mathbf{Mod}^@$$

which with every vertex of D associates a functor and with every arrow associates a natural transformation between corresponding vertex functors. So $i : A \to B$ is mapped to the natural transformation $dia(i) : dia(A) \to dia(B)$.

With these notations, we can define a *semiotic of forms* as follows:

Definition 1 *A* semiotic of forms *is a set map*

$$sem : FORMS \to Types \times Mono\mathbf{Mod}^@ \times Dia(Names/\mathbf{Mod}^@)$$

defined on a subset $FORMS \subset Names$ *with the following properties* (i) to (iv). *To ease language, we use the following notations and terminology:*

- *An element* $F \in FORMS$ *is called a* form name, *and the pair* (F, sem) *a* form *(if sem is clear, the form is identified with its name)*
- $pr_1 \cdot sem(F) = t(F)$ *(= type of F)*
- $pr_2 \cdot sem(F) = id(F)$ *(= identifier of F)*
- $domain(id(F)) = fun(F)$ *(= functor or "space" of F)*
- $codomain(id(F)) = frame(F)$ *(= frame or "frame space" of F)*
- $pr_3 \cdot sem(F) = coord(F)$ *(= coordinator of F)*

Then these properties are required:

(i) *The empty word* $\emptyset$ *is not a member of FORMS*
(ii) *Within the coordinator of F, if* $t(F) \neq$ **Simple**, *the vertices of the diagram are form names, i.e. elements of FORMS*
(iii) *For any vertex X of the coordinator diagram coord(F), we have*

$$coord(F)(X) = fun(X)$$

[2] This is the current extension of the *ASCII* alphabet code to non-European letters.

(iv) *If the type $t(F)$ is given, we have the following for the corresponding frames:*
- *For* **Syn** *and* **Power**, *the coordinator has exactly one vertex G and no arrows, i.e. $coord(F) : G \to fun(G)$, what means that in theses cases, the coordinator is determined by a form name G.*
  Further, for **Syn**, *we have $frame(F) = fun(G)$, and for* **Power**, *we have $frame(F) = \Omega^{fun(G)}$.*
- *For* **Limit** *and* **Colimit**, *the coordinator is any diagram $coord(F)$. For* **Limit**, *we have the frame $frame(F) = lim(coord(F))$, and for* **Colimit**, *we have the frame $frame(F) = colim(coord(F))$.*
- *For type* **Simple**, *the coordinator has the unique vertex $\emptyset$, and a value $coord(F) : \emptyset \to @M$ for a module M, or, in a more sloppy notation: $coord(F) = M$.*

Given a form semiology, we shall denote a form by the symbol

$$F \underset{id(F):fun(F)\to frame(F)}{\longrightarrow} t(F)\,(coord(F))$$

and omit the identifier if it is the identity. We also write

$$F \underset{id(F)}{\longrightarrow} \mathbf{Limit}\,(F_0, F_1, \ldots F_n) \quad \text{and} \quad F \underset{id(F)}{\longrightarrow} \mathbf{Colimit}\,(F_0, F_1, \ldots F_n)$$

if the diagram reduces to the discrete set of forms $F_0, F_1, \ldots F_n$.

Given two forms F, G in a semiotic of forms *sem*, a morphism $f : F \to G$ is just a natural transformation $f : fun(F) \to fun(G)$. Hence every semiotic of forms defines its category $Forms_{sem}$ of forms. Observe that from semiotics can be defined over any topus instead of Mod^G, see [30].

The general problem of existence and size of form semiotics, i.e., the extent of the *FORMS* set, maximal candidates of such sets, gluing such sets together along compatible intersections, etc., is far from being settled. We shall not pursue this interesting and logically essential branch for reasons of space. The least one should say is that *regular* forms, i.e., those forms which are built from simple forms by transfinite recursion, may be supposed to be included in a form semiotics without further danger concerning logical consistency.

Example 1 For non-negative integers m, n, consider the forms

$$OnMod_m \underset{Id}{\longrightarrow} \mathbf{Simple}\,(\mathbb{Z}_m)$$

$$PiMod_n \underset{Id}{\longrightarrow} \mathbf{Simple}\,(\mathbb{Z}_n)$$

$$OnPiMod_{m,n} \underset{Id}{\longrightarrow} \mathbf{Limit}\,(OnMod_m, PiMod_n)$$

$$IntMod_{m,n} \underset{Id}{\longrightarrow} \mathbf{Limit}\,(\mathbb{D})$$

with $\mathbb{Z}$-modules $\mathbb{Z}_m$ and $\mathbb{Z}_n$ as coordinators, and with the diagram $\mathbb{D} =$

$$OnPiMod_{m,n} \xrightarrow{pr_1} OnMod_m \xleftarrow{pr_1} OnPiMod_{m,n}$$

associated with the canonical projection onto the form $OnMod_m$. The name "OnMod" symbolizes "onset modulo ... " whereas "PiMod" symbolizes "pitch modulo ... ", i.e., ordinary pitch classes. We see that the last form's diagram is just the condition that we should take the fiber product over onset, i.e., the simultaneity of two events in pitch and onset; this is a way to encode an interval of simultaneous note events.

But *circular*, i.e., non-regular forms do not exist automatically, nor are they uniquely defined. For example, defining a form

$$F \underset{I}{\longrightarrow} \mathbf{Power}\,(F)$$

is equivalent to selecting any monomorphism $I : G \rightarrowtail \Omega^G$, and setting $fun(F) = G$. To elaborate canonical monomorpisms, consider a set $S \subset A@G$ for a presheaf G. This defines a subfunctor $S^@ \subset @A \times F$ which in the morphism $f : B \to A$ takes the value $f@S^@ = \{f\} \times S.f$. Since we have $Id_M@S^@ = \{Id_M\} \times S$, S is recovered by $S^@$. This defines a presheaf monomorphism

$$?^@ : 2^G \rightarrowtail \Omega^G$$

on the presheaf 2^G of all subsets $2^{A@G}$ at address A. When combined with the singleton monomorphism $sing : G \rightarrowtail Fin(G) : x \mapsto \{x\}$ with the codomain presheaf $Fin(G) \subset 2^G$ of all finite subsets (per address), we have this chain

$$G \rightarrowtail Fin(G) \rightarrowtail 2^G \rightarrowtail \Omega^G$$

of monomorphisms. A number of common circular forms can be constructed by use of the following proposition ([30, Appendix G]):

Proposition 1 *Let H be a presheaf in* $\mathbf{Mod}^@$. *Then there are presheaves X and Y in* $\mathbf{Mod}^@$*such that*

$$X \stackrel{\sim}{\to} Fin(H \times X) \text{ and}$$
$$Y \stackrel{\sim}{\to} H \times Fin(Y)\,.$$

Example 2 It is common to consider sound events which share a specific grouping behavior, for example when dealing with arpeggios, trills or larger groupings such as they are considered in Schenker or in Jackendoff-Lerdahl theory [14]. We want to deal with this phenomenon in defining *MakroEvent* forms. Put generically, let *Basic* be a form which describes a sound event type, for example the above event type $Basic = OnPiMod_{m,n}$. We then set

$$Makro_{Basic} \underset{f:F \stackrel{\sim}{\to} Fin(FK) \rightarrowtail \Omega^{FK}}{\longrightarrow} \mathbf{Power}\,(Knot_{Basic})$$

with $F = fun(Makro_{Basic}), FK = fun(Knot_{Basic})$

and the limit form

$$Knot_{Basic} \underset{Id}{\longrightarrow} \mathbf{Limit}\,(Basic, Makro_{Basic})\,,$$

a form definition which by the above proposition yields existing forms.

The typical situation here is an existing form semiotic *sem* and a bunch of 'equations' $E_{F_1,F_2,\ldots F_n}(F)$ which contain the form names $F_1, F_2, \ldots F_n$ already covered by *sem*, and the new form name F. The equations are just form definitions, using different types and other ingredients which specify forms. The existence of an extended semiotics sem' which fits with these equations is a kind of algebraic field extension which solves the equations E. This type of conceptual Galois theory should answer the question about all possible solutions and their symmetry group, i.e., the automorphisms of sem' over *sem*. No systematic account of these problems has been given to the date, but in view of the central role of circular forms in any field of non-trivial knowledge bases [3], the topic asks for serious research.

The level of forms is still not the substance we are looking for. The substance is what is called a denotator. More precisely, given an address A and a form F, a *denotator* is a quatruple $Name : A \rightsquigarrow F(c)$, consisting of a string D (in *UNICODE*), its name, its address A, its form F, and its coordinates $c \in A@\mathit{fun}(F)$. So a denotator is a kind of substance point, sitting in its form-space, and fixed on a determined address. This approach is really a restatement of Aristotelian principles according to which the real thing is a substance plus its "instanciation" in a determined form space. Restating the above coordinates as a morphism $c : @A \to \mathit{fun}(F)$ on the representable contravariant functor $@A$ of address A by the Yoneda lemma, the "pure substance" concept crystallizes on the representable functor $@A$, the "pure form" on the functor $\mathit{fun}(F)$, and the "real thing" on the morphism between pure substance and pure form.

In classical mathematical music theory [23], denotators were always special zero-addressed objects in the following sense: If M is a non-empty R-module, and if $0 = 0_{\mathbb{Z}}$ is the zero module over the integers, we have the well-known bijection $0@M \xrightarrow{\sim} M$, and the elements of M may be identified with zero-addressed points of M. Therefore, a *local composition* from classical mathematical music theory, i.e., a finite set $K \subset M$, is identified with a denotator $K^* : 0 \rightsquigarrow \mathit{Loc}(M)(K)$, with form

$$\mathit{Loc}(M) \underset{\mathit{Fin}([M]) \rightarrowtail \Omega^{[M]}}{\longrightarrow} \textbf{Power}([M])$$

and $[M] \longrightarrow \textbf{Simple}(M)$.

Evidently, this approach relates to approaches to set theory, such as Aczel's hyperset theory [1] which reconsiders the set theory as developed and published by Finsler[3] in the early twenties of the last century [8,9]. The present setup is a generalization on two levels (besides the functorial setup): It includes circularity on the level of forms and circularity on the level of denotators. For instance, the above circular form named $\mathit{Makro}_{\mathit{Basic}}$ enables

[3] It is not clear whether Aczel is aware of this pioneer who is more known for his works in differential geometry ("Finsler spaces").

denotators which have infinite descent in their knot sets. Similar constructs intervene for frequency modulation denotators, see [30,28].

The denotator approach evidently fails to cover more connotative strata of the complex musical sign system. But it is shown in music semiotics [28, section 1.2.2] that the highly connotative Hjelmslev stratification of music can be construed by successive connotational enrichment around the core system of denotators. This is the reason why the naming "denotator" was chosen: Denotators are the denotative kernel objects.

12.2 Categories of Local and Global Compositions

Although the category of all denotators is defined [30], we shall focus on the classically prominent subcategory of local compositions. These are the denotators $D : A \rightsquigarrow F(x)$ whose form F is of power type. More precisely, we shall consider A-addressed denotators with coordinates $x \subset @A \times \mathit{fun}(S)$, where form S is called the *ambient space*[4] of D. If there is a set $X \subset A@S$ such that $x = X^{@}$, the local composition is said to be *objective*, otherwise, we call it *functorial.*

Given two local compositions $D : A \rightsquigarrow F(x), E : B \rightsquigarrow G(y)$, a morphism $f/\alpha : D \to E$ is a couple $(f : x \to y, \alpha \in A@B)$, consisting of a morphism of presheaves f and an address change α such that there is a form morphism $h : S \to T$ which makes the diagram of presheaves

$$\begin{array}{ccc} x & \longrightarrow & @A \times S \\ {\scriptstyle f}\downarrow & & \downarrow{\scriptstyle @\alpha \times h} \\ y & \longrightarrow & @B \times T \end{array}$$

commute. This defines the *category Loc of local compositions.* If both, D, E are objective with $x = X^{@}, y = Y^{@}$, one may also define morphisms on the sets X, Y by the expressions $f/\alpha : X \to Y$ (forgetting about the names) which means that $f : X \to Y$ is a set map such that there is a form morphism $h : S \to T$ which makes the diagram

$$\begin{array}{ccc} X & \longrightarrow & A@S \\ {\scriptstyle f}\downarrow & & \downarrow{\scriptstyle A@h} \\ Y.\alpha & \longrightarrow & A@T \end{array}$$

of sets commute. This defines the *category ObLoc of objective local compositions.* Every objective morphism $f/\alpha : X \to Y$ induces a functorial morphism $f^{@}/\alpha : x \to y$ in an evident way. This defines a functor

$$?^{@} : \mathit{ObLoc} \to \mathit{Loc}.$$

[4] If no confusion is likely, we identify S with $\mathit{fun}(S)$.

This functor is fully faithful. Moreover, each functorial local composition x (again forgetting about names) gives rise to its *objective trace* $X = x_@$ where $\{Id_A\} \times X = Id_A@x$. If we fix the address A and restrict to the identity $\alpha = Id_A$ as address change, we obtain subcategories $ObLoc_A, Loc_A$ and a corresonding fully faithful embedding $?^@_A : ObLoc_A \to Loc_A$. In this context, the objective trace canonically extends to a left inverse functor $?_{@A}$ of $?^@_A$. Moreover

Proposition 2 *The morphisms* $?_{@A}$ *and* $?^@_A$ *build an adjoint pair* $?^@_A \dashv ?_{@A}$.

So, on a fixed address, objective and associated functorial local compositions are quite the same. But there is a characteristic difference when allowing address change. This relates to universal constructions:

Theorem 1 [30] *The category Loc is finitely complete.*

If we admit general adderess changes, the subcategory of objective local compositions is not finitely complete, there are examples [30] of diagrams $E \to D \leftarrow G$ of objective local compositions whose fiber product $E \times_D G$ is not objective[5]. Therefore address change – which is the portal to the full Yoneda point of view – enforces functorial local compositions if one insists on finite completeness. This latter requirement is however crucial if, for example, Grothendieck topologies must be defined (see below).

The dual situation is less simple: There are no general colimits in *Loc*. This is the reason why global compositions, i.e., 'manifolds' defined by (finite) atlases whose charts are local compositions, have been introduced to mathematical music theory [18,23]. More precisely, given an address A, a *global composition* G^I is a presheaf G which is covered by a finite atlas I of subsheaves G_i which are isomorphic to (the functors of) A-addressed local compositions with transition isomorphisms $f_{i,j}/Id_A$ on the inverse images of the intersections $G_i \cap G_j$. The significant difference of this concept is that the covering $(G_i)_I$ is part of the global composition, i.e., no passage to the limit of atlas refinements is admitted. For music this is a semiotically important information since the covering of a musical composition is a significant part of its understanding [20]. In fact, a typical construction of global compositions starts with a local composition and then covers its functor by a familiy of subfunctors, together with the induced atlas of the canonical restrictions, the result is called an *interpretation*. The absence of colimits in *Loc* can be restated in terms that there are global compositions which are not isomorphic[6] to interpretations.

The theory of global composition is a proper extension of the more local theory of denotators and forms. The fact that local compositions admit arbitrary finite limits implies that the same is true for the category *Glob* of global

---

[5] The right adjointness of the objective trace functor for fixed addresses only guarantees preservation of limits for fixed addresses.

[6] Global composition build a category, morphisms being defined by gluing together local morphisms, see [23,30].

compositions. We may therefore define a Grothendieck (pre)topology on *Glob* via covering families. Its covering families for a global composition G^I are finite collections of morphisms $(H_k^{J_k} \to G^I)_k$ which generate the functor of G^I. Various Čech cohomology groups (in the sense of Verdier [13, exposé V]) can be associated to covering families of this *finite cover* Grothendieck topology [30]. The sheaf of affine function of a global composition is used in the *classification theory of global compositions.* This theory has been worked out for objective global compositions which have locally free addresses of finite rank [30, Chapter 15, Theorem 18], see [23] for an early version of that result concerning the zero address. The result exhibits a *locally projective classifying scheme whose rational points represent isomorphism classes of global compositions.* For ambient spaces which stem from finite modules, combinatorial results regarding the number of isomorphism classes of selected global compositions (such as chords, motives, dodecaphonic series, mosaics) have been obtained by Fripertinger [10], using Pólya's and de Bruijn's enumeration theory [7].

12.3 "Grand Unification" of Harmony and Counterpoint

In this section, we shall shortly illustrate on a concrete musicological situation: harmony and counterpoint, why some of the above general concepts have been introduced. Classically, mathematical music theory worked on the pitch class space $PiMod_{12}$ introduced above. In what follows, we shall slightly adjust it by the "fifth circle" automorphism $.7 : \mathbb{Z}_{12} \xrightarrow{\sim} \mathbb{Z}_{12}$, i.e., we consider the synonymous form

$$FiPiMod_{12} \underset{@.7}{\longrightarrow} \mathbf{Syn}\,(PiMod_{12})$$

which means that pitch denotators are now thought in terms of multiples of fifths, a common point of view in harmony. On this pitch space, two extensions are necessary: extension to intervals and extension to chords. The first one will be realized by a new form space

$$IntMod_{12} \longrightarrow \mathbf{Simple}\,(\mathbb{Z}_{12}[\epsilon])$$

with the module $\mathbb{Z}_{12}[\epsilon]$ of dual numbers over the pitch module $\mathbb{Z}_{12}$. We have the evident form embedding

$$\otimes 1 : FiPiMod_{12} \rightarrowtail IntMod_{12} : x \mapsto x \otimes 1$$

of this extension, where we should pay attention to the interpretation of a zero-addressed interval denotator

$$D : 0 \rightsquigarrow IntMod_{12}\,(a + \epsilon.b)\,.$$

It means that D has cantus firmus pitch a and interval quantity b in terms of multiples of fifths. For example, the interval coordinate $1+\epsilon.5$ denotes the pitch of fifth from the basic pitch (say 'g' if zero corresponds to pitch 'c'), together with the interval of $7.5 = 11$, i.e., the major seventh ('b' in our setup). The set K_ϵ of consonant intervals in counterpoint are then given by the zero-addressed denotators with coordinate $a + \epsilon.k, k \in K = \{0, 1, 3, 4, 8, 9, \}$. The set D_ϵ of dissonant intervals are the remaining denotators $a + \epsilon.d, d \in D = \mathbb{Z}_{12} - K$.

The counterpoint model of mathematical music theory [20] which yields an excellent coincidence of counterpoint rules between this model and Fux' traditional rules [11] is deduced from a unique affine automorphism, the autocomplementary involution $AC = e^2.5$ on the pitch space: we have $AC(K) = D$, $AC(D) = K$. It can be shown [20,32] that this unique involution and the fact that K is a multiplicative monoid uniquely characterize the consonance-dissonance dichotomy among all 924 mathematically possible 6-6-dichotomies. This model's involution has also been recognized by neurophysiological investigations in human depth EEG [26]. Consider the consonance stabilizer $\mathit{Trans}(K_\epsilon, K_\epsilon) \subset \mathbb{Z}_{12}[\epsilon]@\mathbb{Z}_{12}[\epsilon]$. This one is canonically related to Riemann harmony in the following sense.

In his PhD thesis, Noll succeeded in reconstructing Riemann harmony on the basis of "self-addressed chords". This means that pitch denotators

$$D : \mathbb{Z}_{12} \rightsquigarrow \mathit{FiPiMod}_{12}\,(e^y.x)$$

are considered instead of usual zero-address pitch denotators which here appear as those which factor through the zero address change $\alpha : \mathbb{Z}_{12} \to 0$, i.e., the constant pitches. A *self-addressed chord* is defined to be a local composition with ambient space $\mathit{FiPiMod}_{12}$, and Noll's point was to replace zero-addressed chords by self-addressed ones. Figure 12.1 shows a zero-addressed and a self-addressed triad in the pitch form PiMod_{12}.

In Riemann's spirit [34–36], the harmonic "consonance perspective" between the constant dominant triad $\mathit{Dominant} : 0 \rightsquigarrow \mathit{FiPiMod}_{12}\,(1, 5, 2)$ and the constant tonic triad $\mathit{Tonic} : 0 \rightsquigarrow \mathit{FiPiMod}_{12}\,(0, 4, 1)$ is defined by the monoid $\mathit{Trans}(\mathit{Dominant}, \mathit{Tonic}) \subset \mathbb{Z}_{12}@\mathbb{Z}_{12}$, a self-addressed chord generated by the transporter set of all morphisms $u : \mathit{Dominant} \to \mathit{Tonic}$.

This self-addressed chord is related to the above stabilizer as follows: Consider the tensor multiplication embedding

$$\otimes\epsilon : \mathbb{Z}_{12}@\mathbb{Z}_{12} \rightarrowtail \mathbb{Z}_{12}[\epsilon]@\mathbb{Z}_{12}[\epsilon] : e^u.v \mapsto e^{(u+\epsilon.0)}.(v \otimes \mathbb{Z}_{12}[\epsilon]).$$

Then we have a "grand unification" theorem ([32], see also [33] for more details):

Theorem 2 *With the above notations, we have*

$$\mathit{Trans}(\mathit{Dominant}, \mathit{Tonic}) = \otimes\epsilon^{-1}\,\mathit{Trans}(K_\epsilon, K_\epsilon).$$

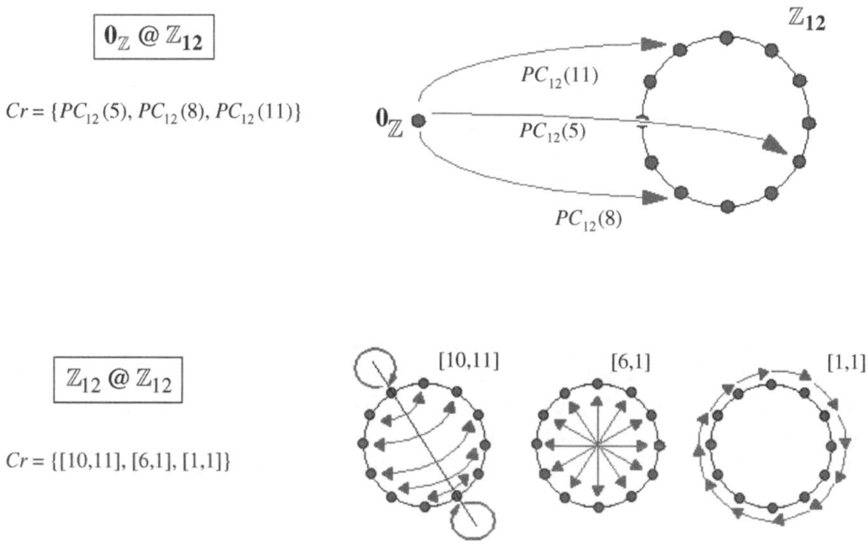

Fig. 12.1. Above, a zero-addressed 12-tempered pitch class triad is shown. The zero address is the domain of three constant affine maps which target at the pitch classes $PC_{12}(5)$ etc. in $\mathbb{Z}_{12}$. Below a self-addressed 12-tempered pitch class triad is shown. Its elements are affine endomorhisms $[10, 11] = e^{10}.11$ etc. of $\mathbb{Z}_{12}$

This means that the Fux and Riemann theories are intimately related by this denotator-theoretic connections. At present, it is not known to what extent this structural relation has been involved in the historical development from contrapuntal polyphony to harmonic homophony.

12.4 Truth and Beauty

So far, the exposé covers a powerful concept framework, including circular space and point concepts, as well as an elaborate theory of categories of local and global music objects, comprising classification by algebro-geometric parametrization in the vein of Grothendieck topologies and associated sheaf topoi. The flavor is however still overly geometric, and logical evaluation has not yet been addressed explicitely.

The logical perspective intervenes when one tries to distinguish between mathematically relevant objects and objects which share musical or musicological facticity. For example, the denotators which parametrize the nine symphonies of Beethoven are facts whereas another denotator of the same form – supposing that we are given a common form called "Symphony", say – which could possibly describe Beethoven's 'Tenth Symphony' is pure mathematical fiction. To grasp this difference between mathematical potentiality and musicological facticity, we introduced so-called textual predicates, concepts which are related to Agawu's work on music semiology [2] which builds on the tradition of Jakobson's [15,16] research in modern poetology.

Textual prediactes are extensional and relate to what Agawu in Jakobson's terminology calls introversive semiosis[7], i.e., production of meaning on the basis of intratextual signs, the "universe of structure". Examples: Schenker's "Ursatz" (beginning/middle/ending), Ratner's model of harmonic functions, and, of course, all elementary signs for metric, rhythmical, motivic, harmonical, etc. structures. Introversive semiosis can be said to be production of textual meaning because the text is the relevant reference level for introversive semiosis.

To control the variety of textual semiosis it is necessary to set up an adequate system of signification mechanisms for facticity. To this end, suppose that a determined category *Den* of denotators has been selected, for example the category *Loc* of local compositions discussed above. Denote by $Den^\infty = \coprod Den^k$ the category which is the union of all positive cartesian products of *Den*. Select a module I of "truth values", and set

$$Val(I) \longrightarrow \mathbf{Simple}\,(I)$$

$$TRUTH(I) \longrightarrow \mathbf{Power}\,(\,Val(I))$$

$$T_I = \text{ set of all denotators over } TRUTH(I)$$

$$Texig(Den)_I = T_I^{Den^\infty}\,, \text{ the set-theoretic exponential set}\,.$$

With this, a *textual semiosis* is a map

$$Sig_{Den} : Tex \to Texig(Den)_I$$

on a subset *Tex* of the name set *Names* which is also called *set of expressions* to distinguish its elements from form or denotator names, and whose elements are called *textual expressions.* If $Ex \in Tex$ is any such textual expression, and if $f. \in Den^k$ is any k-tuple of denotators or morphisms (identifying objects with the identities, as usual) we write $f./Ex$ for the value $Sig_{Den}(Ex)(f.)$.

This evaluation relates to facticity, i.e., "being the case"[8] as follows. If the value $f./Ex$ identifies to $f./Ex : A \rightsquigarrow TRUTH(I)\,(t)$, we have a sieve $t \subset @(A \times I)$.

In the special case of $A = I = Null$, the zero module over the zero ring, the sieve t identifies to a truth arrow $t : 1 \to \Omega$ since the final element 1 identifies to $@Null$, and by Yoneda, $Hom(1, \Omega) \stackrel{\sim}{\to} Null@\Omega \stackrel{\sim}{\to} Null@\Omega^1$. So we have

[7] Agawu's extroversive semiosis relates to what we call paratextual predicates. These ones involve signs which transcend the system of musical signs in the narrow sense of the word. Agawu calls them "the universe of topics". Topics are signs which have a signified beyond the text. We shall not deal with this intensional signification process here.

[8] Recall Wittgenstein's starting proposition: "Die Welt ist alles, was der Fall ist." in his tractatus [39].

a classical truth value from topos theory, including $t = \emptyset = \bot, t = \top = @Null$ as false and true arrows. In the special case of truth module $I = S_1$, the circle group, any half-open interval $[0, e[\subset S_1, 0 \leq e < 1$ defines a truth denotator

$$Fuzzy(e) : Null \rightsquigarrow TRUTH(S_1)\left([0, e[^{@}\right)$$

of ordinary fuzzy logic with false $Fuzzy(0)$ and true $Fuzzy(1)$. But here, we are really approaching objects of the local composition theory. In fact, the ambient space $TRUTH(S_1)$ includes the subspace $PiMod_m$ via the inclusion $\mathbb{Z}_m \overset{\sim}{\rightarrow} \frac{1}{m}\mathbb{Z}/\mathbb{Z} \rightarrowtail S_1$. This reviews zero-addressed chords $Ch : 0 \rightsquigarrow PiMod_m(Ch_1, \ldots, Ch_n)$ as being the objective counterpart of a discrete fuzzy truth value $Ch^{@}$ with respect to the circle group.

As usual, the complete Heyting algebra $A@\Omega^{@I}$ of A-addressed truth values in I recovers the logical operations (negation, implication, conjunction, disjunction, sup and inf for universal quantifiers).

The essential of this perspective is this:

Principle 1 *The truth denotators are ordinary local compositions, and we may therefore embedd them in the general theory of local and global (!) compositions as a very special item of musical objects, i.e., of objects which were meant to describe beauty, not truth.*

The theory of textual predicates which takes off at this point is concerned with construction methods of new predicates from given ones. Basically, *atomic* predicates are defined by (1) mathematical formulas, (2) prima vista predicates (corresponding to proper introversive semiosis), and (3) shifter predicates, defined by arbitrary usage. They are combined by logical and topos-theoretic constructs to build semiotically *motivated predicates*, see [25,28–30] for details. A particular case of this approach is Orlarey's idea of applying lambda calculus to composition software, such as it is realized in *OpenMusic* or *Elody* [38, Ch.2,3]. The connection is that a denotator D may be interpreted as having predicate Ex, $D/Ex = \top$, where $/Ex$ is generated by abstraction *with respect to a specific mathematical formula.* In other words, the idea is that a composer takes a concrete musical object D which is then lifted to a predicate extension and may be varied by taking another object D' with $D'/Ex = \top$. In this way, musical composition and analysis, abstract representation and facticity are on their way to a unified realm of true beauty. A beautiful truth.

References

1. Aczel, P.: *Non-well-founded Sets.* No. 14 in CSLI Lecture Notes. Stanford: Center for the Study of Language and Information 1988
2. Agawu, V.K.: *Playing with Signs.* Princeton: Princeton University Press 1991
3. Barwise, J., Etchemendy, J.: *The Liar: An Essay on Truth and Circularity.* New York: Oxford University Press 1987

4. Beran, J.: *Cirri.* Centaur Records 1991
5. Beran, J., Mazzola, G.: *Immaculate Concept.* Zürich: SToA music 1992
6. Beran, J.: *Śānti.* Bad Wiessee: col-legno 2000
7. de Bruijn, N.G.: Pólya's Theory of Counting. In: Beckenbach, E.F. (ed.): *Applied Combinatorial Mathematics*, Chapt. 5. New York: Wiley 1964
8. Finsler, P.: Über die Grundlegung der Mengenlehre. Erster Teil. Die Mengen und ihre Axiome. Math. Z. **25**, 683–713 (1926)
9. Finsler, P.: *Aufsätze zur Mengenlehre.* Unger, G. (ed.). Darmstadt: Wiss. Buchgesellschaft 1975
10. Fripertinger, H.: *Endliche Gruppenaktionen in Funktionenmengen – Das Lemma von Burnside – Repräsentantenkonstruktionen – Anwendungen in der Musiktheorie.* Doctoral Thesis, Univ. Graz 1993
11. Fux, J.J.: *Gradus ad Parnassum (1725).* Dt. und kommentiert von L. Mitzler. Leipzig 1742
12. Grothendieck, A.: Correspondence with G. Mazzola. April, 1990
13. Grothendieck, A., Dieudonné, J.: *Eléments de Géométrie Algébrique I–IV.* Publ. Math IHES no. 4, 8, 11, 17, 20, 24, 28, 32. Bures-sur-Yvette 1960–1967
14. Jackendoff, R., Lerdahl, F.: *A Generative Theory of Tonal Music.* Cambridge, MA: MIT Press 1983
15. Jakobson, R.: Linguistics and Poetics. In: Seboek, T.A. (ed.): *Style in Language.* New York: Wiley 1960
16. Jakobson, R.: Language in relation to other communication systems. In: *Linguaggi nella società e nella tecnica.* Edizioni di Communità. Milano 1960
17. Mac Lane, S., Moerdijk, I.: *Sheaves in Geometry and Logic.* Berlin, Heidelberg, New York: Springer-Verlag 1994
18. Mazzola, G.: Die gruppentheoretische Methode in der Musik. *Lecture Notes*, Notices by H. Gross, SS 1981. Zürich: Mathematisches Institut der Universität 1981
19. Mazzola, G.: *Gruppen und Kategorien in der Musik.* Berlin: Heldermann 1985
20. Mazzola, G.: *Geometrie der Töne.* Basel: Birkhäuser 1990
21. Mazzola, G.: presto *Software Manual.* Zürich: SToA music 1989–1994
22. Mazzola, G.: *Synthesis.* Zürich: SToA music 1990
23. Mazzola, G.: Mathematische Musiktheorie: Status quo 1990. Jber. d. Dt. Math.-Verein. **93**, 6–29 (1991)
24. Mazzola, G., Zahorka, O.: The RUBATO Performance Workstation on NeXTSTEP. In: ICMA (ed.): Proceedings of the ICMC 94, S. Francisco 1994
25. Mazzola, G., Zahorka, O.: Geometry and Logic of Musical Performance I, II, III. *SNSF Research Reports* (469 pp.), Zürich: Universität Zürich 1993–1995
26. Mazzola, G. et al.: Neuronal Response in Limbic and Neocortical Structures During Perception of Consonances and Dissonances. In: Steinberg, R. (ed.): *Music and the Mind Machine.* Berlin, Heidelberg, New York: Springer-Verlag 1995
27. Mazzola, G., Zahorka, O.: *RUBATO on the Internet.* Univ. Zűrich 1996. http://www.rubato.org
28. Mazzola, G.: Semiotic Aspects of Musicology: Semiotics of Music. In: Posner, R. et al. (eds.): *A Handbook on the Sign-Theoretic Foundations of Nature and Culture.* Berlin, New York: Walter de Gruyter 1998. Preview on http://www.ifi.unizh.ch/mml/musicmedia/publications.php4

29. Mazzola, G.: Music@EncycloSpace. In: Enders, B. (ed.): Proceedings of the klangart congress'98. Universität Osnabrück 1998 Preview on http://www.ifi.unizh.ch/mml/musicmedia/publications.php4
30. Mazzola, G. et al.: *The Topos of Music – Geometric Logic of Concepts, Theory, and Performance.* To appear, Basel: Birkhäuser 2002
31. Nattiez, J.-J.: *Fondements d'une Sémiologie de la Musique.* Paris: Edition 10/18 1975
32. Noll, T.: *Morphologische Grundlagen der abendländischen Harmonik.* Doctoral Thesis, TU Berlin 1995
33. Noll, T.: The Consonance/Dissonance-Dichotomy Considered from a Morphological Point of View. In: Zannos, I. (ed.): *Music and Signs – Semiotic and Cognitive Studies in Music.* Bratislava: ASCO Art & Science 1999
34. Riemann, H.: *Musikalische Logik.* Leipzig 1873
35. Riemann, H.: *Vereinfachte Harmonielehre oder die Lehre von den tonalen Funktionen der Akkorde.* London 1893
36. Riemann, H.: *Handbuch der Harmonielehre.* Leipzig 6/1912
37. Ruwet, N.: *Langage, Musique, Poésie.* Paris: Seuil 1972
38. Vinet, H., Delalande, F. (eds.): Interface homme-machine et création musicale. Paris: Hermès 1999
39. Wittgenstein, L.: *Tractatus Logico-Philosophicus (1918).* Frankfurt/Main: Suhrkamp 1969
40. Zahorka, O.: PrediBase – Controlling Semantics of Symbolic Structures in Music. In: ICMA (ed.): *Proceedings of the ICMC 95.* S. Francisco 1995

13 Computing Musical Sound

Jean-Claude Risset

Summary The links between mathematics and music are ancient and profound. The numerology of musical intervals is an important part of the theory of music: it has also played a significant scientific role. Musical notation seems to have inspired the use of cartesian coordinates.

But the intervention of numbers within the human senses should not be taken for granted. In the Antiquity, while the pythagorician conception viewed harmony as ruled by numbers, Aristoxenus objected that the justification of music was in the ear of the listener rather than in some mathematical reason. Indeed, the mathematical rendering of the score can yield a mechanical and unmusical performance.

With the advent of the computer, it has become possible to produce sounds by calculating numbers. In 1957, Max Mathews could record sounds as strings of numbers, and also synthesize musical sounds with the help of a computer calculating numbers specifying sound waves. Beyond composing with sounds, synthesis permits to compose the sound itself, opening new resources for musicians. Digital sound has been popularized by compact discs, synthesizers, samplers, and also by the activity of institutions such as IRCAM. Mathematics is the pervasive tool of this new craft of musical sound, which permits to imitate acoustic instruments; to demonstrate auditory illusions and paradoxes; to create original textures and novel sound material; to set up new situations for real-time musical performance, thanks to the MIDI protocol of numerical description of musical events. However one must remember Aristoxenus' lesson and take in account the specificities of perception.

Introduction

The first section will recall the role of mathematics in the theory of music – intervals, structures, musical syntax. But the article will deal mostly with digital sound: since 1957, the computer permits to deal with sounds as numbers, which opens possibilities to renew the musical vocabulary, namely the sonic material. The composer Edgard Varèse liked to remind that new materials lead to novel structures, in music as well as in architecture. Mathematics is the pervasive tool of this unprecendented craft of musical sound.

Mathematics and Musical Theory

The links between mathematics and music are ancient and profound. According to Leibniz, "music is a secret calculation done by the soul unaware that it is counting". This line of thought goes back to Pythagoras, who applied arithmetics to the description of natural phenomena in his study of music: on the monocord or on the lyra, privileged musical intervals correspond to simple ratios of string lengths. Hence the pythagorian dogma "numbers rule the world" – the music of heavenly spheres as well as that of sounds: this view has stimulated the scientific study of natural phenomena. In the Middle Ages, the quadrivium, highest level of science education, included arithmetics, geometry, astronomy and music. According to the XIIIth century music theorist Johannes de Garlandia, "music is the science of number related to sound" ("*Musica est scientia de numero relato ad sonos*") (cf. [44]).

Cardan, Kepler, Galileo, Descartes, Gassendi, Huygens, Newton, Leibniz all wrote musical treatises. As Johannes de Garlandia and later Euler remarked, those treatises did not deal with musical practice, but rather with music theory – a theory implying some numerological mysticism going back to the Pythagoras school.

Later, romantic composers have despised mathematics. Yet recent research on hearing – in particular the work of Licklider, Plomp and Terhardt – substantiate Leibniz's conception: the evaluation of musical intervals imply counts and correlations, which amount to unconscious arithmetic operations within the brain of the listener.

Musical notation is close to a time-frequency display – one should rather say time-scale. Two musical intervals judged equivalent by the ear correspond to the same frequency ratios. Hence the use of a logarithmic scale, which has been materialized in keyboards and much earlier in lithophones. By mapping time onto space, notation suggests symmetries which have inspired contrapuntal inversion and retrogradation: such transformations are absent from music transmitted through oral tradition. Ars Nova – condemned by the Pope Giovanni XXII – resorted to these figures in complex combinations. In Webern's work as well as in the *Musical Offer* and the *Art of Fugue* by Bach, one can find palindromic canons.

Equal temperament appears as a mathematical approximation. When one compares a tonal melody played in the equally-tempered scale, the just scale (or Zarlino scale) and the Pythagorean scale, the difference can be heard clearly for the IIIe et VIe degree, lower for Zarlino and higher for Pythagoras.

Rameau's theory of the fundamental bass is often considered as the theoretical foundation of tonal music: it is based on the work of Mersenne, Sauveur et d'Alembert[1]. As stressed by the composer and philosopher Hugues Dufourt, here is "a global move of rationality uniting mathematics and music toward the achievement of a common functional goal".

---

[1] In addition to arithmetic considerations, the theory also implies a psychophysical assumption according to which the more common harmonics between tones, the less dissonant their combination. This is confirmed by recent experiments by Plomp and Terhardt.

Musical automata resort to a stored program – probably for the first time: long before Descartes, they code music according to a cartesian representation. According to the British historian Geoffroy Hindley, western musical notation has inspired cartesian coordinates – which played a considerable role in the blooming of western science, with the developement of mathematical analysis, the translation of Newton's law into differential equations, the computation of trajectories and the Laplacian notion of determinism.

In his book *Symmetry*, Hermann Weyl describes the early use of groups of transformations in western musical composition. As stressed by the eminent mathematician Yves Hellegouarch, also an outstanding cello player, a number of other concepts were also implemented implicitly in music before being clearly formulated in mathematics, for instance the notions of logarithm, of arithmetic modulo. Hellegouarch reminds us that Farey has developed his series in the course of a study of the numerology of musical intervals. According to Hellegouarch, the consideration of music could stimulate mathematicians and suggest a different approach, less discursive, more intuitive, holistic and even romantic[2].

The concept of artificial intelligence was first clearly articulated around 1840 by Lady Lovelace, who collaborated with Charles Babbage to implement the Analytical Engine, an ambitious calculating machine which foreshadowed the digital computer: Lady Lovelace understood that this machine could be used to deal with other objects besides music, and she gives the example of musical composition. Fifty years before Chomsky, the Vienna musicologist Heinrich Schenker developed a theory of tonal music introducing the notion of generative grammar.

While the romantic ideology rejected sciences, XXth century music has come back closer to mathematics. Serial music aims at the search of a radical novelty refusing the heritage: its syntax is combinatorial. Mathematics has been a strong source of poetic suggestion for Edgard Varèse and György Ligeti. Iannis Xenakis resorted to mathematical models: his "stochastic" music calls for the statistical control of musical parameters, and so do compositions by James Tenney, Göttfried-Michael Koenig and Denis Lorrain. Such control was implemented earlier in experiments by Pierce, Fucks et Hiller, who attempted to create music and to imitate existing compositions from statistical analysis. Today David Cope has extended this approach of musical style by "analysis-by-synthesis". Research on computer-assisted composition is actively pursued: it calls for techniques such as object and logic programming. However, while one can speak of geometrical art (illustrated by names

[2] In 1970, Hellegouarch has himself shown the way which has recently lead to the demonstration of the Fermat theorem: relating it to elliptic curves which would have implausible properties if the theorem were wrong. Such an imaginative link may have to do with Hellegouarch's own itinerary: Paris Conservatory for cello first prize, he never followed high school.

such as Mondrian, Albers, Vasarely, Herbin ...), it seems fair to say that there is no strong musical school based on mathematics.

Most of the above examples deal with theory rather than with musical practice. Warnings are in order against the risk to reduce compositional and perceptual processes to musical operations. Jean-Toussaint Desanti has insisted on the specificity of mathematical procedures, which can only be verified "from inside". Similarly music has its specificities, which I stressed in a 1977 article. Aristoxenus already declared that the justification of music lies in the ear rather than in some mathematical reason. Frequency ratios such as 2 or 3/2 are no longer perceived as musical octaves or fifths above 5000 Hz (cf. [3]): musical practice takes this in account by limiting the tessitura of instruments to a lower frequency. Also the mathematical rendering of the score can yield a mechanical and unmusical performance. The characteristics of musical performance have been studied by both analysis and synthesis by researchers such as Alf Gabrielsson, Johan Sundberg, Erich Clarke, Carol Palmer, Bruno Repp. These studies have shown that performers deviate from mathematically accurate parameters in systematic ways, in order to underline specific musical structure or articulations.

Digital Sound

With the advent of the computer, it has become possible to produce sounds by calculating numbers. In 1957, Max Mathews could record sounds as strings of numbers, and also synthesize musical sounds with the help of a computer calculating numbers which specify sound waves. Digital synthesis is a novel source of musical material, which permits to perform "integral composition", to compose the sound itself: it has aroused the interest of musicians, as evidenced by the activity of institutions such as IRCAM in Paris. Digital sound has been popularized by compact disks, synthesizers, samplers, transmission of sounds on the world-wide web.

Mathematics is the pervasive tool of this new craft of musical sound[3], which permits to imitate acoustic instruments; to demonstrate auditory illusions and paradoxes; to create original textures and novel sound material; to set up new situations for real-time musical performance, thanks to the MIDI protocol of numerical description of musical events.

Digital representation of continuous signals is an important chapter of discrete mathematics. The simplest coding process is called sampling: a continuous wave $p(t)$ can be represented by a string of numbers $p(t)\partial(t-nT)$, representing the successive values of $p(t)$ at closely spaced time intervals. T, the *sampling period*, is the interval of time separating two successive samples; the

[3] According to Gaston Bachelard, mathematics is the language and the tool of all sciences rather than a science of its own. The present book presents a number of applications of mathematics to sound, for instance modeling, recognition, restoration or compression.

sampling rate is $F = 1/T$ ($\partial(x)$ is 0 if x is non-zero and 1 for $x = 0$). If the frequency spectrum of $p(t)$ does not contain any component above a certain limit $f_{\max}$, the process of sampling – replacing $p(t)$ by $p(t)\partial(t-nT)$ – does not lead to any information loss insofar as $F > 2f_{\max}$. Thus one can sample the audible component of sound – limited to approximately 20 000 Hz – provided one uses sampling rates higher than 40 000 Hz. This condition is fulfilled in compact disks, which use a sampling rate of 44 100 Hz: its validity has been demonstrated by Claude Shannon in his 1947 *Mathematical Theory of Communication.* Actually it had been established earlier by a number of authors: around 1930 by Nyquist, Kupfmüller, Kolmogorov, and as early as 1917 by Whittaker as an interpolation theorem using function $\sin(x)/x$; according by Bernard Picinbono, Cauchy was already aware of the proper conditions for loss-less sampling.

Programming digital computers permits to compute samples in varied and flexible ways. Synthesis can thus yield very diverse sonic structures, with an unprecedented precision and reproducibility. The musician can, so to say, compose the sound itself, directly, freed from the need to build specific mechanical vibrating systems which impose their constraints and their idiosyncrasies. For instance, one can only produce acoustically harmonic sustained sounds – comprising components with frequencies proportional to $1, 2, 3, \ldots$ – since they can only be obtained by forcing quasi-periodic vibrations in mechanical systems: in contradistinction, various algorithms such as additive synthesis or audio frequency modulation permit to realize sustained sounds of arbitrary length with inharmonic spectral content, and to compose such timbres as chords.

However the exploitation of a mathematical property or formula does not necessarily lead to interesting results – most often it does not. It is fair to say that the main foundations of digital signal processing have been established in the context of the musical exploration of the possibilities of digital sound: but one must add that this exploration raises the crucial problem of the auditory perception of sound structures. The pythagorean conception – numbers rule music – must be appraised to take in account the already mentioned objection of Aristoxenus – the justification of music lies in the ear of the listener rather than some a mathematical reason[4].

Indeed, it my be hard to predict the auditory effect of even a simple sound structure. I like to illustrate this with the case of a synthetic sound combining 10 periodic tones (my example is described in detail in [27]). All tones comprise the 10 first harmonics with an equal amplitude, but they have a slightly different fundamental frequency –55 Hz, 55 Hz + 1/20 Hz, 55 Hz+2/20 Hz, 55 Hz+3/20 Hz, ..., 55 Hz+9/20 Hz. Heard alone, the tone of 55 Hz frequency sounds as an ordinary low A. Most listeners do not anticipate

[4] As was mentioned above, the relation between the frequency ratio and the heard interval collapses above 5000 Hz, and so does our capacity to evaluate musical intervals. One cannot dismiss the specific workings of perception.

correctly the way the combination of so-close frequencies sounds like (1/20 Hz corresponds to 1/60th of a semi-tone): one hears a "song of harmonics" with frequencies 55 Hz multiplied by 1, by 2, ..., by 10: each harmonic disappears and reappears at rates porportional to its rank (every 2 seconds – 20/10 – for harmonic 10, every 20/9 s, that is, a little less frequently, for harmonic 9, etc.

We have here a complex phenomenon of beats, according to the trigonometric identity

$$\cos 2\pi f_1 t + \cos 2\pi f_2 t = 2 \cos\{2\pi (f_1 + f_2) t/2\} \cos\{2\pi (f_1 - f_2) t/2\} .$$

The pattern is made clearer by the multiplicity of interferences. One could conclude that this is merely a physico-mathematical phenomenon which could be anticipated. However it is auditory perception which determines whether a vibration of frequency f is perceived as a tone with a pitch (audio domain) or not (ultra-sound or infra-sound domain), hence whether the superposition of two periodic tones is perceived as such (first member of the above identity) or as a single tone with a sub-audio frequency modulation (second member of the identity). Aristoxenus' objection to Pythagoras is inescapable: as Pierre Schaeffer said, "music is meant to be heard". The musical project can be betrayed if one merely stipulates numerical relations between physical parameters without checking that the intended relations are preserved in the sonic realization which carries them to the ear and brain of the listener. In the section "Illusions, paradoxes" below, I shall illustrate this in a more striking way.

The example I just discussed shows that even with relatively simple sound structures, the auditory effect is not always easy to predict. It also shows that the extreme design precision afforded by digital sound permits leads to effects which can be musically interesting. Beyond "harmonic song", I modified the previous example to "animate" arbitrary chords through the interference of periodic sounds with a defective harmonic series: one can choose a low-enough fundamental and harmonics of high-enough rank to correspond to the desired chord, since rational numbers can approximate real numbers with an arbitrary precision.

Synthesis Programs

Thanks to digital-to-analog converters, one can avoid completely the use of mechanical vibrating systems and directly design a sound from blueprints. The diversity of possible sounds results from the rich ressources of programming. Mathews has designed general programs for sound synthesis, called Musicn – those which have been used the most are Music4, Music5 and their variants Cmusic and Csound. These are open and powerful programs. The data particularizing the program – called the "score" for synthesis – act like recipes for sound production: at the same time, they provide thorough descriptions of the sound structures, which help to disseminate the synthesis know-how. In 1969, I have published a catalog of synthesis sounds, collecting the recordings and the Music5 scores of a certain number of musical synthesis experiments: the descriptions included permit to replicate the syntheses later, in other centres, with other programs or with sound synthesizers, and they are still valid and useful.

Programs such as Music5 permit the user to resort to varied protocols and to obtain sounds of diverse morphologies by varying parameters and synthesis models. These models can be as complex and elaborate as desired: but it is useful to briefly describe some basic models.

Additive, Substractive and Non-linear Synthesis Models

Musicians did not wait for Fourier to pile up harmonic partials: since the XVI-th century, organ builders have resort to so-called mutation stops, in which one key opens several pipes tuned to the harmonics of the fundamental frequency – for instance thirds, in the so-called stops, or fifths, in stops called "nasard" ou "larigot". Such stops compose various timbres according to the process of "additive synthesis"; appropriate superpositions of harmonics can evoke certain instrumental timbres, such as the "cornet" stop.

Building a sound via additive synthesis is akin to building a wall by piling up bricks. In Fourier synthesis, the basic bricks for sound are sine waves. Ohm realized that the ear is insensitive to the relative phases of the various harmonics, even though changing them can upset the waveshape: it is sufficient to specify the spectrum, that is, the respective weight of the harmonics, to simulate a periodic tone. To approximate instrumental sounds, one must modulate the amplitude of the components by appropriate functions of time (called envelopes). To evoke gongs, bells or drums, on must choose non-harmonic components. One can also resort to other basic "bricks". Additive synthesis is the most general and intuitive method: but it requires much computer power and a wealth of data, since all the details of the desired sounds must be specified.

In substractive synthesis, the sound is manufactured by eliminating the unwanted components of a rich and complex sound through filtering – similarly to a sculptor who extracts his work from the stone by eliminating the guangue which surrounds it. Digital filtering techniques have considerably progressed since the inception of digital sound synthesis. A process known as predictive coding determines the characteristics of a variable filter so that it can simulate a given sound: this technique is well adapted to the sounds of speech, produced through the evolving filtering, by the vocal tract, of the sound of the vocal chords.

A third process consists to globally transforming a sound, to warp, bend, distort it like clay. Thus distorting a sine wave yields a complex sound. If the distorsion process is such that this sound remains periodic, it can be

Fourier-analyzed into since waves, and the effect of the distorsion is to add harmonics.

Daniel Arfib (1979) has shown that one can distort a sine wave so as to produce an arbitray harmonic spectrum. For the Tchebytcheff polynomial of order k, $Tk(\cos wt) = \cos kwt$: the product of the distorsion of a sine wave of frequency f and amplitude 1 by the Tchebytcheff polynomial of order k is a sine wave of frequency kf. To obtain a Fourier spectrum with amplitudes $A1, A2, \ldots, Ak$, one can distort a sine wave of amplitude 1 by the transfer function $\sum A_k T_k$. To illustrate this process, Arfib has realized sound examples in which a sine wave with an amplitude growing from 0 to 1 is distorted by the successive Tchebytcheff polynomials. When the amplitude is close to 0, the spectrum is close to one containing only harmonic 1 (the fundamental). When the amplitude increases, the spectrum gets richer and includes higher components: when it reaches 1, it is reduced to the kth harmonic for the Tchebytcheff polynomial of order k. If k is even, the sound begins at frequency $2f$, since

$$\cos^2 x = \{\cos 2x + 1\}/2 .$$

In these examples, the components introduced by distorsion are clearly audible: so to say, one can hear the trigonometric transformations realized by the Tchebytcheff polynomials.

Non-linear distorsion also permits to produce non-harmonic tones: if one adds amplitude modulation, the spectral lines are shifted in frequency, since

$$\cos a \cos b = 1/2 \{\cos(a+b) + \cos(a-b)\} .$$

Amplitude modulation produces a periodic sound if the carrier and modulating frequencies are commensurable. If the frequency ratio is rational, one produces aperiodic sounds with a inharmonic spectral content (like bell sounds).

Frequency modulation can be used as a method to produce complex spectra by non-linearity. In radio broadcasting, the carrier frequency – for instance 94.20 MHz – is much larger than the modulating frequencies – which are audiofrequencies, hence lower than 20 kHz – and the radio receiver perform demodulation to restore the audiofrequencies. John Chowning (1973) has implemented frequency modulation to synthesize complex spectra by giving similar values – both audio – to the carrier and the modulating frequencies: this process, "FM", is a powerful and economical method, which provides a very effective control over spectra and spectral width. With this method, Chowning could synthesize a great variety of timbres, instrument-like or not. The computation of spectra produced by frequency modulation calls for Bessel functions. Scanning the so-called modulation index – the amplitude ratio between modulating and carrier frequency – gives a sonic rendition of Bessel functions.

I briefly mention here physical modelling, a significant synthesis method, which has been developed in particular by Claude Cadoz, Annie Luciani et

Jean-Loup Florens: here the computer resolution of the differential equations governing a vibrating system yields a sound signal and an image as well. This method presents the advantage of being intrinsically multi-media, and it produces very "physical" sounds with a strong identity.

Imitation of Instruments

One has long believed that the signature of musical instruments – the physical correlate of their timbre – was their frequency spectrum. This view is valid for quasi-periodic tones. But attempts to simulate the tones of musical instruments by reproducing their Fourier spectrum as given by Acoustics treatises often failed: certain instruments could not be evoked this way, even if one introduces attack and decay transients.

In the sixties only, with computer synthesis, has it become possible to exert complete control over the synthesis of complex sounds. This allows to isolate the physical correlates of timbre and to understand the cues for the identity of such or such instrument. The genesis of acoustic sounds is indirect: it requires the excitation of a vibrating system, and the relatively stable structure of this system ensures the stability of the timbre, which is the signature of the sound origin. In contradistinction, computer synthesis requires the specification of the physical parameters: it forces the user to have some knowledge of the cues for timbre. Synthesis makes it possible to check the auditory relevance of the features extracted by analysis: one can speak of analysis by synthesis.

Using the methodology of analysis by synthesis, I have shown that the signature of "brassy" tones (trumpet, trombone, horn) is not a characteristic Fourier spectrum, but a relationship between spectrum and intensity: when the amplitude increases, the spectrum gets richer in high frequencies. This is also true during the attack phase, which lasts only a thirtieth of a second or so: the ear cannot analyse this short phase to be aware of the non-synchrony of the components, but it recognizes it as a characteristic pattern ([55]). Hence it is impossible to simulate a brassy tone by a tone with a fixed spectrum, even if one carefully controls the durations of the attack and decay transients. But Chowning realized an elegant simulation of brassy tones with his FM method: since the modulation index determines the spectral width, one only has to gang the modulation index to the amplitude envelope of the carrier sine wave. The research by Chowning and others has developed an extended know-how for FM synthesis: this technique was implemented in the popular DX7 synthesizers of Yamaha, which benefitted from this research to bring an unprecedented synthesis quality and diversity.

The main signature of bowed strings (violon, viola, cello) is also a relation, in this case between fundamental frequency and spectrum: this relation results from the frequency response of the instrument body, which displays a number of resonances. This explains the very specific quality of the vibrato: the quasi-periodic frequency modulation is accompanied by a complex synchronous spectral modulation. Such a vibrato cannot be imitated by the frequency modulation of a fixed spectrum.

Instrumental sustained tones are quasi-periodic, with a harmonic spectrum. Percussion instruments have a inharmonic spectrum. The frequencies of the components influences timbre: thus piano tones are close to harmonicity, the first components of a bell tone are close to the first frequencies of a harmonic series, unlike gong, cymbal or drum sounds. The characteristics of the decay – shape, rate, spectral variation – also influence the timbre.

The final goal of digital synthesis is not the production of ersatz: but the stage of imitative synthesis was decisive to understand which physical features determine the identity and the internal life of certain tones. Moreover it opens novel musical possibilities, for instance in the so-called "mixed" pieces associating musical instruments live with synthesis sounds: the latter can come close to instrumental sounds but also diverge from them. If one is able to simulate various instrumental sounds by assigning different parameters to the same synthesis model, it is easy to interpolate between the parameter values to metamorphose an instrumental timbre into another one – a process called morphing in the visual domain – without fade-in/fade-out.

Composition of Textures

The issues of the exploration of musical timbre are far-reaching, beyond scientific understanding: at stake is the creation of novel timbres and the renewal of the sonic vocabulary available for music. Beyond the instrumental domain, rich but loaded with connotations, many musicians are keen on conquering new territories of artificial timbres. In the sixties, Max Mathews, Dick Moore and myself at Bell Laboratories, John Chowning at Stanford University, Barry Vercoe at MIT explored the possibilities of sound synthesis in scientific research institutions. In the seventies, Pierre Boulez has created IRCAM (*Institut de Recherche et de Coordination Acoustique Musique*), where digital techniques are pervasive. These techniques are nowadays available on personal computers.

The ressources of digital synthesis have been exploited to create novel sound textures – to compose the sound itself, disposing of time within sounds, and not to merely disposing sounds in time.

Dissonance relates to the roughness of encounters between close spectral components. It is possible to control the frequency composition of inharmonic tones so as to provoke consonance for other intervals than the octave, the fifth or the third: this was suggested by John Pierce and realized by John Chowning in his piece *Stria*.

In my works *Mutations*, *Inharmonique* and *Songes*, I have set up relations between harmony and timbre by composing sound textures just as chords, and by transforming inharmonic sounds with synchronous attacks (sounding bell-like) into bell-like textures: the dispersion in time of spectral components helps hearing the inside of the sounds, just as a prism disperses the spectral components of white light.

Illusions, Paradoxes

Illusory and paradoxical sounds clearly demonstrate the specific possibilities of digital synthesis.

Shepard has produced twelve tones which form a chromatic scale and which seem to go up endlessly when they are repeated: this evidences the circularity and the non-transitivity of judments of pitch. I have gone beyond this to synthesize continuoulsy gliding sounds that seem to go up or down indefinitely, or to go down the scale, yet which after a while are much higher than where they started. I have produced a tone which seems to go down in pitch when all its frequencies are doubled. This demonstrates that pitch, a perceptual attribute of the auditory experience, is not isomorphous to frequency, a physical parameter that can be measured objectively. It is essential to be aware of the complexity of the pyschoacoustic relation between the physical structure of the sounds and their auditory effect: otherwise the composer could stipulate relations between parameters which the listener would not be perceive as expected by the composer.

This is also true in the rhythmical domain. I have produced a succession of beats which seem to slow down when one doubles the tape speed of the tape recorder on which it is played. The perceived relations can be contrary to the physically programmed ones.

I have used these paradoxical sounds in compositions such as *Little Boy*, *Mutations* or *Contre nature*. They are not mere tricks: as Purkinje wrote, illusions are errors of the senses but truths of perception.

Intimate Transformations and Analysis-Synthesis

Instead of limiting oneself to synthesis sounds, one can take advantage of the ressources of programming to process recorded sounds digitally. One has thus access to a corpus of rich and varied sounds with a rich identity – but these sounds are less ductile than synthesis sounds, the parameters of which can be modified independently. This is a problem for samplers, which reproduce digitally recorded sounds (mostly instrumental sounds).

To recover the malleability of the sound material, one can resort to processes of analysis-synthesis. This raises the problem of representing sound waves in terms of a base of functions – for instance sine waves, but also Walsh–Hadamard functions, predictive coding ... At the Laboratoire de Mécanique et d'Acoustique of Marseille, Daniel Arfib and Richard Kronland-Martinet, collaborating with Alex Grossmann, have applied the decomposition into Gabor grains and Morlet wavelets to analysis-synthesis. In his software Sound

Mutations, Arfib has used Gabor grains to perform intimate transformations such as time stretching without frequency alteration, or cross-synthesis leading to sonic hybrids. Such effects can be heard in his piece *Fragments complets*, and in my pieces *Attracteurs étranges* and *Invisible.*

Real-Time Piano-Computer Interaction

MIDI stands for Musical Instrument Digital Interface: it is a standard of digital description of musical events, introduced around 1980. It does not focus on the details which specify the sound wave, but rather on the parameters of instrumental control, those which a pianist acts upon when he or she plays piano: for each note, which key is played, the instants of striking and of releasing the key, the velocity of the key. MIDI permits to connect keyboards, sound synthesizers and computers. Playing a keyboard generates control signals for synthesis. This standard thus facilitates the instrumental use of digital synthesizers, exploited for instance in pieces by Philippe Manoury (Jupiter, Pluton) or Pierre Boulez (Explosante-fixe, Répons). Max Mathews has developed a radio-baton, which is in fact a gesture controller, to perform in real-time digital music elaborated in advance – not like a performer who specifies all the notes, but rather like a conductor who controls tempo and nuances.

The Yamaha "Disklavier" is a special acoustic piano, equipped with MIDI inputs and outputs. On this piano, each key can be played from the keyboard, but also triggered by a MIDI signal which controls the motors to lower or release the keys. Each time a key is depressed, it sends a MIDI signal indicating when and at what intensity. I took advantage of this instrument to implement a *Duet for one pianist*: the pianist plays his part, and a second part is added on the same piano by a computer program which follows the playing of the pianist. This "accompaniment" depends upon what the pianist plays and how he plays: the pianist dialogues with an invisible, virtual partner, which performs in a programmed but sensitive way. A computer receives the MIDI data indicating what the pianist plays, and sends back to the piano the MIDI signals specifying the accompaniment: the programming determines in what way the computer part depends on what the pianist plays.

I implemented this real-time piano-computer interaction in 1989 at the Media Lab of MIT, with the help of Scott Van Duyne, then at the Laboratoire de Mécanique et d'Acoustique in Marseille, using the graphic environment MAX by Miller Puckette, developed at IRCAM: this powerful musical software permits to specify a variety of real-time interactions. The reaction to the gesture of the pianist is perceived as instantaneous: its speed is actually limited by the mechanical inertia of the piano.

I have explored various kinds of real-time relation between the pianist and the computer. In particular, I have implemented simple mathematical transformations in the time-frequency plane: a translation in this plane corresponds to a musical transposition, possibly with a delay, as in an accompaniment, a canon or a fugue; a symmetry around a given time value corresponds to a "retrogradation"; a symmetry around a given frequency corresponds to an inversion of intervals. This latter transformation permits to program the accompaniment so as to play the second *Variation opus 27* by Webern with one hand: in this variation, each note is followed by the symmetrical note with respect ot the central A of the keyboard, with a delay of a eighth note; it is easy to program the computer to accomplish this mirror operation. Such simple transformations involve the Klein group, which governs transformations used in counterpoint.

One can implement a multiple translation which "fractalises" a melody by reproducing it in different octaves. When the interval of transposition is one octave plus a semi-tone, the melodies played by the pianist are strangely distorted: thus an octave jump upwards is heard as a semi-tone descent. This can be understood with a similar tonal combination on the piano, by playing C, C# one octave higher, D one octave higher D# one octave higher, then this same set of tones transposed one octave higher: the ear performs local rather than global pitch comparisons.

As transformation, one can also use an affinity, which corresponds to a melodic and/or rhythmical enlargement or contraction. The implementation on the piano gives a lively illustration of these transformations at work.

Beyond those note-to-note transformations, one can trigger generative processes, for instance instruct the program to respond to a given note with arppegioes. The speed of the arpeggioes can be specified by the tempo adopted by the pianist, or by other means, for instance by making the arpeggioes get faster when the pianist plays louder. The latter interaction is sensitive, reactive and playful – it is completely different from the relation a performer can have with another pianist. One can also use the memory of the computer to accompany the pianist according a pre-established score: the program can play its part by following the pianist as an accompanist follows a singer. And the computer, even though it is a deterministic machine, can simulate randomness: one can imagine to program quite different relations between the pianist and his programmed clone.

Conclusion

I wish to conclude with three observations.

First, the computation of sound has opened new musical possibilities, illustrated in the works by John Chowning and myself, followed by many others. It has also drastically changed our understanding of musical sound and its perception. The exploration of the possibilities of digital sound synthesis and processing takes advantage of discrete mathematics and stimulates it: sonic and musical demands raise interesting problems for mathematics and computer science as well as music, as shown by a number of chapters in this book. However one must remember Aristoxenus' lesson and take in account the specificities of perception.

Second, visual representation is extensively used in many areas of mathematics. It could also be fruitful to represent them as sounds, to "auralize" them, as can be done for Bessel functions or Tchebycheff polynomials with John Chowning's audio frequency modulation of Daniel Arfib's non-linear distorsion. Perceptual organization evidences a genuine intelligence. The auditory sense resorts to specific ways of processing and apprehension, capable to perform highly selective identifications. The ear – actually the auditory brain – can recognize patterns such as words, voices, timbres, much more effectively than any artificial intelligence program. It can unravel a mixture of conversations, a fugue with eight parts, a complex orchestral jumble, by grouping components with "common fate" – in synchrony, coherence or co-modulation. One can easily differentiate between two sounds arriving at the ear with the same physical energy, one emitted loud by a remote source and the other coming from a close and soft source: no machine or program can accomplish such a feat. It could be revealing to submit certain mathematical structures to tests of listening: unsuspected aspects might appear.

Last, one may notice that mathematics does not seem to shed light on the specificity of time, the stuff music is made of or plays with. The irreversibility of time appears in thermodynamics – and of course in biology. It is sometimes speculated that this "temporal disability" might relate with certain a prioris in mathematics: it may also have to do with the limits of mathematics in the field of music.

References

1. Arfib, D.: Digital synthesis of complex spectra by means of multiplication of non-linear distorted sine waves. Journal of the Audio Engineering Society **27**, 757–768 (1979)
2. Arfib, D.: Analysis, transformation, and resynthesis of musical sounds with the help of a time-frequency representation. In: De Poli et al. (Cf. ci-dessous), pp. 87–118. 1991
3. Bachem, A.: Chroma fixations at the ends of the musical frequency scale. Journal of the Acoustical Society of America **20**, 704–705 (1948)
4. Barbaud, P.: *La musique, discipline scientifique.* Dunod 1968
5. Barrière, J.B. (ed.): *Le timbre – une métaphore pour la composition.* Paris: C. Bourgois & IRCAM 1991
6. Cadoz, C.: *Les réalités virtuelles.* Collection Dominos. Paris: Flammarion 1994
7. Cadoz, C., Luciani, A., Florens, J.L.: Responsive input devices and sound synthesis by simulation of instrumental mechanisms. Computer Music Journal **14**(2), 47–51 (1984)
8. Charbonneau, G., Risset, J.C.: Circularité de hauteur sonore. Comptes Rendus de l'Académie des Sciences **277** (série B), 623–626 (1973)
9. Charbonneau, G., Risset, J.C.: Jugements relatifs de hauteur: schémas linéaires et hélicoïdaux. Comptes Rendus de l'Académie des Sciences **281** (série B), 289–292 (1975)

10. Chemillier, M., Duchamp, G.: Recherches et Gazette des mathématiciens I: **81**, 27–39; II: **82**, 26–30 (1999)
11. Chemillier, M., Pachet, F. (sous la direction de): *Recherches et applications en informatique musicale.* Paris: Hermès 1998
12. Chowning, J.: The synthesis of audio spectra by means of frequency modulation. Journal of the Audio Engineering Society **21**, 526–534 (1973)
13. Chowning, J.: Music from machines: perceptual fusion and auditory perspective. In: *Für György Ligeti – Die Referate des Ligeti-Kongresses Hamburg 1988.* Laaber-Verlag 1991
14. Collectif: *Nuova Atlantide – il continente della musica elettronica 1900–1986.* Biennale di Venezia 1986
15. Cook, P. (ed.): *Music cognition and computerized sound.* Cambridge, Mass.: MIT Press 1999
16. Cope, D.: *Experiments in musical intelligence.* Computer music and digital audio series, Vol. 12. Madison, WI: A-R Editions 1996
17. De Poli, G., Piccialli, A., Roads, C. (ed.): *Representations of musical signals.* Cambridge, Mass.: MIT Press 1991
18. Dodge, C., Bahn, C.: *Musical fractals.* Byte, june 1986, pp. 185–196. 1986
19. Dufourt, H.: *Musique, pouvoir, écriture.* Paris: C. Bourgois 1991
20. Dufourt, H.: Musique, mathésis et crises de l'antiquité à l'âge classique. In: Loi, M. (sous la direction de): *Mathématiques et art*, pp. 153–183. Paris: Hermann 1995
21. Escot, P.: *The poetics of simple mathematics in music.* Cambridge, Mass.: Publication Contact International 1999
22. Euler, L.: *Tentamen novae theoriae musicae.* Publications of St Petersburg Academy of Sciences 1739
23. Feichtinger, H., Dörfler, M. (ed.): *Diderot Forum on Mathematics and Music.* Wien: Österreichische Computer Gesellschaft 1999
24. Fucks, W.: Music analysis by mathematics, random sequences, music and accident. Gravesaner Blätter **23/24**(6), 146–168 (1962)
25. Genevois, H., Orlarey, Y. (sous la direction de): *Musique et mathématiques.* Lyon: Grame/Aléas 1997
26. Grossmann, A., Morlet, J.: Decomposition of Hardy functions into square integrable wavelets of constant shape. SIAM Journal of Mathematical Analysis **15**, 723–736 (1984)
27. Hartmann, W.H.: The frequency-domain grating. J. Acoust. Soc. Am. **78**, 1421–1425 (1985)
28. Hellegouarch, Y.: *Scales.* IV, 05 & V, 02. C. R. Soc. Roy. Math. Canada 1982/1983
29. Hellegouarch, Y.: L'"essai d'une nouvelle théorie de la musique" de Leonhard Euler (1); Le romantisme des mathématiques, ou un regard oblique sur les mathématiques du XIXème siècle (2). In: *Destins de l'art, desseins de la science*, pp. 47–88 (1); pp. 371–390 (2). Université de Caen/CNRS (1991)
30. Hellegouarch, Y.: Gammes naturelles. Gazette des mathématiciens I: **81**, 27–39; II: **82**, 13–26 (1999)
31. Hiller, L.A., Isaacson, L.M.: *Experimental Music.* McGraw Hill 1959
32. Kircher, A.: *Musurgia Universalis.* New York: Olms 1650/1970
33. Licklider, J.C.R.: A duplex theory of pitch perception. Experientia (Suisse) **7**, 128–133L (1951)

34. Lusson, P.: Entendre le formel, comprendre la musique. In: Loi, M. (sous la direction de): *Mathématiques et art*, pp. 197–204. Paris: Hermann 1995
35. Loi, M. (sous la direction de): *Mathématiques et art.* Paris: Hermann 1995
36. Mathews, M.V.: *The technology of computer music.* Cambridge, Mass.: MIT Press 1969
37. Mathews, M.V., Pierce, J.R.: *Current directions in computer music research.* (Avec un disque compact d'exemples sonores). Cambridge, Mass.: MIT Press 1989
38. Moorer, J.A.: Signal processing aspects of computer music. Proceedings of the Institute of Electric and Electronic Engineers **65**, 1108–1135 (1977)
39. Parzysz, B.: *Musique et mathématique* (avec "Gammes naturelles" par Yves Hellegouarch). Publication n° 53 de l'APMEP (Association des Professeurs de Mathématiques de l'Enseignement Public; 13 rue du Jura, 75013 Paris). 1983
40. Pascal, R.: Structure mathématique de groupe dans la composition musicale. In: *Destins de l'art, desseins de la science* – Actes du Colloque ADERHEM, Université de Caen, 24–29 octobre 1986 (publiés avec le concours du CNRS). 1991
41. Pierce, J.R.: *Science, Art, and Communication.* New York: Clarkson N. Potter 1968
42. Pierce, J.R.: *Le son musical* (avec exemples sonores). Paris: Pour la Science/ Belin 1984
43. Plomp, R.: *Aspects of tone sensation.* New York: Academic Press 1976
44. Reimer, E.: *Johannes de Garlandia: De mensurabili musica.* Beihefte zum Archiv für Musikwissenschaft 10, Vol. 1, p. 4. Wiesbaden 1972
45. Riotte, A.: Ecriture intuitive ou conception consciente? Création, formalismes, modèles, technologies. Musurgia **2**(2), Paris (1995)
46. Risset, J.C.: *An introductory catalog of computer-synthesized sounds* (Réédité avec le disque compact Wergo "History of computer music", 1992). Bell Laboratories 1969
47. Risset, J.C.: Musique, calcul secret? Critique **359** (numéro spécial "Mathématiques: heur et malheur"), 414–429 (1977)
48. Risset, J.C.: Stochastic processes in music and art. In: *Stochastic processes in quantum theory ans statistical physics.* New York: Springer 1982
49. Risset, J.C.: Pitch and rhythm paradoxes. Journal of the Acoustical Society of America **80**, 961–962 (1986)
50. Risset, J.C.: Symétrie et arts sonores. In: *"La symétrie aujourd'hui"* (entretiens avec Emile Noël), pp. 188–203. Paris: Editions du Seuil 1989
51. Risset, J.C.: Hasard et arts sonores. In: *"Le hasard aujourd'hui"* (entretiens avec Emile Noël), pp. 81–93. Paris: Editions du Seuil 1991
52. Risset, J.C.: Timbre analysis by synthesis: representations, imitations, and variants for musical composition. In: De Poli et al. (Cf. ci-dessus), pp. 7–43. 1991
53. Risset, J.C.: Moments newtoniens. Alliages **10** (1991), 38–41 (1992)
54. Risset, J.C.: Aujourd'hui le son musical se calcule. In: Loi, M. (sous la direction de): *Mathématiques et art*, pp. 210–233. Paris: Hermann 1995
55. Risset, J.C., Mathews, M.V.: Analysis of musical instrument tones. Physics Today **22**(2), 23–30 (1969)
56. Risset, J.C., Wessel, D.L.: Exploration of timbre by analysis and synthesis. In: Deutsch, D. (ed.): *The psychology of music*, pp. 113–169. Academic Press 1999

57. Shepard, R.: Circularity of judgments of relative pitch. Journal of the Acoustical Society of America **36**, 2346–2353 (1964)
58. Sundberg, J. (ed.): *Studies of music performance.* Stockholm: Royal Academy of Music, Publication n° 39, 1983
59. Terhardt, E.: Pitch, consonance, and harmony. Journal of the Acoustical Society of America **55**, 1061–1069 (1974)
60. Wessel, D.L.: Timbre space as a musical control structure. Computer Music Journal **3**(2), 45–52 (1979)
61. Wessel, D.L., Risset, J.C.: Les illusions auditives. *Universalia* (Encyclopedia Universalis), 167–171 (1979)
62. Weyl, H.: *Symmetry.* Princeton University Press 1952
63. Xenakis, I.: Stochastic music. Gravesaner Blätter **23/24**(6), 169–185 (1962)
64. Xenakis, I.: *Formalized music.* Bloomington, Indiana: Indiana University Press 1971
65. Youngblood, J.: Style as information. J. of Music Theory **2**, 24–29 (1958)

14 The Mathematics of Tuning Musical Instruments – a Simple Toolkit for Experiments

Erich Neuwirth

This paper gives an overview of the (rather simple) mathematics underlying the theory of tuning musical instruments. Besides demonstrating the fundamental problems and discussing the different solutions (only on an introductory level), we also give Mathematica code that makes it possible to listen to the constructed scales and chords. To really get a "feeling" for the contents of this paper it is very important to hear the tones and intervals that are mentioned. The paper also has the purpose of giving the reader a Mathematica toolkit to experiment with different tunings.

More than 250 years ago Johann Sebastian Bach composed "Das wohl temperirte Clavier" (the well tempered piano) to celebrate an achievement combining music, mathematics, and science. Finally, a method of tuning musical instruments had been devised which allowed playing pieces in all 12 major and all 12 minor scales on the same instrument without retuning. Nowadays, we are so used to this fact that we almost lack an understanding for the kind of problems musicians were facing for a few hundred years.

The appendix of this paper contains some Mathematica code. This code allows us to play scales and chords with given frequencies. Using these functions, we will be able to listen to the musical facts we are describing in a mathematical way. (Warning: On slower machines this code may take some time to create the sounds.) The waveform used for this sound is not a sine wave. For musicians, sine waves sound very bad. Therefore, we are using a more complicated waveform, which has been described in [4] and is heavily used in [3]. Our code defines three Mathematica functions, `PlayScale`, `PlayChord`, and `PlayStereoScale`, and we will explain the use of our examples later in the paper. The code will run on any computer with a sound device and a Mathematica version supporting the `Play` function on this platform. In particular, it will run on PCs with any 32-bit version of Microsoft Windows.

Now let us start historically.

The ancient Greeks, and especially the Pythagoreans, noticed that the length of strings (of equal) tension and the musical intervals they produced showed some interesting relationships. Using more modern knowledge from physics we know that the length of strings and the frequency of the tone are inversely proportional. So in the context of this paper we will study the relation between frequencies, frequency ratios, and musical relationships like consonance.

The first fact we note in this respect is that when we compare two tones and the frequency of the second tone is double the frequency of the first tone, we feel that this is "the same tone on a higher level". The interval created this way is called an octave, and it seems to be a universal musical constant in the sense that an octave is perceived as a consonance in every musical culture.

To listen to this phenomenon, we can execute the command `PlayScale[{220,440},1.5]`. To listen to these two tones played as a chord, we execute `PlayChord[{220,440},1]`. We also can hear that this fact only depends on the ration and not on the absolute frequencies by playing `PlayChord[330*{1,2},1.5]` or `PlayChord[264*{1,2},1.5]`.

Since we noticed that doubling the frequency produces something musical, we also might be interested in listening to a series of tones consisting of the first few integer multiples of a base frequency, for example the sequence 220, 440, 660, 880, 1100, 1320, 1540, 1760.

In our code, we use `PlayScale[220*{1,2,3,4,5,6,7,8},1.5]`.

Listening to this sequence, between the consecutive tones we hear many intervals, which in Western music are considered to be consonant. In musical terms, we hear 7 intervals, and the first five are octave, fifth, fourth, major third, and minor third. The last 2 intervals normally are not used in Western music. Especially the tone with sevenfold base frequency is not used in Western music, but it is used in Jazz.

Considering integer multiples of a base frequency is not just "mathematical aesthetics", valveless fixed length wind instruments like historical horns and trumpets only can produces tones with exactly this property. So asking about the kind of music possible under these restrictions in not just academic, but connected with real wind instruments.

The 2 musically most important intervals in our sequence are the major third and the fifth. From our series we see and hear that the fifth corresponds to $3/2$ and the major third corresponds to $5/4$. Playing a base frequency and these two intervals at the same time produces a major triad, probably the most used chord in Western music.

Defining `MajorTriad={1, 5/4, 3/2}` we can do `PlayChord[264*MajorTriad,1.5]` and hear that is sounds very consonant.

The sequence of tones having integer multiple frequencies of a base frequency is often called overtone series.

Now let us try to construct a major scale by using only intervals we found between neighboring tones in the overtone series we just studied. Using a piano keyboard as our visual aid for constructing a scale we see that we immediately can create the base tone and tones for the third, the fourth and the fifth and, of course, for the octave.

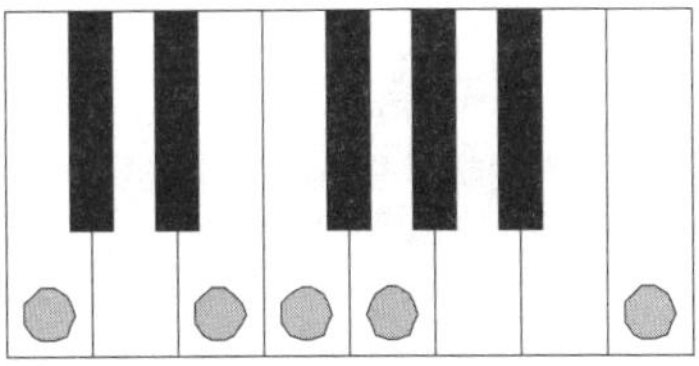

Fig. 14.1.

Defining `PartialScale1={1,5/4,4/3,3/2,2}` we can listen to `PlayScale[264*PartialScale1,1.5]`.

So we still are missing the second, the sixth, and the seventh. To find the corresponding frequency ratios, we look an the following picture:

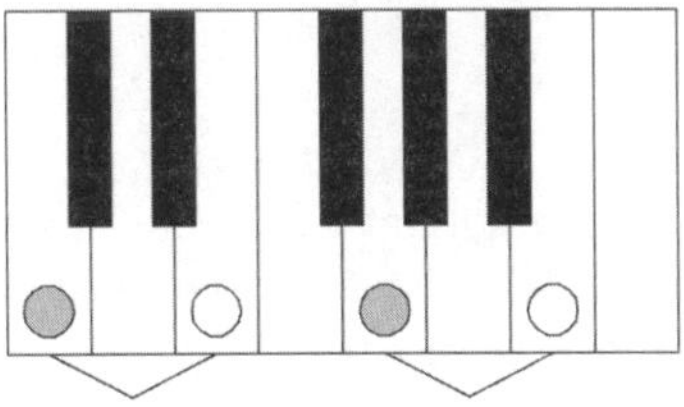

Fig. 14.2.

We see that the the lower interval marked by one dark and one light circle are similar intervals. We know that the lower interval, as a major third, corresponds to a frequency ratio of $5/4$. The lower tone of the upper interval has a frequency ratio of $3/2$ to the base tone. Therefore, we use a frequency ration of $(3/2).(5/4) = 15/8$ for the seventh.

We can listen to these intervals:
`Third={1,5/4}`. `PlayScale[264*Third,1.5]` and `PlayScale[264*3/2*Third,1.5]`. We also can listen to these intervals as chords: `PlayChord[264*Third,1.5]` and `PlayChord[264*3/2*Third,1.5]`. Using this we can define `PartialScale2={1,5/4,4/3,3/2,15/8,2}` and do `PlayScale[264*PartialScale2,1.5]`.

So we have been able to fill one of the holes in our scale. Similarly, we can construct the sixth by noting that the sixth is one third above the fourth:

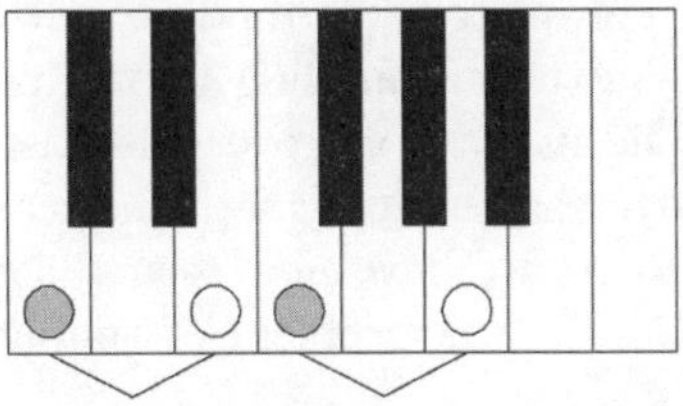

Fig. 14.3.

So the frequency ratio we need for this tome is $(4/3).(5/4) = 5/3$.

We can check the musical quality of these intervals with `PlayChord[264*Third,1.5]` and `PlayChord[264*4/3*Third,1.5]`. To extend our scale, we define `PartialScale3={1,5/4,4/3,3/2,5/3,15/8,2}` and do `PlayScale[264*PartialScale3,1.5]`.

Summarizing we see that we have almost all the tones we need for a major scale:

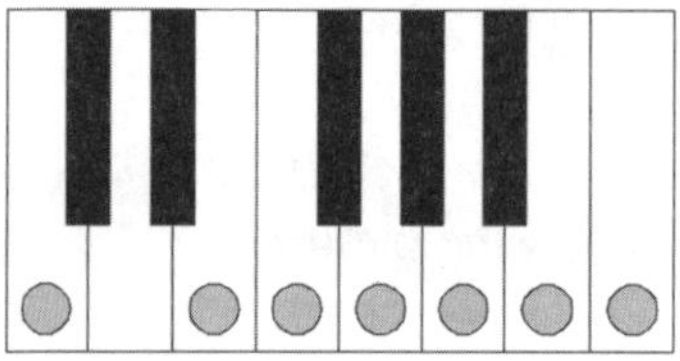

Fig. 14.4.

The only tone we are missing is the second. We cannot get the second from the overtone series up to the eightfold multiple of the base tone (i.e. within a range of 3 octaves of the base tone). But we can note that by extending the keyboard a little bit and going up 2 fifths:

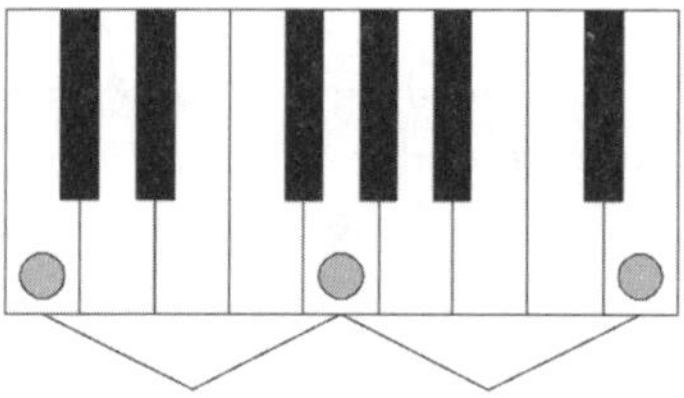

Fig. 14.5.

we get the tone one octave above the second. Just going down one octave (i.e. multiplying with 1/2) we see that we can construct the second as $(3/2).(3/2).(1/2) = 9/8$.

So now we have completed our scale, `PureMajorScale={1,9/8,5/4,4/3,3/2,5/3,15/8,2}` and we can listen to the scale with `PlayScale[264*PureMajorScale,1.5]`.

Now we have a musically pleasing scale represented by fractions with rather small numerators and denominators. The tuning building upon this scale is called pure tuning (or just tuning). The major triad over the base tone is the one we already discussed, it has a very simple mathematical description, and it sounds very harmonic. So what we have now seems like a mathematically and musically perfect solution to the problem of tuning instruments. To test the musical qualities of our scale, let us try a few other chords consisting just of thirds and fifths taken from our scale.

`PlayChord[264*{3/2,15/8,2*9/8},1.5]` is the major triad based on the fifth of our scale, and it sounds musically pleasing also.

This chord, has the same frequency rations as the major triad on the base tone: $(15/8)/(3/2) = 5/4$ and $(9/4)/(3/2) = 3/2$.

Now let us look at the triad based on the second of our scale (it is a minor triad).

`PlayChord[264*{9/8,4/3,5/3},1.5]` does not sound musically pleasing.

Let us look at the internal frequency ratios of this chord: $(4/3)/(9/8) = 32/27$, and $(5/3)/(9/8) = 40/27$. These ratios are not related to the intervals we derived from the overtone series. For the "upper" interval, however, we have $(5/3)/(4/3) = 5/4$, and this is a pure major third. We would expect a pure fifth for the ratio between the lowest tomne and the highest tone in our triad, so instead of $40/27$ we would need $3/2$. If we try to change the lowest tone of the triad such that we get the pure fifth, we have to take a $(5/3)/(3/2) = 10/9$ for the ratio between the base one and the second.

Listening to a chord containing this tone,
`PlayChord[264*{10/9,4/3,5/3},1.5]` gives us a consonant musical experience again.

We see that for a pure triad upon the fifth we need a second of $9/8$ and for a pure triad on the second, we need a second of $10/9$. So the problem is that when we try to play different chords with the tones taken from one scale, we are getting into musical trouble.

The frequency ratio between the two different seconds we need is $(9/8)/(10/9) = 81/80$, and it is called the syntonic comma. It also occurs in a different problem. Musically speaking, when he go up 4 fifths and then go down 2 octaves, we should arrive at the third above the base tone. Up 4 fifths and down 2 octaves corresponds to $(3/2).(3/2).(3/2).(3/2)/4 = 81/64$, one third corresponds to $5/4 = 80/64$, so the ratio occurring here also is the syntonic comma of $81/80 = 1.0125$. We can say that the syntonic comma is the degree of incompatibility between the pure third and the pure fifth.

Musically speaking, we would like to have compatible fifths and thirds. Since the fifth is the simplest interval in the overtone series (except the octave, of course), we try to keep the value for the fifth and use a third, which is "compatible" in the sense that one third and 4 fifth essentially produce the same tone. To achieve this, we have to use a frequency ratio of $81/64$ for the third. Since in our construction of the scale we used the third in 3 places, for the third, for the sixth, and for the seventh, we have to change the definition of the corresponding intervals in our scale.

So we define `PythagoreanMajorScale=`
`{1,9/8,81/64,4/3,3/2,4/3*81/64,3/2*81/64,2}`
and we play `PlayScale[264*PythagorealMajorScale,1.5]`. To hear the difference between the two different thirds (the pure third and the Pythagorean

third) we do `PlayScale[264*{5/4,81/64},1.5]`. We also can listen to the two thirds played on the two stereo channels:
`PlayStereoScale[264*{5/4},264*{81/64},1.5]`. The audible beats demonstrate that the difference between these two tones really matters musically. Finally, we can listen to the Pure scale and the Pythagorean scale played simultaneously on the two stereo channels: `PlayStereoScale [264*PureMajorScale,264*PythagoreanMajorScale,1.5]`.

We already noted that the Pythagorean third does not sound too harmonic when played as a constituent of a major triad.

To demonstrate this, try `PlayChord[264*{1,81/64,3/2},1.5]`.

The Pure triad sounds much better.

To demonstrate this, try `PlayChord[264*{1,5/4,3/2},1.5]`.

So, music played in this temperament (temperament in musical context is just another word for tuning) probably should avoid the major triad on the base tone. Generally speaking, also the chords on the fourth and the fifth do not sound too well in Pythagorean tuning. As long as music is played monophonically, this does not really matter that much, but when playing chords, tuning becomes a serious issue. This fact explains why interest in tuning grew in the 16th century: keyboard instruments became popular. Keyboard instruments are used to play chords, and it is rater difficult to change tuning. Therefore, there was the need for a tuning method which would allow to play many different chords reasonably well. As we have seen, Pythagorean tuning and pure tuning both have serious problems with some very basic chords. So people tried to find alternative methods of tuning. One of the basic problems to be solved was incompatibility between the third and the fifth. Pythagorean tuning had enlarged the third to make it compatible with the pure fifth. The alternative is to reduce the fifth to make it compatible with the pure third. This implies that the fifths are not pure any more. To make the fifth compatible with the pure third we note that one third and 2 octaves together produce a frequency ratio of 5. Therefore, to create a fifth with the property that 4 fifths produce the same interval as one third and 2 octaves, the fifth has to have a frequency ratio of the fourth root of 5, $\sqrt[4]{5}$. The frequency ration for the fourth then is $2/\sqrt[4]{5}$. Using the building principles we applied to create the pure tuning we can create the scale for this new tuning called meantone tuning.

`MeantoneMajorScale={1,5^(1/2)/2,5/4,2/5^(1/4),5^(1/4), 5^(1/4)*5^(1/2)/2,5^(1/4)*5/4,2}`. Using this definition, we can do `PlayScale[264*MeantoneMajorScale, 1.5]`. We also can compare the meantone scale to the pure scale,
`PlayStereoScale[264*MeantoneMajorScale, 264*PureMajorScale, 1.5]`, and to the Pythagorean scale `PlayStereoScale[264*MeantoneMajorScale, 264*PythagoreanMajorScale, 1.5]`.

Major triads in this tuning sound acceptable.

To hear this, we do `PlayChord[264*{1,5/4,5^(1/4)},1.5]`.

Comparing this with the with the major triad in Pythagorean tuning really makes an audible difference.

To hear this, we do `PlayChord[264*{1,81/64,3/2},1.5]`.

There is another problem we have not coped with until now, the circle of fifths. 12 consecutive fifths would bring us back to the original tone, or, in other word, should be the same as 7 consecutive octaves. If this were true for pure tuning, we should have $(3/2)^{12} = 2^7$, which of course is not true. The frequency ratio $(3/2)^{12}/2^7 = 3^{12}/2^{19} = 531441/524288 = 1.01364$ is called the Pythagorean comma and it is the measure of incompatibility between the pure fifth and the octave. To make these intervals compatible, we could either make the fifth smaller or make the octave larger. Since the factor 2 for the octave is an almost universal constant, we will change the fifth to be compatible with the octave. To accomplish that, we need a fifth with a frequency ratio of $2^{7/12} = \left(\sqrt[12]{2}\right)^7 = 1.49831$. Since we also want a third compatible with this fifth, we need a third of $\left(2^{(7/12)}\right)^4/4 = 2^{1/3} = 1.25992$

Using these values for the fifth and the third, we can construct a new tuning for a scale. For reasons we will mention briefly later this temperament is called equal temperament.

After some easy algebraic transformation this scale can be defined as follows:

```
EqualMajorScale={1,2^(1/6),2^(1/3),2^(5/12),2^(7/12),2^(3/4),
2^(11/12),2}.
```

We can play this scale, `PlayScale[264*EqualMajorScale, 1.5]`. Like with our previous scale examples, we can also compare it with other scales using stereo sounds, e.g.

`PlayStereoScale[264*EqualMajorScale,264*PureMajorScale,1.5]`.

Chords in equal temperament sound reasonably well, we do not have very bad dissonances. On the other hand, we do not have any pure interval except the octave.

This is illustrated by comparing

`PlayChord[264*{1,2^(1/3),2^(7/12)},1.5]` and

`PlayChord[264*{1,5/4,3/2},1.5]`.

Our Mathematica toolkit also contains a function Triad allowing to select a triad with a given base note from a give scale. `Triad[264*PureMajorScale,2]` will construct the frequencies for the triad constructed from the second, the fourth and the sixth of a scale in pure tuning with a frequency of 264 Hz for the base tone of the scale. Using this toolkit, the reader can do extensive comparisons of chord in different tunings and learn how differently notationally identical chords sound in different tunings.

We have seen that the fundamental numbers for tuning are the frequency ratios for the third and the fifth, all the other frequencies ratios are derived from these two.

So let us compare these ratios for all the tunings we have studied in a table:

	Pure	Pyth.	Meantone	Equal
third	1.2500	1.2656	1.2500	1.2599
fifth	1.5000	1.5000	1.4953	1.4983

We see that the that the largest difference occurs for the pure and the Pythagorean third. This at least partially explains why triads in Pythagorean tuning sound very bad. Meantone tuning for the two basic intervals is very similar to pure tuning, therefore the basic major triad sounds rather well in meantone tuning. The third in equal tuning is also quite different from the pure third; therefore the musical characteristics of the major triad in equal temperament are quite different from pure tuning.

In the framework of thismpaper we could only discuss some of the mathematical problems for tuning musical instruments. Especially, we only studied diatonic scales (i.e. scales without using the black keys on keyboards). The problems get much more complicated when tunings are extended to chromatic scales. Detailed discussions of these problem can be found in [1] and [2]. For studying chords and scales along the lines we have described here, [3] gives a very large set of almost 400 sound examples.

References

1. Blackwood, E.: *The Structure of Recognizable Diatonic Tunings.* Princeton University Press 1985
2. Lindley, M., Turner-Smith, R.: *Mathematical Models of Musical Scales.* Bonn: Verlag für systematische Musikwissenschaft 1993
3. Neuwirth, E.: *Musical Temperaments.* Transl. from the German by Rita Steblin. With CD-ROM for Windows. (English) Vienna: Springer-Verlag 1997
4. Neuwirth, E.: Designing a Pleasing Sound Mathematically. Mathematics Magazine. **74**, 2001, pp. 91–98

Appendix: Mathematica code

```
PlayChord[FreqList_, Duration_] :=
  Play[Min[1, 20*t, 20*(Duration - t)]*
    Sum[Sum[(0.6)^k*Sin[2*Pi*t*k*FreqList[[i]]],
        {i, 1, Length[FreqList]}],
      {k, 1, 10}],
   {t, 0, Duration - 0.001},
   SampleRate → 22050]
```

```
PlayScale[FreqList_, Duration_] :=
  Play[Min[1, Abs[20*(t - Duration*Floor[t/Duration])],
      Abs[20*(t - Duration*(1 + Floor[t/Duration]))]]*
    Sum[(0.6)^k*Sin[2*Pi*t*k*FreqList[[1 + Floor[t/Duration]]]],
       {k, 1, 10}],
   {t, 0, Duration*Length[FreqList] - 0.001},
   SampleRate → 22050]
```

```
PlayStereoScale[FreqListLeft_, FreqListRight_, Duration_]:=
  Play[{Min[1, Abs[20*(t - Duration*Floor[t/Duration])],
        Abs[20*(t - Duration*(1 + Floor[t/Duration]))]]*
          Sum[(0.6)^k*Sin[2*Pi*t*k*
                     FreqListLeft[[1 + Floor[t/Duration]]]],
          {k, 1, 10}],
      Min[1, Abs[20*(t - Duration*Floor[t/Duration])],
        Abs[20*(t - Duration*(1 + Floor[t/Duration]))]]*
      Sum[(0.6)^k*Sin[2*Pi*t*k*
                     FreqListRight[[1 + Floor[t/Duration]]]],
          {k, 1, 10}]},
    {t, 0, Duration*Length[FreqListLeft] - 0.001},
    SampleRate → 22050]
```

```
Triad[Scale_, BaseTone_] :=
  List[Scale[[BaseTone]],
    If[BaseTone + 2 > 8, 2*Scale[[BaseTone - 5]],
                         Scale[[BaseTone + 2]]],
    If[BaseTone + 4 > 8, 2*Scale[[BaseTone - 3]],
                         Scale[[BaseTone + 4]]]]
```

15 The Musical Communication Chain and its Modeling

Xavier Serra

15.1 Introduction

We know that Music is a complex phenomenon impossible to approach from any single point of view; thus, most scientific attempts to understand music will not make justice to its amazing richness. However, this cannot prevent us from trying to explain and formalize some aspects. In this way we also contribute to its understanding and develop new tools to create and enjoy new musical artwork. This in turn can be seen as a form of further enrichment of that marvelous complexity.

In his book *Elements of Computer Music* ([25]), Richard Moore presents a view of the musical communication chain that despite being based on a traditional conception of music, it is a very useful starting point for discussing many relevant issues involving computers and music. Starting from that view, in this article we will mention some of the active research areas in Computer Music and present topics that are still very much open to be looked into. No attempt is made to present a comprehensive overview of the Computer Music field and its related disciplines. A part from the book by Moore there are a few other books that can give the reader an overview of the field ([27,12]).

15.2 The Communication Chain

The musical communication chain proposed by Moore, and shown in the figure, is a loop of interconnected signals (data) and processes (transformations of the signals) that encompass all the elements involved in the making, transmission and reception of music. As shown in the diagram, this loop is the encounter of two knowledge bases, the musical one, basically mental and fundamented on our cultural tradition, and a physical base, in which the laws of physics have a much greater influence.

Starting from the top we can see that from the perceptual inputs and the personal musical background, the composer is able to create a symbolic representation that expresses a musical idea. From such symbolic representation and by sharing some common musical background, a performer is then capable of producing gestures, or temporal controls, to drive a musical instrument. Such an instrument is a crafted physical object that can produce an air vibration, sound source, from the performers gestures. The sound

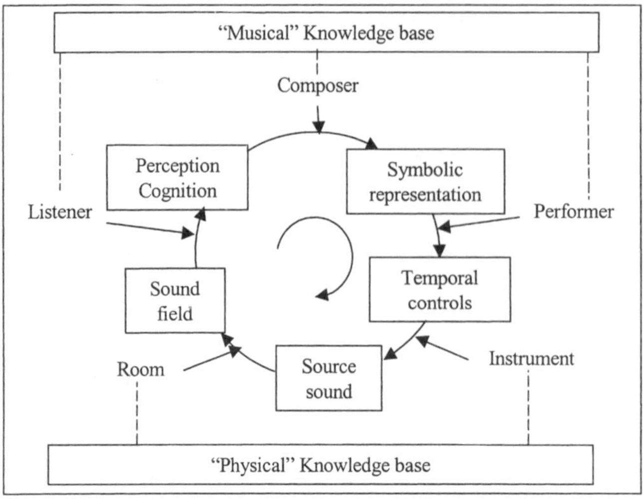

Fig. 15.1. Diagram of the Musical Communication chain proposed by Moore

source produced by the instrument is then propagated in a room and a sound field is created which a listener then perceives. Finally, based on previous perceptual experiences the listener processes the acoustic signal that enters the ear, thus combining perceptual and cognitive experiences. The composer closes the loop by using his/her own perceptual and cognitive experiences in the musical creative decisions.

This description is a traditional view of the music communication chain and it is clear that the scientific and technological developments of the second half of the 20th century have had a great impact in this chain. An important alteration caused by these developments has been the flexibilization of some of the processes, since now they can not only be undertaken by human beings or mechanical devices, but also by electronic machines or even by software simulations. For example, a room might be completely skipped in the communication loop, since we can listen to a performance on headphones, or a performer might also be leaved out since the composer can produce sounds directly with a computer without the traditional physical interaction performer-instrument. The most important overall impact on the musical communication chain has been the incorporation of new creative possibilities at different levels, for the composer, the performer and the instrument builder. Many of these improvements are based on our enhanced capability to create mathematical models of the various steps in that chain and to perform computer simulations on this basis, e.g. in order to tune parameters, or open up new possibilities which could not be realized with classical instruments, for example.

Despite all the possible shortcomings and limitations of the loop, its discussion will give us useful insights that are very much applicable to the current developments in the area of Computer Music. In fact, we are particularly interested in discussing this chain in the context of the current scientific and technological developments, viewing in order to investigate how they affect the communication chain. Let us now go into some detail through the different parts of the chain.

15.3 Composer

Traditional music composition is the process of producing symbolic representation of musical thought. Even though musical thought is difficult, or even impossible, to formalize, there is a long tradition of formalization of the compositional process and of its symbolic output. Thus it has become a natural step to use the computer to help in the compositional process and to develop formal composition algorithms; modeling some specific compositional tasks and devising tools for helping the composer ([9,19,21,29,36]).

The idea of automated, or algorithmic, composition is as old as music composition itself. It generally refers to the application of rigid, well-defined rules to the process of composing music. Since composers always follow some rigid rules and structures, we could say that classical compositions are also algorithmic compositions. However, we normally restrict the term of algorithmic composition to the situations when there is minimal human intervention. The techniques used in the automated composition programs ([6]) cover a wide spectrum of approaches coming from such different domains as law (rules), mathematics (mathematical functions), psycho neurology (connectionism), and biology (generative processes).

- *Rule-based.* The most common of the early algorithmic systems were those which applied rules, of counterpoint for example, to musical choices. These systems rely on the "wisdom" of those who design the rules. Rule-based systems can be extremely complex but always rely on the specification of musical "heuristics" by the programmer. Rules can be derived from analysis of previous musical works or from other formal structures such as linguistic grammars, and mathematical formulae.
- *Mathematical Functions.* The direct use of mathematical functions has also been a fruitful source of inspiration for the development of algorithmic system. For example using probability functions, when choices between various options are weighted, or more recently with the use of equations derived from the Chaos theory or Fractal structures.
- *Connectionist.* A connectionist system, such as a neural network, is "trained" by being exposed to existing music, at which time it alters the characteristics of the connections between its nodes (neurons) so they reflect the patterns in the input. After training the connectionist system it can be seeded and then produces an output based on the acquired patterns.
- *Generative Processes.* Generative systems are based on theories of genetic evolution. The basic idea is that a melody, or a more complex music

structure, can "grow" similarly to the evolutionary development of a life form. The oldest form of evolutionist computer algorithms are "cellular automata". Another common class of evolutionary algorithms are those based on "genetic algorithms", which model the splitting, mutation, and recombination of genes. For example, within a gene pool of notes new children can be created by taking the pitch from one parent and the duration of another. Mutations are created by randomly changing the values of the data and a melody is generated by selecting notes from the pool of notes.

Every single composer has a particular way of thinking about music and thus it has been difficult to come up with software applications that can be used by more than a few composers or compositional styles. The most successful systems are either the very specific one, which are useful for precise tasks, or those making very little assumption on the aesthetic or mental process and give a lot of freedom to the composer to make his/her own music model. This last group of systems are, essentially, computer languages that support basic music and sound constructs by means of which the user can express his/her own musical thought.

15.4 Symbolic Representation

The traditional symbolic representation of classical western music, Common Practice Notation, CPN, is essentially a highly encoded abstract representation of music that lies somewhere between instructions for performance and representation of the sound. It has been an excellent way to communicate between composers and performers for several centuries and it assumes that performers and composers share a common musical tradition, which enables performers to infer many of the non-written performance instructions from that tradition. The usefulness of this representation breaks down as soon as the shared tradition does not exists, as in some contemporary music, or the performer is a computer program, which has not had the "appropriate" musical training. In these situations a more detailed representation, or set of instructions, is required ([30]).

Each musical usage has a different set of requirements and thus a different representation is needed for it. For example there are representations for controlling digital synthesizers and representations for analyzing the musical structure of a piece of music with a computer. Given the currently developed computer applications we could group the used representations into three categories: (1) sound-related codes, (2) music notation codes, and (3) music data for analysis.

MIDI (Musical Instrument Data Interface) has been for some years, and still is, the most prevalent representation of music for computer applications. It was designed as a universal interconnection scheme for instrument controllers and synthesizers, thus in the category of sound-related codes, but through the years its usage has been extended to many other applications, even though it is not the most appropriate one. A part from MIDI, there are other representations at the sound level, such as Csound ([2]) that are also used for making music with software programs. But there is no definite solution to the music representation problem at the sound control level, a representation that should be able to express music at different levels of detail and be useful at the same time for the control of all the available synthesis techniques.

Music notation codes are mainly used for computer representation and printing of traditional music, thus they are used by all the notation programs currently available. Looking at the quality obtained by these programs it is clear that the basic problems have mostly been solved. This is not the case for the codes used for music analysis, there is very little done in this area and it is not a simple problem to be solved. We need a representation from which structural analysis can be done, thus it has to show the relationships between the different music elements at different abstraction levels.

15.5 Performer

Although the performer a key element in the music production chain, there has been little formalization, or scientific studies, of what his actual contribution to that chain is. Performance skills are learned in an intuitive way and thus they are quite elusive to scientific analysis. However, in the past few years it has turned into a very active and fruitful field of research (Dannenberg, De Poli, 1998; Gabrielsson, 1999) with already some very promising results.

What makes a piece of music come alive is the performer's understanding of the structure and "meaning" of a piece of music, and his/her expression of this understanding via expressive performance. Hence the key research issue is to explain and quantify the principles that govern expressive performance. There are several strategies, complementary to each other, to approach such a problem:

- *Measuring performances.* We compare the parameters defined in a musical score with those obtained from a recording of the same score, either audio recording or some measured gesture data (normally MIDI). We can then try to formalize the differences with a set of performance rules or models. Such an approach requires analysis techniques from which to extract the musical parameters of the recordings and ways to compare them with the musical score data.
- *By an analysis-by-synthesis method.* We formalize the knowledge given by experimented performers using the perceptual feed-back of a sound synthesis system. Thus we can implement and test the validity of the formalization, which is generally a symbolic rules system. This has been the approach carried out by the KTH group in Stockholm ([5,14]).

- *Devising performance models with automatic methods.* The goal is to develop AI methods to come up with quantitative models of expressive performances. For example, machine learning algorithms are devised that search for systematic connections between structural aspects of the music and typical expression patterns ([34]).

Current results on performance analysis give us some light into expressive aspects such as tempo and dynamics, but we are far from understanding most of what a performer really brings into the musical communication process.

15.6 Temporal Controls

One of the problems in studying performance issues is that it is very hard to differentiate the output of the performer from the output of the instrument being played. There is a very strong coupling between the two, and in fact, in most cases, the two processes form a feed back loop impossible to understand one without the other. Generally, we can only record the sound output of the instrument, from which we will then have to differentiate the two types of data.

By placing sensors on the instrument it is possible to measure the action of the performer, such as pressure values of the fingers on a string. For example, in the case of the piano it is quite easy to measure the action of every single key of the keyboard and convert that to MIDI values. This is one of the reasons why most research on performance issues is being done on MIDI data extracted from piano performances. In self-sustained instruments, like bowed string or wind instruments, is much harder to measure all the performance gestures and very little work has been done on developing representations for this type of performance data.

In the context of electronic instruments it is feasible to separate the controlling aspect of a musical instrument from its sound producing capabilities. We can build controllers and interfaces to capture performance gestures and sound modules to produce sounds. With this division a staggering range of possibilities become available ([28,35]). Since the invention of the first electronic instruments there has been considerable research on developing new controllers with which to explore new creative possibilities ([7]) and communication protocols to interface the controllers with the sound generation devices ([30]). Thus the concept of performance takes a new meaning and with it the concept of instrument. There is a huge open ground for research by considering the performer-instrument interface in the general framework of human-computer interaction.

15.7 Instrument

This is one of the best-defined elements of the chain and at the same time yields the clearest scientific and technological problems. The understanding of the acoustics of the musical instruments is a problem already introduced by the Greeks, with still fascinating challenges to be solved, for example we still do not know why a Stradivarius sounds the way it does. In the Computer Music context we are also very much interested in inventing digital instruments with the use of computer models, or in modifying the sound of existing acoustical instruments by digital means, thus extending the creative possibilities of the composer, performer, and instrument builder.

The way to create digital instruments is by using synthesis techniques ([27]) and implementing them either in hardware or software. Traditionally, these techniques have been classified into: additive synthesis, subtractive synthesis, and non-linear synthesis. Additive synthesis is based on the sum of elementary sounds, each of which is generated by an oscillator. Subtractive synthesis is based on a complementary idea of filtering out parts of a complex sound. The last group, non-linear synthesis, is a jumble in which a great number of techniques based on mathematical equations with non-linear behavior are included.

After many years trying to come up with "yet a new synthesis technique" the main research efforts in sound synthesis and instrument modeling are currently centered around two clear modeling problems. Either we want to model the sound source (physical modeling approach) or we want to model the perceived signal (spectral modeling approach).

With the physical modeling approach we generate sounds describing the behavior of the elements that make up a musical instrument, such as strings, reeds, lips, tubes, membranes and resonant cavities. All these elements, mechanically stimulated, vibrate and produce disturbances, generally periodic, in the air that surrounds them. It is this disturbance that arrives to our hearing system and is perceived as sound. Historically, physical models have been carried out by means of very complex algorithms that can hardly work in real time with current technology. These implementations have been based on numerical integration of the equation that describes wave propagation in a fluid. Recently, more efficient solutions have been found for this problem and systems have begun to appear with interest for musicians ([33]).

Spectral models are based on the description of sound characteristics that the listener perceives. To obtain the sound of a string, instead of specifying the physical properties, we describe the timbre or spectral characteristics of the string sound. Then, sound generation is carried out from these perceptual data, thanks to diverse mathematical procedures developed in the last few decades. One advantage of these models is that techniques exist for analyzing sounds and obtaining the corresponding perceptual parameters. That is to say, by analyzing a specific sound we can extract its perceptual parameters. From the analysis, it is possible to synthesize the original sound again and the parameters can be modified in the process so that the resulting sound is new but maintains aspects of the sound analyzed ([31]).

15.8 Source Sound

The output of the instrument can easily be recorded with a microphone and processed with an analog to digital converter. Thus we get a direct representation of the sound as a one-dimensional pressure wave, represented digitally as a signal regularly sampled in time. There are many standard ways to store this information and also ways to visualize it and analyze it. This is necessary in order to study the important physical attributes of the sound, such as its frequency, amplitude and timbre.

Another useful sound representation is obtained by decomposing the time waveform into its frequency components, thus obtaining a spectral representation. Such a representation is based on the assumption that the waveform is stable during a certain length of time and it is generally expressed in polar coordinates as a magnitude and a phase values. To capture the time-varying characteristics of the sound we use a frame-based approach, thus the representation is a sequence of spectra. There are many different techniques for obtaining this representation from the time domain waveform, each one having a different set of compromises and being used in different applications. However, it is fair to say that the Fourier transform is the most important single technique, used in one form or another at least as a component of more complex systems (e.g. wavelet or time-frequency analysis methods).

There is a lot of research to come up with compressed representations of the signal, in order to reduce the memory needed for its storage or the bandwidth required for its transmission. But basically, it is a solved problem the way to represent the direct output of a musical instrument.

15.9 Room

The space within which the music is played may have the same impact on the listener's experience as the instrument used. Nevertheless, despite the fact that the science of room accoustics is quite well known – the complexity of real spaces is so huge that we are still far from being able to design the "perfect" concert hall.

In the context of Computer Music we are especially interested in simulating spaces with computer models and at the same time being able to control the location of our sound sources, or instruments, inside these spaces ([1]). We can characterize the reverberation of a space by its impulse response, or from a parameterization of that signal. From this characterization we can then create digital reverberators using different signal processing strategies based on digital filters and delay lines. But given the complexity of real spaces, the creation of natural reverberations is still a great challenge. At the same time the simulation of localization and movement of sound through space brings many interesting problems. Depending on the sound reproduction situation the simulation of localization and movement cues is done in different way. The possible strategies and the problems to be solved will be completely different from each other, depending on the situation: we may have a standard stereo system, a multichannel system, or different types of headphones, for example. Much of the work on sound spatialization is based on psychoacoustics' studies, thus reproducing the space perceptual cues.

15.10 Sound Field

Many of the recent improvements made to the commercial sound systems have been incorporated in order to have the possibility to preserve into a sound recording a given sound field and to be able to reproduce it ([1]). Thus, giving the listener the sensation of being in the "original" acoustic space, a 3D sound environment. From the old monophonic recordings to the current surround systems there has been an enormous evolution. Some of the landmarks have been: Two-channel recordings, Three-channel systems, Four-channel surround systems, Binaural systems, Ambisonics ([23]), and Motion Picture Sound (ex. Dolby Stereo).

15.11 Listener

From a medical point of view, the physiology of the human hearing is quite well understood, but the relationship between acoustics, musical structure, and emotion is still a subject of ongoing scientific investigation and far from being well understood. There are many different steps involved in this process, starting from the entrance of the sound waveform into our ear and ending with the musical sensation we get from that sound. We understand some of the low level (hearing system) issues but we are far from understanding the cognition issues. There are many areas involved, related and interrelated to the different qualities and aspects from sound to music, some of which are: basic auditory processes, low-high grouping mechanisms, timbre, pitch, time and rhythm perception ([11]).

In auditory modeling the aim is to find mathematical models that represent some physiological behavior or some perceptual aspects of human hearing ([17]). Thus with a good model we can analyze audio signals in a way similar to the brain.

The computational models of the ear ([22]) generally pay particular attention the behavior of the cochlea, the most important part of the inner ear, that acts essentially as a non-linear filter bank. But one of the better known issues is the filtering effect of the head and the outer ear, which is critical for the spatial perception of sound ([3]). This processing effect is generally modeled by the Head Related Transfer Functions, HRTF, which are represented by pole/zero models, series expansions or structural models.

Beyond the low level physiological aspects there are many research topics related to disciplines such as psychoacoustics, psychology or cognitive musicology. For example, to deal with the sound environment the auditory system has to extract the essential sound components from the composite sound signal reaching the ear and to construct an auditory world. This function of the auditory system is known as auditory scene analysis ([4]). We are still far from understanding this process, but considerable work is done on its simulation on computers.

Another active field is related to our ability to distinguish and categorize sounds. The classification of musical sound into different timbre groups has received much attention in the perceptual literature in the past, but the principal components of timbre still remain elusive. Existing theories of timbre are derived from perceptual experiments, such as the well known multidimensional scaling experiments of Grey ([16]) in which the cognitive relationships amongst a group of sounds are represented geometrically. Current alternative approaches are based on machine learning algorithms or statistical analysis ([18]).

At the level of musical cognition several theories have been proposed, such as Narmour's implication/realization model ([26]). It proposes a theory of cognition of melodies based on simple characterize pattern of melodic implications, which constitute the basic units of the listener perception. Lerdahl and Jackendoff's proposed a generative theory of tonal music ([20]) which offers an alternative approach to understanding melodies based on a hierarchical structure of musical cognition.

15.12 Perception and Cognition

A perceptually meaningful sound representation should be based on log-frequency in order to reflect the relative salience of low frequency components with respect to the high frequencies. There must be also temporal information that gives precedence to the transient portion of the signal. The standard Fourier representations do not fulfill these requirements and thus there has been considerable research into alternative representations. We could just mention a few:

- The correlogram ([32]) allows us to see where energy is located in a log frequency, but also the value of the autocorrelation lag for which the signals of the cochlear channels have the same periodicity.
- Mel-warping of spectra is commonly used in cepstral-based speech processing and since the Mel scale was derived from human psychophysics, the resulting frequency scale is cochlear like.
- The cochleogram is the direct output of a cochlear filterbank model.
- The field of Computational Auditory Scene Analysis is emerging with new, non FFT-based, representations of audio with a view to solving difficult

auditory scene analysis problems. Frequency components are grouped using gestalt principles such as synchrony of onset and temporal proximity as well as psycho-acoustic principles such as harmonicity and critical band masking effects.

15.13 Conclusions

The music communication chain is too complex to be fully explained by any single discipline. Throughout this short article we have referenced contributions from such disciplines as: Music, Computer Science, Electrical Engineering, Psychology, Physics. This is precisely one of the virtues of the Computer Music field and what Moore ([25]) tries to express in the diagram shown in Fig. 15.2.

The number of open problems is still vast and we have only explained the most important ones. However it is clear that the interdisciplinary approach expressed in this article is the only way to tackle most of them.

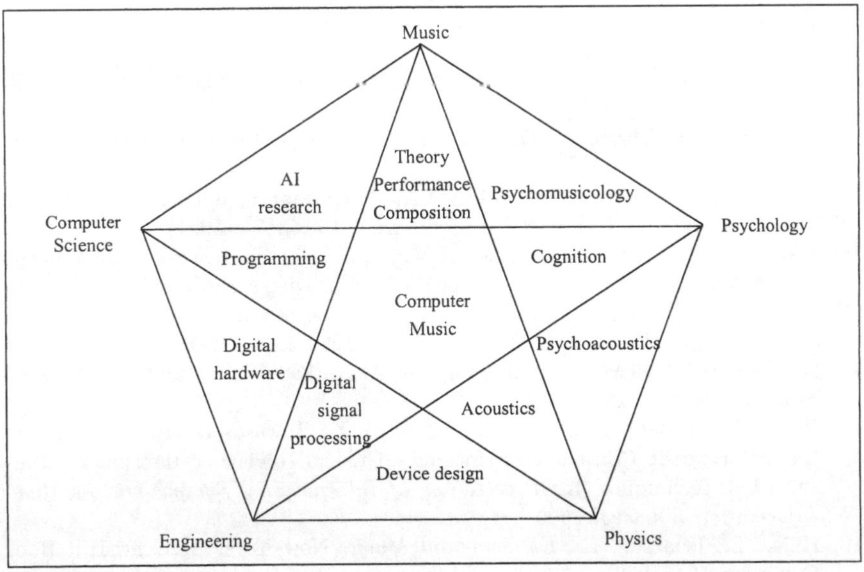

Fig. 15.2. Diagram of the interdisciplinariety of the Computer Music field as proposed by Moore

References

1. Begault, D.R.: *3-D Sound For Virtual Reality and Multimedia.* Academic Press 1994
2. Boulanger, R.C.: *The Csound Book: Perspectives in Software Synthesis, Sound Design, Signal Processing, and Programming.* Cambridge: MIT Press 2000
3. Blauert, J.: *Spatial Hearing.* Cambridge: MIT Press 1983
4. Bregman, A.S.: *Auditory Scene Analysis: The Perceptual Organization of Sound.* Cambridge: MIT Press 1990
5. Bresin, R.: *Virtual Virtuosity: Studies in Automatic Music Performance.* Ph.D. Thesis. KTH, Stockholm, Sweden 2000
6. Brown, A.R.: "Your Friend, the Algorithm". Music Forum **4**(6), 26–29 (1998)
7. Chadabe, J.: *Electric Sound: The Past and Promise of Electronic Music.* New Jersey: Prentice Hall 1997
8. Cook, P.R.: *Music Cognition, and Computerized Sound: An Introduction to Psychoacoustics.* Cambridge: MIT Press 1999
9. Cope, D.: *Computers and Musical Style* (The Computer Music and Digital Audio Series, Vol. 6). Madisonsconsin: A-R Editions 1991
10. Dannenberg, R., De Poli, G.: "Synthesis of Performance Nuance". Special issue of Journal of New Music Research **27**(3), (1998)
11. Deutsch, D.: *The Psychology of Music.* 2nd edn. Academic Press Series in Cognition and Perception 1998
12. Dodge, C., Jerse, T.A.: *Computer Music.* 2nd edn. New York: Schirmer Books 1996
13. Fletcher, N.H., Rossing, T.D.: *The Physics of Musical Instruments.* New York: Springer 1991
14. Friberg, A.: "Generative Rules for Music Performance: A Formal Description of a Rule System." Computer Music Journal **15**, 56–71 (1991)
15. Gabrielsson, A.: The Performance of Music, in Deutsch, D. (Ed.) *The Psychology of Music* [second edition]. pp. 501–602, San Diego: Academic Press 1999
16. Grey, J.M.: "Multidimensional perceptual scaling of musical timbres." Journal of the Acoustical Society of America **61**(5), 1270–1277 (1976)
17. Hawkins, H.L., McMullen, T.A., Popper, A. N., Fay, R.R.: *Auditory Computation.* New York: Springer 1996
18. Herrera, P, Amatriain, X., Batlle, E., Serra, X.: Towards Instrument Segmentation for Music Content Description: a Critical Review of Instrument Classification Techniques. In: *Proceedings of International Symposium on Music Information Retrieval* 2000
19. Hiller, L., Isaacson, L.: *Experimental Music.* New York: McGraw-Hill Book Company, Inc. 1959
20. Lerdahl, F., Jackendoff, R.: An overview of hierarchical structure in music. In: Scwanaver, S.M., Levitt, D.A., (Eds.) *Machine Models of Music.* Reproduced from Music Perception 1993
21. Loy, G.: "Composing with Computers: A survey of Some Compositional Formalisms and Music Programming Languages". In: Mathews, M.V., Pierce, J.R. (eds.): *Current Directions in Computer Music Research.* Cambridge: MIT Press 1989
22. Lyon, R.F.: "A Computational Model of Filtering, Detection, and Compression in the Cochlea", In: *Proceedings of IEEE-ICASSP-82*, pp. 1282–1285. 1982

23. Malham, D.G., Myatt, A.: 3-D Sound Spatialization using Ambisonic Techniques. Computer Music Journal **19**(4), 58–70 (1995)
24. McAdams, S.: "Audition: Cognitive Psychology of Music". In: Llinas, R., Churchland, P. (Eds.): *The Mind-Brain Continuum*, pp. 251–279. Cambridge: MIT Press 1996
25. Moore, F.R.: *Elements of Computer Music.* New Jersey: Prentice-Hall 1990
26. Narmour, E: *The analysis and cognition of basic melodic structures: the implication-realization model.* University of Chicago Press 1990
27. Roads, C.: *The Computer Music Tutorial.* Cambridge: MIT Press 1996
28. Rowe, R.: *Interactive Music Systems – Machine Listening and Composing.* Cambridge: MIT Press 1994
29. Schillinger, J.: *The Mathematical Basis of the Arts.* New York: The Philosophical Library 1948
30. Selfridge-Field, E.: *Beyond MIDI: The Handbook of Musical Codes.* Cambridge: MIT Press 1997
31. Serra, X.: "Musical Sound Modeling with Sinusoids plus Noise". In: Poli, G.D., Picialli, A., Pope. S.T., Roads, C. (Eds.): *Musical Signal Processing* Swets & Zeitlinger Publishers 1997
32. Slaney, M., Lyon, R.F.: "On the Importance of Time: A Temporal Representation of Sound". In: Cooke, M., Beet, S., Crawford, M. (Eds.): *Visual Representations of Speech Signals*, pp. 95–115. Chichester: Wiley & Sons 1993
33 Smith, J.O.: Physical modeling using digital waveguides. Computer Music Journal **16**, 74–87 (1992)
34. Widmer, G.: "Machine Learning and Expressive Music Performance". In: *AI Communications* (2001)
35. Winkler, T.: *Composing interactive music: techniques and ideas using MAX.* Cambridge: MIT Press 1998
36. Xenakis, I.: *Formalized Music.* Bloomington: Indiana University Press 1971

16 Computational Models for Musical Sound Sources

Giovanni De Poli and Davide Rocchesso

Abstract. As a result of the progress in information technologies, algorithms for sound generation and transformation are now ubiquitous in multimedia systems, even though their performance and quality is rarely satisfactory. For the specific needs of music production and multimedia art, sound models are needed which are versatile, responsive to user's expectations, and having high audio quality. Moreover, for human-machine interaction model flexibility is a major issue. We will review some of the most important computational models that are being used in musical sound production, and we will see that models based on the physics of actual or virtual objects can meet most of the requirements, thus allowing the user to rely on high-level descriptions of the sounding entities.

16.1 Introduction

In our everyday experience, musical sounds are increasingly listened to by means of loudspeakers. On the one hand, it is desirable to achieve a faithful reproduction of the sound of acoustic instruments in high-quality auditoria. On the other hand, the possibilities offered by digital technologies should be exploited to approach sound-related phenomena in a creative way. Both of these needs call for mathematical and computational models of sound generation and processing.

The sound produced by acoustic musical instruments is caused by the physical vibration of a certain resonating structure. This vibration can be described by signals that correspond to the time-evolution of the acoustic pressure associated to it. The fact that the sound can be characterized by a set of signals suggests quite naturally that some computing equipment could be successfully employed for generating sounds, for either the imitation of acoustic instruments or the creation of new sounds with novel timbral properties.

The focus of this chapter is on computational models of sounds, especially on those models that are directly based on physical descriptions of sounding objects. The general framework of sound modeling is explained in Sects. 16.2 and 16.3. Section 16.4 proposes an organization of sound manipulations into generative and processing models. We will see how different modeling paradigms can be used for both categories, and we will divide these paradigms into signal models and physics-based models. Sections 16.5

and 16.6 form the kernel of the chapter and illustrate physically-based modeling with some detail. Several techniques are presented for modeling sound sources and general, linear and nonlinear acoustic systems. Finally, Sect. 16.7 is a physics-based view of sound processing models, such as reverberation and spatialization techniques.

16.2 Computational Models as Musical Instruments

In order to generate, manipulate, and think about sounds, it is useful to organize our intuitive sound abstractions into objects, in the same way as abstract categories are needed for defining visual objects. The first extensive investigation and systematization of sound objects from a perceptual viewpoint was done by Pierre Shaeffer in the fifties [67]. Nowadays, a common terminology is available for describing sound objects both from a phenomenological or a referential viewpoint, and for describing collections of such objects (i.e. *soundscapes*) [38,50,75].

For effective generation and manipulation of the sound objects it is necessary to define models for sound synthesis, processing, and composition. Identifying models, either visual or acoustic, is equivalent to making high-level constructive interpretations, built up from the zero level (i.e. pixels or sound samples). It is important for the model to be associated with a semantic interpretation, in such a way that an intuitive action on model parameters becomes possible.

A sound model is implemented by means of sound synthesis and processing techniques. A wide variety of sound synthesis algorithms is currently available either commercially or in the literature. Each one of them exhibits some peculiar characteristics that could make it preferable to others, depending on goals and needs. Technological progress has made enormous steps forward in the past few years as far as the computational power that can be made available at low cost is concerned. At the same time, sound synthesis methods have become more and more computationally efficient and the user interface has become friendlier and friendlier. As a consequence, musicians can nowadays access a wide collection of synthesis techniques (all available at low cost in their full functionality), and concentrate on their timbral properties.

Each sound synthesis algorithm can be thought of as a computational model for the sound itself. Though this observation may seem quite obvious, its meaning for sound synthesis is not so straightforward. As a matter of fact, modeling sounds is much more than just generating them, as a computational model can be used for representing and generating a whole class of sounds, depending on the choice of control parameters. The idea of associating a class of sounds to a digital sound model is in complete accordance with the way we tend to classify natural musical instruments according to their sound generation mechanism. For example, strings and woodwinds are normally seen as timbral classes of acoustic instruments characterized by their sound generation mechanism. It should be clear that the degree of compactness of a class of sounds is determined, on one hand, by the sensitivity of the digital model to parameter variations and, on the other hand, the amount of control that is necessary to obtain a certain desired sound. As an extreme example we may think of a situation in which a musician is required to generate sounds sample by sample, while the task of the computing equipment is just that of playing the samples. In this case the control signal is represented by the sound itself, therefore the class of sounds that can be produced is unlimited but the instrument is impossible for a musician to control and play. An opposite extremal situation is that in which the synthesis technique is actually the model of an acoustic musical instrument. In this case the class of sounds that can be produced is much more limited (it is characteristic of the mechanism that is being modeled by the algorithm), but the degree of difficulty involved in generating the control parameters is quite modest, as it corresponds to physical parameters that have an intuitive counterpart in the experience of the musician.

An interesting conclusion that could be already drawn in the light of what we stated above is that the generality of the class of sounds associated to a sound synthesis algorithm is somehow in contrast with the "playability" of the algorithm itself. One should remember that the "playability" is of crucial importance for the success of a specific sound synthesis algorithm as, in order for a sound synthesis algorithm to be suitable for musical purposes, the musician needs an intuitive and easy access to its control parameters during both the sound design process and the performance. Such requirements often represents the reason why a certain synthesis technique is preferred to others.

From a mathematical viewpoint, the musical use of sound models opens some interesting issues: description of a class of models that are suitable for the representation of musically-relevant acoustic phenomena; description of efficient and versatile algorithms that realize the models; mapping between meaningful acoustic and musical parameters and numerical parameters of the models; analysis of sound signals that produces estimates of model parameters and control signals; approximation and simplification of the models based on the perceptual relevance of their features; generalization of computational structures and models in order to enhance versatility.

16.3 Sound Modeling

In the music sound domain, we define *generative models* as those models which give computational form to abstract objects, thus representing a sound generation mechanism. Sound fruition requires a further processing step, which accounts for sound propagation in enclosures and for the listener position relative to the sound source. Modifications such as these, which intervene on the attributes of sound objects, are controllable by means of *space models*.

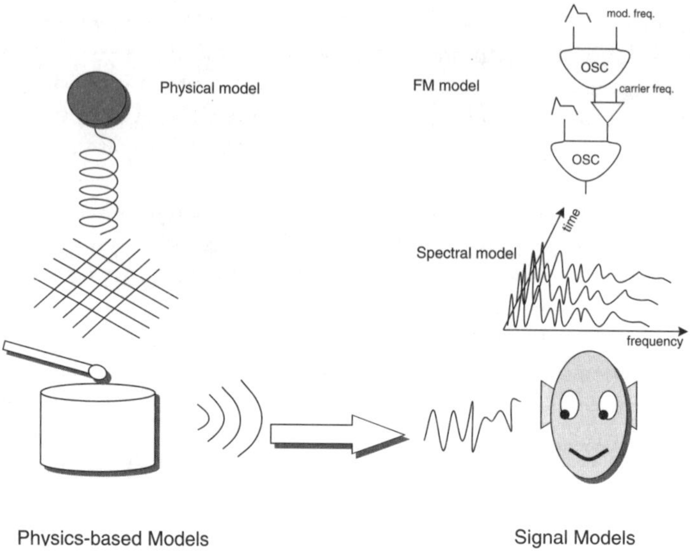

Fig. 16.1. Physics-based Models and Signal Models

Generative models can represent the dynamics of real or virtual generating objects (*physics-based models*), or they can represent the physical quantities as they arrive to human senses (*signal models*) [71] (see Fig. 16.1). In our terminology, signal models are models of signals as they are emitted from loudspeakers or arrive to the ears. The connection with human perception is better understood when considering the evaluation criteria of the generative models. The evaluation of a signal model should be done according to certain perceptual cues. On the contrary, physics-based models are better evaluated according to the physical behaviors involved in the sound production process. Space models can be as well classified with respect to their commitment to model the causes or the effects of sound propagation from the source to the ears. For example, a reverberation system can be built from an abstract signal processing algorithm where its parameters are mapped to perceptual cues (e.g. warmth or brilliance) or to physical attributes (e.g. wall absorption or diffusion).

In classic sound synthesis, signal models dominated the scene, due to the availability of very efficient and widely applicable algorithms (e.g. frequency modulation). Moreover, signal models allow to design sounds as objects *per se* without having to rely on actual pieces of material which act as a sound source. However, many people are becoming convinced of the fact that physics-based models are closer to the users/designers' needs of interacting with sound objects. The semantic power of these models seems to make them preferable for this purpose. The computational complexity of physically-based algorithms is becoming affordable with nowadays technology, even for real-time applications. We keep in mind that the advantage we gain in model expressivity comes to the expense of the flexibility of several general-purpose signal models. For this reason, signal models keep being the model of choice in many applications, especially for music composition.

In the perspective of a multisensorial unification under common models, physics-based models offer an evident advantage over signal models. In fact, the mechanisms of perception for sight and hearing are very different, and a unification at this level looks difficult. Even though analogies based on perception are possible, an authentic sensorial coherence seems to be ensured only by physics-based models. The interaction among various perceptions can be an essential feature if we want to maximize the amount of information conveyed to the spectator/actor. The unification of visual and aural cues is more properly done at the level of abstractions, where the cultural and experience aspects become fundamental. Thus, building models closer to the abstract object, as it is conceived by the designer, is a fundamental step in the direction of this unification.

16.4 Classic Signal Models

Here we will briefly overview the most important signal models for musical sounds. A more extensive presentation can be found in several tutorial articles and books on sound synthesis techniques [21,22,31,42,52,53]. Complementary to this, Sect. 16.5 will cover the most relevant paradigms in physically-based sound modeling.

16.4.1 Spectral Models

Since the human ear acts as a particular spectrum analyser, a first class of synthesis models aims at modeling and generating sound spectra. The Short Time Fourier Transform and other time-frequency representations provide powerful sound analysis tools for computing the time-varying spectrum of a given sound.

Sinusoidal model. When we analyze a pitched sound, we find that its spectral energy is mainly concentrated at a few discrete (slowly time-varying) frequencies f_i. These frequency lines correspond to different sinusoidal components called partials. If the sound is almost periodic, the frequencies of partials are approximately multiple of the fundamental frequency f_0, ie. $f_i(t) \simeq i\, f_0(t)$. The amplitude a_i of each partial is not constant and its time-variation is critical for timbre characterization. If there is a good degree of correlation among the frequency and amplitude variations of different partials, these are perceived as fused to give a unique sound with its timbre identity.

The sinusoidal model assumes that the sound can be modeled as a sum of sinusoidal oscillators whose amplitude a_i and frequency f_i are slowly time-varying

$$s_s(t) = \sum_i a_i(t) \cos[\phi_i(t)] , \tag{16.1}$$

$$\phi_i(t) = \int_0^t 2\pi f_i(\tau) d\tau + \phi_i(0) , \tag{16.2}$$

or, digitally,

$$s_s(n) = \sum_i a_i(n) \cos[\phi_i(n)] , \tag{16.3}$$

$$\phi_i(n) = 2\pi\, f_i(n) T_s + \phi_i(n-1) , \tag{16.4}$$

where T_s is the sampling period. Equations (16.1) and (16.2) are a generalization of the Fourier theorem, that states that a periodic sound of frequency f_0 can be decomposed as a sum of harmonically related sinusoids $s_s(t) = \sum_i a_i \cos(2\pi i f_0 t + \phi_i)$.

This model is also capable of reproducing aperiodic and inharmonic sounds, as long as their spectral energy is concentrated near discrete frequencies (spectral lines). In computer music this model is called *additive synthesis* and is widely used in music composition. Notice that the idea behind this method is not new. As a matter of fact, additive synthesis has been used for centuries in some traditional instruments such as organs. Organ pipes, in fact, produce relatively simple sounds that, combined together, contribute to the richer spectrum of some registers. Particularly rich registers are created by using many pipes of different pitch at the same time. Moreover this method, developed for simulating natural sounds, has become the "metaphorical" foundation of a compositional methodology based on the expansion of the time scale and the reinterpretation of the spectrum in harmonic structures.

Random noise models. The spread part of the spectrum is perceived as random noise. The basic noise generation algorithm is the congruential method

$$s_n = [a s_n(n-1) + b] \bmod M . \tag{16.5}$$

With a suitable choice of the coefficients a and b it produces pseudorandom sequences with flat spectral density magnitude (white noise). Different spectral shapes can be obtained using white noise as input to a filter.

Filters. Some sources can be modeled as an exciter, characterized by a spectrally rich signal, and a resonator, described by a linear system, connected in a feed-forward relationship. An example is the voice, where the periodic pulses or random fluctuations produced by the vocal folds are filtered by the vocal tract, that shapes the spectral envelope. The vowel quality and the voice color greatly depends on the resonance regions of the filter, called formants.

If the system is linear and time-invariant, it can be described by the filter $H(z) = B(z)/A(z)$ that can be computed by a difference equation

$$s_f(n) = \sum_i b_i u(n-i) - \sum_k a_k s_f(n-k) . \tag{16.6}$$

where a_k resp. b_i are the filter coefficients and $u(n)$ resp. $s_f(n)$ are input and output signals. The model is also represented by the convolution of the source $u(n)$ with the impulse response of the filter

$$s_f(n) = (u * h)(n) \triangleq \sum_k h(n-k)u(k) . \tag{16.7}$$

Digital signal processing theory gives us the tools to design the filter structure and to estimate the filter coefficients in order to obtain a desired frequency response. This model combines the spectral fine structure (spectral lines, broadband or narrowband noise, etc.) of the input with the spectral envelope shaping properties of the filter: $S_f(f) = U(f)\,H(f)$. Therefore, it is possible to control and modify separately the pitch from the formant structure of a speech sound. In computer music this model is called *subtractive synthesis*. If the filter is static, the temporal features of the input signal are maintained. If, conversely, the filter coefficients are varied, the frequency response changes. As a consequence, the output will be a combination of temporal variations of the input and of the filter (*cross-synthesis*).

If we make some simplifying hypothesis about the input, it is possible to estimate both the parameters of the source and the filter of a given sound. The most common procedure is linear predictive coding (LPC) which assumes that the source is either a periodic impulse train or white noise, and that the filter is all pole (i.e., no zeros) [30]. LPC is widely used for speech synthesis and modification.

A special case is when the filter features a long delay as in

$$s_f(n) = \beta u(n) - \alpha s_f(n - N_p) . \tag{16.8}$$

This is a comb type filter featuring frequency resonances multiple of a fundamental $f_p = F_s/N_p$, where $F_s = 1/T_s$ is the sampling rate. If initial values are set for the whole delay line, for example random values, all the frequency components that do not coincide with resonance frequencies are progressively filtered out until a harmonic sound is left. If there is attenuation ($\alpha < 1$) the sound will have a decreasing envelope. Substituting α and/or β with filters, the sound decay time will depend on frequency. For example if α is smaller at higher frequencies, the upper harmonics will decay faster than the lower ones. We can thus obtain simple sound simulations of the plucked strings [34,33], where the delay line serves to establish oscillations. This method is suitable to model sounds produced by a brief excitation of a resonator, where the latter establishes the periodicity, and the interaction between exciter and resonator can be assumed to be feedforward. This method is called *long-term prediction* or *Karplus-Strong* synthesis. More general musical oscillators will be discussed in Sect. 16.6.

16.4.2 Time Domain Models

When the sound characteristics are rapidly varying, as during attacks or non stationary sounds, spectral models tend to present artifacts, due to low time-frequency resolution or to the increase of the amount of data used in the representation. To overcome these difficulties, time domain models were proposed. A first class, called *sampling* or *wavetable*, stores the waveforms of musical sounds or sound fragments in a database. During synthesis, a waveform is selected and reproduced with simple modifications, such as looping of the periodic part, or sample interpolation for pitch shifting. The same idea is used for simple oscillators, that repeats a waveform stored in a table (*table-lookup oscillator*).

Granular models. More creative is the *granular synthesis* model. The basic idea is that a sound can be considered as a sequence, possibly with overlaps, of elementary and short acoustic elements called *grains*. Additive synthesis starts from the idea of dividing the sound in the frequency domain into a number of simpler elements (sinusoidal). Granular synthesis, instead, starts from the idea of dividing the sound in the time domain into a sequence of short elements called "grains". The parameters of this technique are the waveform of the grain $g_k(\cdot)$, its temporal location l_k and amplitude a_k

$$s_g(n) = \sum_k a_k g_k(n - l_k) \; . \tag{16.9}$$

A complex and dynamic acoustic event can be constructed starting from a large quantity of grains. The features of the grains and their temporal locations determine the sound timbre. We can see it as being similar to cinema, where a rapid sequence of static images gives the impression of objects in movement. The initial idea of granular synthesis dates back to Gabor [20], while in music it arises from early experiences of tape electronic music. The choice of parameters can be via various criteria, at the base of which, for each one, there is an interpretation model of the sound. In general, granular synthesis is not a single synthesis model but a way of realizing many different models using waveforms that are locally defined. The choice of the interpretation model implies operational processes that may affect the sonic material in various ways.

The most important and classic type of granular synthesis (*asynchronous granular synthesis*) distributes grains irregularly on the time-frequency plane in form of clouds [51]. The grain waveform is

$$g_k(i) = w_d(i) \cos(2\pi f_k T_s i) \ , \tag{16.10}$$

where $w_d(i)$ is a window of length d samples, that controls the time span and the spectral bandwidth around f_k. For example, randomly scattered grains within a mask, which delimits a particular frequency/amplitude/time region, result in a sound cloud or musical texture that varies over time. The density of the grains within the mask can be controlled. As a result, articulated sounds can be modeled and, wherever there is no interest in controlling the microstructure exactly, problems involving the detailed control of the temporal characteristics of the grains can be avoided. Another peculiarity of granular synthesis is that it eases the design of sound events as parts of a larger temporal architecture. For composers, this means a unification of compositional metaphors on different scales and, as a consequence, the control over a time continuum ranging from the milliseconds to the tens of seconds. There are psychoacoustic effects that can be easily experimented by using this algorithm, for example crumbling effects and waveform fusions, which have the corresponding counterpart in the effects of separation and fusion of tones.

16.4.3 Hybrid Models

Different models can be combined in order to have a more flexible and effective sound generation. One approach is *Spectral Modeling Synthesis* (SMS) [70] that considers sounds as composed by a sinusoidal part $s_s(t)$ (see Eq. 16.1), corresponding to the main system modes of vibration, and a residual $r(t)$, modeled as the convolution of white noise with a time-varying frequency shaping filter (see Eq. 16.7)

$$s_{sr}(t) = s_s(t) + r(t) \ . \tag{16.11}$$

The residual comprises the energy produced in the excitation mechanism which is not transformed into stationary vibrations, plus any other energy contribution that is not sinusoidal in nature. By using the short time Fourier transform and a peak detection algorithm, it is possible to separate the two parts at the analysis stage, and to estimate the time varying parameters of these models. The main advantage of this model is that it is quite robust to sound transformations that are musically relevant, such as time stretching, pitch shifting, and spectral morphing.

In the SMS model, transients and rapid signal variations are not well represented. Verma et al. [81] proposed an extension of SMS that includes a third component due to transients. Their method is called Sinusoids+Transients+Noise (S+T+N) and is expressed by

$$s_{STN}(t) = s_s(t) + s_g(t) + r(t) \ , \tag{16.12}$$

where $s_g(t)$ is a granular term representing the signal transients. This term is automatically extracted from the SMS residual using the Discrete Cosine Transform, followed by a second SMS analysis in the frequency domain.

16.4.4 Abstract Models: Frequency Modulation

Another class of sound synthesis algorithms is neither derived from physical mechanisms of sound production, nor from any sound analysis techniques. These are algorithms derived from the mathematical properties of a formula. The most important of these algorithms is the so called synthesis by *Frequency Modulation* (FM) [17]. The technique works as an instantaneous modulation of the phase or frequency of a sinusoidal carrier according to the behavior of another signal (modulator), which is usually sinusoidal. The basic scheme can be expressed as follows:

$$s(t) = \sin\left[2\pi f_c t + I \sin\left(2\pi f_m t\right)\right] = \sum_{k=-\infty}^{\infty} J_k(I) \sin\left[2\pi\left(f_c + k f_m\right) t\right] \quad (16.13)$$

where $J_k(I)$ is the Bessel function of order k. The resulting spectrum presents lines at frequencies $|f_c \pm k f_m|$. The ratio f_c/f_m determines the spectral content of sounds, and is directly linked to some important features, like the absence of even components, or the inharmonicity. The parameter I (modulation Index) controls the spectral bandwidth around f_c, and is usually associated with a time curve (the so called envelope), in such a way that time evolution of the spectrum is similar to that of traditional instruments. For instance, a high value of the modulation index determines a wide frequency bandwidth, as it is found during the attack of typical instrumental sounds. On the other hand, the gradual decrease of the modulation index determines a natural shrinking of the frequency bandwidth during the decay phase. From the basic scheme, other variants can be derived, such as parallel modulators and feedback modulation. So far, however, no general algorithm has been found for deriving the parameters of an FM model from the analysis of a given sound, and no intuitive interpretation can be given to the parameter choice, as this synthesis technique does not evoke any previous musical experience of the performer. The main qualities of FM, i.e. great timbre dynamics with just a few parameters and a low computational cost, are progressively losing importance within modern digital systems. Other synthesis techniques, though more expensive, can be controlled in a more natural and intuitive fashion. The FM synthesis, however, still preserves the attractiveness of its own peculiar timbre space and, although it is not particularly suitable for the simulation of natural sounds, it offers a wide range of original synthetic sounds that are of considerable interest for computer musicians.

16.5 Physics-based Models

In the family of physics-based models we put all the algorithms generating sounds as a side effect of a more general process of simulation of a physical phenomenon. Physics-based models can be classified according to the way of representing, simulating and discretizing the physical reality. Hence, we can talk about cellular, finite-difference, and waveguide models, thus intending that these categories are not disjoint but, in some cases, they represent different viewpoints on the same computational mechanism. Moreover, physics-based models have not necessarily to be based on the physics of the real world, but they can, more generally, gain inspiration from it; in this case we will talk about pseudo-physical models.

In this chapter, the approach to physically-based synthesis is carried on with particular reference to real-time applications, therefore the time complexity of algorithms plays a key role. We can summarize the general objective of the presentation saying that we want to obtain models for large families of sounding objects, and these models have to provide a satisfactory representation of the acoustic behavior with the minimum computational effort.

16.5.1 Functional Blocks

In real objects we can often outline functionally distinct parts, and express the overall behavior of the system as the interaction of these parts. Outlining functional blocks helps the task of modeling, because for each block a different representation strategy can be chosen. In addition, the range of parameters can be better specified in isolated blocks, and the gain in semantic clearness is evident. Our analysis stems from musical instruments, and this is justified by the fact that the same generative mechanisms can be found in many other physical objects. In fact, we find it difficult to think about a physical process producing sound and having no analogy in some musical instrument. For instance, friction can be found in bowed string instruments, striking in percussion instruments, air turbulences in jet-driven instruments, etc. . Generally speaking, we can think of musical instruments as a specialization of natural dynamics for artistic purposes. Musical instruments are important for the whole area of sonification in multimedia environments because they constitute a testbed where the various simulation techniques can easily show their merits and pitfalls.

The first level of conceptual decomposition that we can devise for musical instruments is represented by the interaction scheme of Fig. 16.2, where two functional blocks are outlined: a resonator and an exciter. The resonator sustains and controls the oscillation, and is related with sound attributes like pitch and spectral envelope. The exciter is the place where energy is injected into the instrument, and it strongly affects the attack transient of sound, which is fundamental for timbre identification. The interaction of exciter and resonator is the main source of richness and variety of nuances that can be obtained from a musical instruments. When translating the conceptual decomposition into a model, two dynamic systems are found [8]: the excitation block, which is strongly non-linear, and the resonator, supposed to be linear to a great extent. The player controls the performance by means of inputs to the two blocks. The interaction can be "feedforward", when the exciter doesn't receive any information from the resonator, or "feedback", when the two blocks exert a mutual information exchange. In this conceptual scheme, the radiating element (bell, resonating body, etc.) is implicitly enclosed within the resonator. In a clarinet, for instance, we have a feedback structure where the reed is the exciter and the bore with its bell acts as a resonator. The player exert exciting actions such as controlling the mouth pressure and the embouchure, as well as modulating actions such as changing the bore effective length by opening and closing the holes. In a plucked string instrument, such as a guitar, the excitation is provided by plucking the string, the resonator is given by the strings and the body, and modulating actions take the form of fingering. The interaction is only weakly feedback, so that a feedforward scheme can be adopted as a good approximation: the excitation imposes the initial conditions and the resonator is then left free to vibrate.

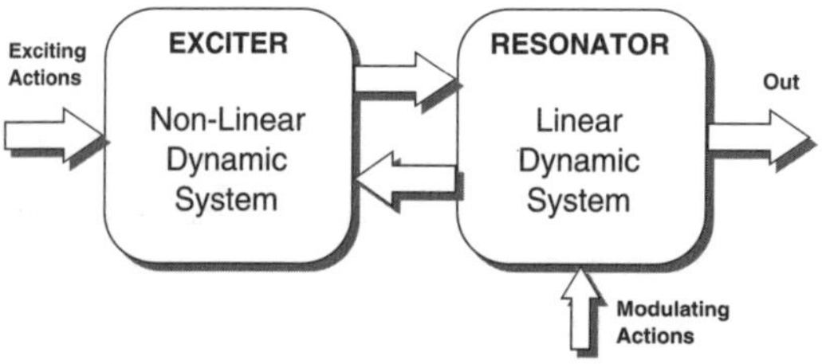

Fig. 16.2. Exciter-Resonator Interaction Scheme

In practical physical modeling the block decomposition can be extended to finer levels of detail, as both the exciter and the resonator can be further decomposed into simpler functional components, e.g. the holes and the bell of a clarinet as a refinement of the resonator. At each stage of model decomposition, we are faced with the choice of expanding the blocks further (white-box modeling), or just considering the input-output behavior of the basic components (black-box modeling). In particular, it is very tempting to model just the input-output behavior of linear blocks, because in this case the problem reduces to filter design. However, such an approach provides structures whose parameters are difficult to interpret and, therefore, to control. In any case, when the decomposition of an instrument into blocks corresponds to a similar decomposition in digital structures, a premium in efficiency and versatility is likely to be obtained. In fact, we can focus on functionally distinct parts and try to obtain the best results from each before coupling them together [7].

In digital implementations, in between the two blocks exciter and resonator, a third block is often found. This is an interaction block and it can convert the variables used in the exciter to the variables used in the resonator, or avoid possible anomalies introduced by the discretization process. The idea is to have a sort of adaptor for connecting different blocks in a modular way. This adaptor might also serve to compensate the simplifications introduced by the modeling process. To this end, a residual signal might be introduced in this block in order to improve the sound realism. The limits of a detailed physical simulation are also found when we try to model the behavior of a complex linear vibrating structure, such as a soundboard; in such cases it can be useful to record its impulse response and include it in the excitation signal as it is provided to a feedforward interaction scheme. Such a method is called *commuted synthesis*, since it makes use of commutativity of linear, time-invariant blocks [73,76].

It is interesting to notice that the integration of sampled noises or impulse responses into physical models is analogous to texture mapping in computer graphics [5]. In both cases the realism of a synthetic scene is increased by insertion of snapshots of textures (either visual or aural) taken from actual objects and projected onto the model.

16.5.2 Cellular Models

A possible approach to simulation of complex dynamical systems is their decomposition into a multitude of interacting particles. The dynamics of each of these particles are discretized and quantized in some way to produce a finite-state automaton (a cell), suitable for implementation on a processing element of a parallel computer. The discrete dynamical system consisting of a regular lattice of elementary cells is called a cellular automaton [82,85]. The state of any cell is updated by a transition rule which is applied to the previous-step state of its neighborhood. When the cellular automaton comes from the discretization of a homogeneous and isotropic medium it is natural to assume functional homogeneity and isotropy, i.e. all the cells behave according to the same rules and are connected to all their immediate neighbors in the same way [82]. If the cellular automaton has to be representative of a physical system, the state of cells must be characterized by values of selected physical parameters, e.g. displacement, velocity, force.

Several approaches to physically-based sound modeling can be recast in terms of cellular automata, the most notable being the CORDIS-ANIMA system introduced by Cadoz and his co-workers [26,11,32], who came up with cells as discrete-time models of small mass-spring-damper systems, with the possible introduction of nonlinearities. The main goal of the CORDIS-ANIMA project was to achieve high degrees of modularity and parallelism, and to provide a unified formalism for rigid and flexible bodies. The technique is very expensive for an accurate sequential simulation of wide vibrating objects, but is probably the only effective way in the case of a multiplicity of micro-objects (e.g. sand grains) or for very irregular media, since it allows an embedding of the material characteristics (viscosity, etc.). An example of CORDIS-ANIMA network discretizing a membrane is shown in Fig. 16.3, where we have surrounded by triangles the equal cells which provide output variables depending on the internal state and on input variables from neighboring cells. Even though the CORDIS-ANIMA system uses heterogeneous elements such as matter points or visco-elastic links, Fig. 16.3 shows how a network can be restated in terms of a cellular automaton showing functional homogeneity and isotropy.

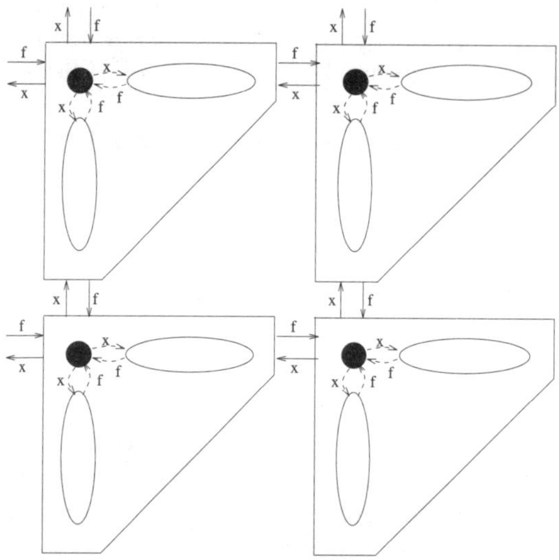

Fig. 16.3. A CORDIS-ANIMA network (a piece of a rectangular mesh) restated as a cellular automaton. Black dots indicate mass points and white ovals indicate link elements, such as a visco-elastic connection. x represents a position variable and f represents a force variable

A cellular automaton is inherently parallel, and its implementation on a parallel computer shows excellent scalability. Moreover, in the case of the multiplicity of micro-objects, it has shown good effectiveness for joint production of audio and video simulations [12]. It might be possible to show that a two-dimensional cellular automaton can implement the model of a membrane as it is expressed by a waveguide mesh. However, as we will see in Sects. 16.5.3 and 16.5.4, when the system to be modeled is the medium where waves propagate, the natural approach is to start from the wave equation and to discretize it or its solutions. In the fields of finite-difference methods or waveguide modeling, theoretical tools do exist for assessing the correctness of these discretizations. On the other hand, only qualitative criteria seem to be applicable to cellular automata in their general formulation.

16.5.3 Finite-difference Models

When modeling vibrations of real-world objects, it can be useful to consider them as rigid bodies connected by lumped, idealized elements (e.g. dashpots, springs, geometric constraints, etc.) or, alternatively, to treat them as flexible bodies where forces and matter are distributed over a continuous space (e.g. a string, a membrane, etc.). In the two cases the physical behavior can be represented by ordinary or partial differential equations, whose form can be learned from physics textbooks [25] and whose coefficient values can be obtained from physicists' investigations or from direct measurements. These differential equations often give only a crude approximation of reality, as the objects being modeled are just too complicated. Moreover, as we try to solve the equations by numerical means, a further amount of approximation is added to the simulated behavior, so that the final result can be quite far from the real behavior.

One of the most popular ways of solving differential equations is finite differencing, where a grid is constructed in the spatial and time variables, and derivatives are replaced by linear combinations of the values on this grid. Two are the main problems to be faced when designing a finite-difference scheme for a partial differential equation: numerical losses and numerical dispersion. There is a standard technique [49,74] for evaluating the performance of a finite-difference scheme in contrasting these problems: the von Neumann analysis. It can be quickly explained on the simple case of the ideal string (or the ideal acoustic tube), whose wave equation is [45]

$$\frac{\partial^2 p(x,t)}{\partial t^2} = c^2 \frac{\partial^2 p(x,t)}{\partial x^2} , \tag{16.14}$$

where c is the wave velocity of propagation, t and x are the time and space variables, and p is the string displacement (or acoustic pressure). By replacing the second derivatives by central second-order differences, the explicit updating scheme for the i-th spatial sample of displacement (or pressure) is:

$$\begin{aligned} p(i,n+1) = 2\left(1 - \frac{c^2 \Delta t^2}{\Delta x^2}\right) p(i,n) - p(i,n-1) \\ + \frac{c^2 \Delta t^2}{\Delta x^2} \left[p(i+1,n) + p(i-1,n)\right] , \end{aligned} \tag{16.15}$$

where Δt and Δx are the time and space grid steps. The von Neumann analysis assumes that the equation parameters are locally constant and checks the time evolution of a spatial Fourier transform of (16.15). In this way a spectral amplification factor is found whose deviations from unit magnitude and linear phase give respectively the numerical loss (or amplification) and dispersion errors. For the scheme (16.15) it can be shown that a unit-magnitude amplification factor is ensured as long as the Courant-Friedrichs-Lewy condition [49]

$$\frac{c\Delta t}{\Delta x} \leq 1 \tag{16.16}$$

is satisfied, and that no numerical dispersion is found if equality applies in (16.16). A first consequence of (16.16) is that only strings having length which is an integer number of $c\Delta t$ are exactly simulated. Moreover, when the string deviates from ideality and higher spatial derivatives appear (physical dispersion), the simulation becomes always approximate. In these cases, the resort to implicit schemes can allow the tuning of the discrete algorithm to the amount of physical dispersion, in such a way that as many partials as possible are reproduced in the band of interest [14].

It is worth noting that if c in (16.14) is a function of time and space, the finite difference method retains its validity because it is based on a local (in time and space) discretization of the wave equation. Another advantage of finite differencing over other modeling techniques is that the medium is accessible at all the points of the time-space grid, thus maximizing the possibilities of interaction with other objects.

When the objects being simulated are rigid bodies, they can be described by ordinary differential equations

$$\begin{aligned} \dot{\mathbf{y}}(t) &= \mathbf{F}\left[\mathbf{y}(t), \mathbf{u}(t), t\right] , \\ \mathbf{y}(0) &= \mathbf{y_0} , \end{aligned} \tag{16.17}$$

being $\mathbf{u}(t)$ the vector describing the set of input signals and $\mathbf{y_0}$ initial conditions. Numerical analysis developed a plethora of techniques for their integration [49] transforming Eq. (16.17) into difference equations

$$\mathbf{y}(n+1) = \mathbf{F_d}\left[\mathbf{y}(n), \mathbf{u}(n), n\right] . \tag{16.18}$$

However, attention must be paid to stability issues and to the correct reproduction of important physical attributes. These issues are strongly dependent on the numerical integration technique and on the sampling rate which are to be used. In most of the cases, there is no better method than trying several techniques and comparing the results, but the task is often facilitated by the fact that the strong nonlinearities are lumped. For example, in [29], the dynamics of a clarinet reed is discretized by using a fourth-order Runge–Kutta method, Euler differencing, and bilinear transformation [46]. The Runge–Kutta method turns out to be unstable for low sampling rates, while Euler differencing shows a poor reproduction of the characteristic resonance of the reed, due to numerical losses. For that specific case, the best choice seems to be the bilinear transform, which corresponds to a trapezoidal integration of the differential equations, possibly with some warping of the frequency axis [44] for adjusting the resonance central frequency. The discretization by impulse invariance [46] is also a reliable tool when aliasing can be neglected, and its performance is often preferable to bilinear transformation in acoustic modeling because it is free of frequency warping and artificial damping. Other discretization methods have recently been compared using the clarinet reed as a testbed [1]. As a result, it seems that a technique that employs polynomial interpolation of the input signals [83] gives the best reproduction of the reed resonance at an affordable cost. This latter technique can be interpreted as an extension of the impulse invariance that includes some antialias filtering.

Further studies are needed to establish the most suitable discretization techniques for the many kinds of lumped dynamics, with special attention to be paid to the looped connection of lumped non-linear elements with memory-less nonlinearities and distributed resonators. Several techniques from signal processing and numerical analysis are yet to be experimented, while some general methodologies are just being proposed. In this respect, Sect. 16.5.4 will show how, switching to a wave-variable representation of the physical quantities, it is possible to apply the paradigms of Wave Digital filters and Waveguide networks to the lumped and distributed elements respectively.

16.5.4 Wave Models

When discretizing physical systems a key role is played by the efficiency and accuracy of the discretization technique. Namely, we would like to be able to simulate simple vibrating structures and exciters with no artifacts (e.g. aliasing, or non-computable dependencies) and with low computational complexity. Due to its good properties with respect to these two criteria, one of the most popular ways of approaching physical modeling of acoustic systems makes use of wave variables instead of absolute physical quantities. Given the dual physical variables p and u (let us call them pressure and velocity), the pressure waves are defined as

$$\begin{aligned} p^+ &= (p + Z_0 u)/2 \, , \\ p^- &= (p - Z_0 u)/2 \, , \end{aligned} \tag{16.19}$$

where Z_0 is an arbitrary reference impedance.

When wave variables are adopted in the digital domain for representing lumped components this approach is called Wave Digital Filtering [24] It is possible to show that a lumped component having impedance $Z(z) = P(z)/U(z)$ can be represented in pressure waves by

$$R(z) = \frac{P^-(z)}{P^+(z)} = \frac{Z(z) - Z_0}{Z(z) + Z_0} \, . \tag{16.20}$$

The reference impedance Z_0 is chosen in such a way that there is at least one delay element in any signal path connecting p^+ with p^-. The complete wave network is derived by applying the Kirchhoff principles [3] to junctions of components derived by the previous steps (wave-scattering formulation of the network) and to abrupt changes of the characteristic impedance.

On the other hand, when the components to be modeled are distributed wave-propagating media, Digital Waveguide Networks [72] can be used to simulate them. In these models the physical variables are decomposed into their respective wave variables and their propagation is simulated by means of delay lines. Low-pass and all-pass filters are added to simulate dissipative and dispersive effects in the medium.

As opposed to finite differencing, which discretize the wave equation (see Eqs. (16.14) and (16.15)), waveguide models come from discretization of the solution of the wave equation. The solution to the one-dimensional wave equation (16.14) was found by D'Alembert in 1747 in terms of traveling waves:

$$p(x,t) = p^{+}(t - x/c) + p^{-}(t + x/c) \, . \tag{16.21}$$

Equation (16.21) shows that the physical quantity p (e.g. string displacement or acoustic pressure) can be expressed as the sum of two wave quantities traveling in opposite directions. In waveguide models waves are sampled in space and time in such a way that equality holds in (16.16). If propagation along a one-dimensional medium, such as a cylinder, is ideal, i.e. linear, non-dissipative and non-dispersive, wave propagation is represented in the discrete-time domain by a couple of digital delay lines (Fig. 16.4), which propagates wave variables, as defined in (16.19) with Z_0 characteristic impedance of the medium. As a slight generalization, it can be seen that the wave equation in a cone is identical to the wave equation in a cylinder (Eq. (16.14), except that $p(x,t)$ is replaced by $x\,p(x,t)$ where x is the radial position along the cone axis. Thus, the solution is a superposition of left- and right-going traveling wave components, scaled by $1/x$ and can still be implemented by a couple of delay lines.

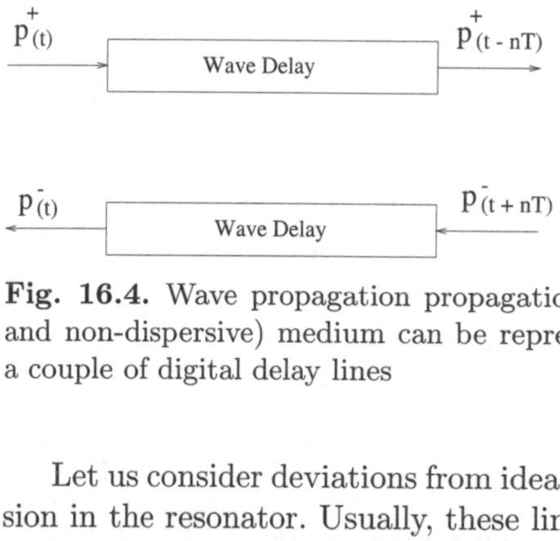

Fig. 16.4. Wave propagation propagation in a ideal (i.e. linear, non-dissipative and non-dispersive) medium can be represented, in the discrete-time domain, by a couple of digital delay lines

Let us consider deviations from ideal propagation due to losses and dispersion in the resonator. Usually, these linear effects are lumped and simulated with a few filters which are cascaded with the delay lines. Losses due to terminations, internal frictions, etc., give rise to gentle low pass filters, whose parameters can be identified from measurements [76]. Wave dispersion, which is often due to medium stiffness, is simulated by means of allpass filters whose effect is to produce a frequency-dependent propagation velocity [59].

In order to increase the computational efficiency, delay lines and filters should be lumped into as few processing blocks as possible. However, when considering the interaction with an exciter or signal pick-up from certain points of the resonator, the process of commuting and lumping linear blocks must be done with care. If the excitation is a velocity signal injected into a string, it will produce two velocity waves outgoing from the excitation point, and therefore at least two delay lines will be needed to represent propagation. The process of commuting and lumping must maintain the semantics at the observation points, while in the other points of the structure it is not necessary to have a strict correspondence with the physical reality.

Another aspect that we like to mention is that of simulating fractional delays. This is necessary when modeling musical instruments, since the proper tuning usually requires a space discretization much finer than dictated by the sample rate. More generally, fractional delay lengths are needed whenever time-varying acoustic objects (such as a string which is varying its length) are being modeled by digital waveguide networks. For this purpose, allpass filters or Lagrange interpolators of various orders can be used [37], the former suffering from phase distortion in high frequency, the latter suffering from both phase and amplitude distortion. However, low-order filters of both families can be used satisfactorily in most practical cases [37]. In other cases, the problem of designing a tuning filter is superseded by the more general problem of modeling wave dispersion [58]. When the physical medium is changing its internal properties during vibration (e.g. a string exhibiting tension modulation), we should use a digital delay line that allows a continuous variation of the spatial sampling rate [57], or integrate the effects of the internal changes along the whole resonator length so that it becomes possible to treat them in a lumped fashion as length modulations [77].

So far, we have talked about one-dimensional resonators, but many musical instruments (e.g. percussions) and most of the real-world objects are subject to deformation along several dimensions. The algorithms presented so far can be adapted to the case of multidimensional propagation of waves, even though new problems of efficiency and accuracy arise. All the models grow in computational complexity with the increase of dimensionality, and for any of them, the choice of the right discretization grid is critical. For example, a rectangular waveguide mesh can be effective for simulating a vibrating flexible membrane, but the simulation of wave propagation turns out to be exact only along the diagonals of the mesh, while elsewhere it is affected by a dispersive phenomenon due to the fact that we are simulating circular waves by portions of plane waves. Waveguide meshes are shown to be equivalent to special kinds of finite-difference schemes [79], so that the von Neumann analysis can be used to evaluate the numerical properties of the algorithms. Special attention has to be paid to the dependence of the dispersion factor on frequency, direction, and mesh geometry, because this influences the distribution of resonances, and therefore affects the tone color and intonation. For membrane simulation, one of the most accurate yet efficient meshes is the triangular mesh [27,28]. Interpolated waveguide meshes have recently been introduced for improving the accuracy while using simple geometries [65]. For three-dimensional wave propagation, the tetrahedral mesh is very attractive because its junctions can be implemented without any multiplication [80]. As compared to more conventional numerical techniques, such as finite difference schemes, the waveguide meshes offer the advantage that all the signals in the discrete-time system have a physical meaning, so that it is relatively straightforward to augment the model with some extra load (e.g. the air load for a membrane) or with nonlinearities.

16.6 Non Linear Musical Oscillators

Nonlinearities assume a great importance in acoustic systems, especially where a wave-propagation medium is excited. A good model for these non-linear mechanisms is essential for timbral quality, and is the real kernel of a physical model, which could otherwise be reduced to linear postprocessing of an excitation signal. Since the area where the excitation takes place is usually small, it makes sense to use lumped models for the excitation non-linearity. In some cases, physical measurements provide a representation of the relation among some physical variables involved in excitation, and this relation can be directly implemented in the simulation. For example, for a simplified bowed string the transversal velocity as a function of force can be found in the literature [39] for different values of bow pressure and velocity (which are control parameters). The instantaneous non-linear function can be approximated analytically or sampled and stored in a lookup table, which is in general multidimensional [61].

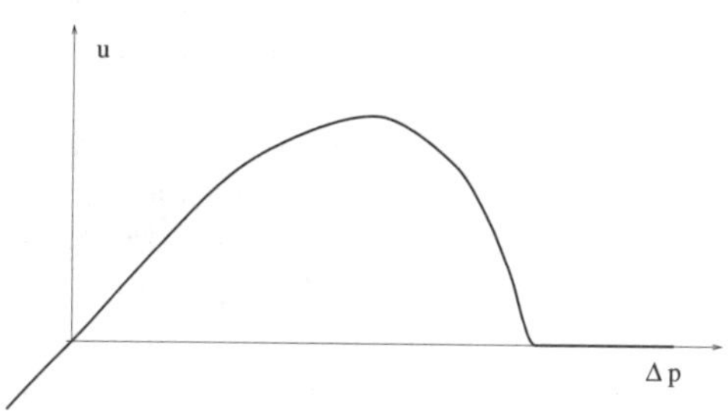

Fig. 16.5. Nonlinear relation between pressure difference Δp (inside pressure – mouth pressure) and particle velocity u of the air entering into a clarinet

As an example, let us consider a rough model of clarinet. The non-linear block representing the reed can simply be an instantaneous non-linear map (Fig. 16.5) relating the particle velocity u_r to the pressure difference Δp:

$$u_r = F_r(\Delta p) = F_r(p_m - p) , \tag{16.22}$$

where p_m is the player's mouth pressure and p is the pressure inside the bore at the excitation point [39]. Let us assume that there is, right after the exciter, a tract of constant-section tube having the characteristic impedance Z_0, and terminated by the radiation impedance $Z(s)$ of the bell. In order to simplify the analysis, we can see the tube as a lossless transmission line for the dual quantities pressure and flow, so that the D'Alembert decomposition (16.21) does hold. This use of instantaneous nonlinearities gives rise to a general scheme of nonlinear oscillator, depicted in Fig. 16.6, which is composed of an instantaneous map $y = G(x, u)$, possibly dependent on input parameters or signals u, a delay-line section, which determines the periodicity, and a linear filter $R(z)$ (see Eq. (16.7)) which can be tuned to give the desired spectral dynamics. In the case of the simplified clarinet $x \doteq p^-$, $y \doteq p^+$ and $u \doteq p_m$. If the filter $R(z)$ is reduced to a constant r, and the input signal u is constant, the system evolution is described by the iterated map

$$y(n) = G[ry(n-m), u] = F[y(n-m)] \tag{16.23}$$

where m is the total length of the delay lines. This formulation permits us to introduce qualitative reasoning about the conditions for establishing oscillations and the periodic, multiperiodic, or chaotic nature of the oscillations themselves [62,63]. Notice that Eq. (16.23) is the non linear generalization of Eq. (16.8).

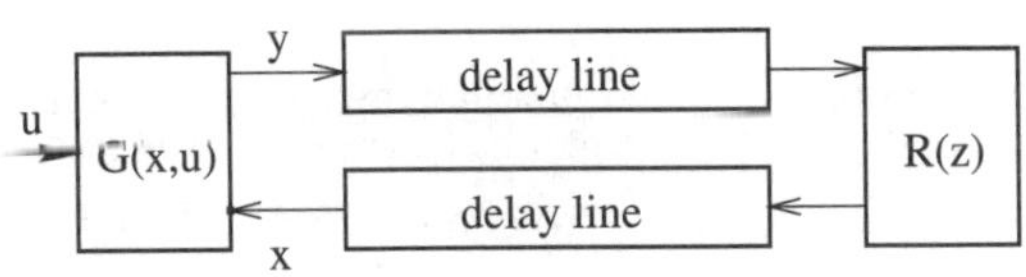

Fig. 16.6. A computational model for non linear oscillators, useful for musical sounds

In order to achieve a satisfactory audio quality of physics-based models, it is often necessary to use structures which are far more complex than that of the simplified clarinet. A first improvement over instantaneous non-linear excitation is considering the dynamics of the exciter: this implies the introduction of a state (i.e., memory) inside the non-linear block. When physicists study the behavior of musical instruments, they often use dynamic models of the exciter. A non-trivial task is the translation of these models into efficient computational schemes for real-time sound synthesis. A general structure has been found for good simulations of wide instrumental families [8,54]. In Fig. 16.7, this structure is schematically depicted. The block NL is a non-linear instantaneous function of several variables, while the block L is a linear dynamic system enclosing the exciter memory. The computational model is described by

$$\begin{aligned} \mathbf{y}(n) &= \mathbf{F_{NL}}\left[\mathbf{x}(n), \mathbf{u}(n), \mathbf{u_E}(n)\right] \\ \mathbf{x}(n+1) &= \mathbf{F_L}\left[\mathbf{x}(n), \mathbf{u}(n), \mathbf{u_E}(n), \mathbf{y}(n)\right] \end{aligned}$$

where $\mathbf{x}$ is the linear system state, $\mathbf{u}$ and $\mathbf{u_E}$ are respectively the exciting actions and the signals coming from the resonator, $\mathbf{y}$ is the output. Studies

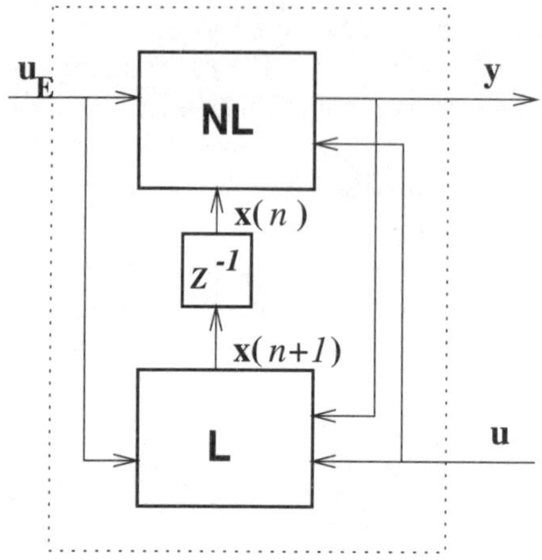

Fig. 16.7. Scheme for a Dynamic Exciter

and simulations have shown that reeds, air jets, bows and percussions, can all be represented by this scheme to a certain extent.

If a lumped modeling of dynamics is used, it has to be represented by discretization of some differential equations. To this end, one approach is to apply a finite difference scheme to these equations and introduce the instantaneous nonlinearity in the resulting signal flowgraph [15,78]. This procedure must be tackled carefully because it is easy to come up with delay-free (noncomputable) loops: in some cases these inconsistencies can be overcome by introducing fictitious delay elements, but at the sampling rates that we can afford it is likely that these delays are not acceptable, so that other techniques are needed. A notable example is found in the piano hammer-string interaction [6], whose model can be applied to several percussive sound sources. The computable discretization scheme of the non-linear hammer is obtained from the straightforward finite difference approximation of the dynamics, from which a term dependent on previous and known terms is separated from an instantaneous and unknown term. Finally, the instantaneous nonlinearity is recast into new variables in such a way that the delay-free loop is canceled [9]. Figure 16.8 compares the force signals obtained by the hammer-string model with and without the correct elimination of delay-free loops. It can be seen that the simple introduction of a decoupling unit delay in the delay-free loop leads to instabilities.

Another approach to the representation of the exciter dynamics is to resort to a model based on lumped linear or non-linear circuit components. The circuit components can be translated into the discrete-time domain using the wave digital filter method briefly explained in Sect. 16.5.4 [24]. This method has recently been extended to cover non-linear elements without [40] or with [64] memory, i.e., resistances or reactances.

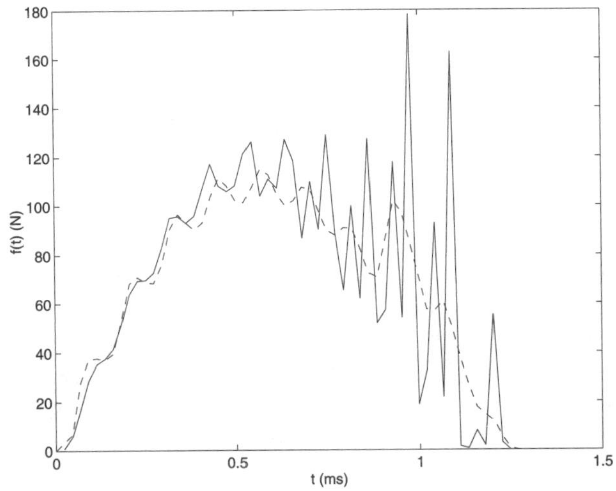

Fig. 16.8. Force in hammer-string interaction computed adding a non physical delay (continous line) and solving the noncomputability problem (dashed line)

When the need is not for a special-case model, but for a model which is adaptable to several sound sources, or when it is desirable to tune the model from sampled sounds, other representations have to be chosen for the nonlinear exciter. It has been proposed to express one-dimensional memoryless nonlinearities as a polynomial function whose coefficients can be identified by Kalman filtering [19], or adaptively by the LMS algorithm [66].

Alternatively, some general-purpose approximation network might be used as a generalized nonlinearity. For example, a Radial-Basis-Function network [48] built with a small number of gaussian kernels has shown its effectiveness in representing the severe nonlinearities found in wind and string instruments [23]. The parameters of these networks have to be identified by some global optimization technique, such as the genetic algorithm, applied to solve a spectral or waveform matching problem.

Nonlinearities might occur in the resonator as well, especially when it is driven into large vibrations. These nonlinearities are usually mild, so that simple saturation characteristics introduced somewhere in the resonator work just fine. However, for some sound sources the resonator nonlinearity is more critical, and special techniques must be devised in order to ensure that energy is not introduced or lost improperly. For instance, a time-variant one-pole allpass filter has been proposed for reproducing the kind of nonlinearity found in the bridge of the sitar [47]. Similar nonlinearities are found distributed in two-dimensional resonators, such as plates, and it is not yet clear how to simulate them by means of a small number of filters, properly placed in the computational structure.

16.7 Models for Sound and Space

In visualization or sonification there is an essential process, which appears basically unvaried in the two domains of image and sound. It is the process of passing from the space of objects, thought of as entities provided by the model, to the space of images or sounds. In computer graphics, this passage is governed by the laws of perspective, illumination and visibility. In sound computing these concepts are respectively replaced by sound localization, radiation/diffusion and masking. Most of the aforementioned models consider sounds as monodimensional functions of time picked up at some point close to the source. While loudspeakers and voice can be considered as point-like sources, in many musical instruments sound is emitted from many points or from a radiating surface. Then, sound waves are diffused in the environment and they reach the ears of the listener, that we can consider as point-like pickups.

This section presents models that account for the effects introduced on pressure waves by propagation in air, interaction with surrounding objects, and enclosing surfaces. They are based on perceptual descriptions of acoustic scenes (e.g., reverberation, spaciousness), or on physical descriptions of the environment (e.g., geometry of the room, position of the sound source). The section is not intended to be an exhaustive survey of techniques (for an extensive treatment of the topic the reader might look at the "Spatial Effects" chapter in the book [86]), but rather a brief description of some models that allow composers and sound designers to control the spatial attributes of sounds effectively.

16.7.1 Sound Spatialization

Any human is capable of detecting the direction where a certain sound is coming from with a precision of a few degrees [4]. Three are the necessary (although not always sufficient) parameters which are needed to detect the direction where a sound source is located: the Interaural Time Difference (ITD), Interaural Intensity Difference (IID) and Head-Related Transfer Functions (HRTF) [35]. The ITD is a time shift between the two signals arriving at the ears, and it is due to the different lengths of the direct paths going from the source to the left and right ears. The IID is due to the different attenuation the waves are subject to along these paths. These two quantities alone do not solve the localization problem in an unambiguous manner, they can only discriminate among source positions in a hemiplane. For a three-dimensional localization we have to deal with HRTFs, which are different filtering patterns offered by the head to signals coming from different directions [2]. These three components have to be simulated with good accuracy if a realistic three-dimensional acoustic field has to be provided through headphones. Moreover, HRTFs should be individualized (earprints). The use of standard HRTFs introduces perceptual distortions such as front-back confusion or poor source externalization.

For simulations of immersive environments the ITD, IID, and HRTF should be updated in real-time in order to follow the movements of the source and of the user's head. Head tracking is needed when the user is not constrained by the visual display to keep the head in a steady position. A great deal of efforts have been devoted to the efficient yet accurate implementation of time-varying filters for sound spatialization. The introduction of physical attributes of the enclosure can add other cues for having a correct spatial impression of the source. For example, the ratio between the intensities of the direct and reverberated sound is often sufficient to control the perceived distance from the source [16]. We will see in Sect. 16.7.2 how a physically-based model of room acoustics can give convincing reverb. The techniques for sound spatialization that we have mentioned so far are based on psychological and physiological attributes of the listener, and on the position of the source with respect to the ears. These techniques do not rely on the physics of wave propagation in an enclosed space, and depend heavily on the use of headphones.

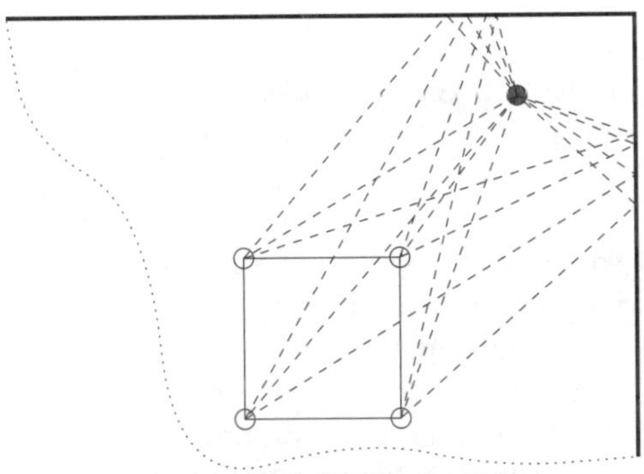

Fig. 16.9. Moore's two-room model for spatialisation. In this model the actual listening space is embedded inside the larger virtual room that we intend to simulate. The model considers direct and first reflection from the source (black circle) to loudspeakers (white circles)

The adoption of loudspeakers for acoustic display forces to change the model of sound spatialization. A model which seems to be suited to a set of loudspeakers is the two-room model (Fig. 16.9) by Moore [41]. In this model the actual listening space is embedded inside the larger virtual room that we intend to simulate. We assume that sound is coming into the inner room from the outer room by holes in the walls, and these holes correspond to the actual positions of the loudspeakers. The model consists of a prescribed number of delay lines, each of them corresponding to a path connecting the source with a loudspeaker. The number of paths depends on the number of wall reflections we are considering. For a convincing perception of space it is often sufficient to consider only the direct paths and the first reflections, leaving to a diffuse reverberation model the simulation of the sound tail produced by further reflections.

Spatialization models should take into account a couple of important physical attributes of a sounding object, namely its size and its radiation properties. The perceived size of an object can be controlled by the crosscorrelation of signals fed to different channels in a multiple-loudspeaker display. For simulating the radiation properties of actual sounding objects, all we have to do is to specify an intensity function dependent on direction. This radiation pattern is inserted in the localization model, and rays departing from the source are weighted by this direction-dependent pattern. A more complex problem is constructing loudspeaker systems that can reproduce a specific radiation pattern as found in a given sounding object [84,13].

16.7.2 Room Modeling and Reverberation

The environment participates to sound propagation by direct modification of sound waves when they hit objects and enclosing surfaces. A real surface typically shows a behavior between a perfect mirror and a perfect (Lambertian) diffusor, and its reflecting properties are frequency dependent [36].

The simulation of the diffusive properties of all the objects involved in sound propagation in a given environment is a heavy task. The problem is similar to that encountered in illumination of scenes [18], with the additional problem that propagation time of sound in air can not be neglected. Even recasting the problem in terms of interaction of objects by force fields, the computational time is far beyond a possible implementation in real time. Somehow luckily, the human ear, for a satisfactory perception of ambience, requires only a rough simulation of the room reverberation properties. Classically, reverberation was obtained by recursive filters having little resemblance of the physical reality of the target room. On the contrary, they only reproduced some statistical qualities of the reverberated sound [68,69,43]. This implied a difficult parametrization of reverberators and, almost always, the user relied on some sets of magic numbers, with an evident sacrifice in versatility.

One of the authors introduced a sound processing model called "the Ball within the Box" (BaBo) which is based on three different perspectives on wave propagation in an enclosure [55]. In proximity of the sound source a time-domain perspective best describes wave propagation, and ray tracing or the image method [10] can be used to come up with a tapped delay which accounts for the direct signals and the early reflections. Whenever a sufficiently long distance has been traveled by waves from the sound source, we can adopt the simpler plane-wave description, which is easily interpreted in the frequency domain in terms of normal modes. This level is well described by a bunch of recirculating delay lines which can be exactly parameterized from the geometrical description of a rectangular box. Each of the delay lines corresponds to a plane-wave closed path. The third level of description takes into account the fine-grain phenomena, such as diffusion, which are missed by the other perspectives. These effects are averaged out and lumped in a single scattering object (i.e. the Ball) inserted in a perfectly-reflecting enclosure (i.e. the Box). From a computational viewpoint the scattering object is represented by a feedback matrix which redistributes energy among the delay lines. Figure 16.10 shows the computational structure of the BaBo model, where additional filters T, G, and A have been inserted for simulating losses and, possibly, dispersion. The coefficients b and c are used to control the location of zeros of the structure: while the poles are independent on source and listener positions, the zeros are not. The feedback matrix is chosen to be circulant because of the good properties of the resulting recursive structure. Such a structure is called a Feedback Delay Network [60,56].

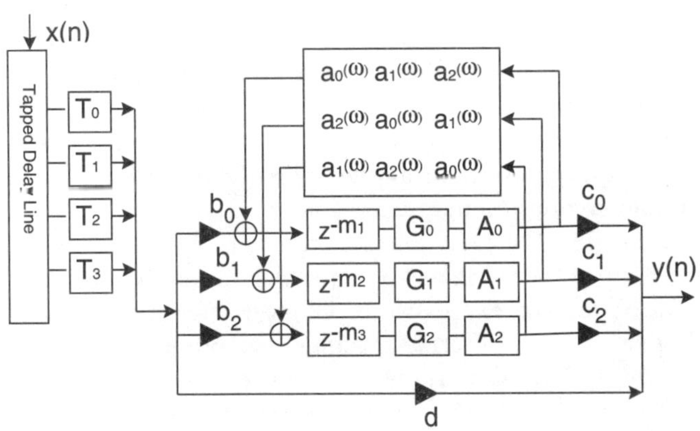

Fig. 16.10. Computational Structure of the BaBo Model

The physical representation offered by the BaBo model, even though simple, allows the determination of the delay lengths based on physical distances and not on number-theoretic considerations, as it was done with classical structures. Moreover, the object properties have a close correspondence with the coefficients of the feedback matrix, thus allowing a tuning of the computational structure based on the diffusive properties we intend to obtain.

16.8 Conclusions

We have presented some of the main modeling approaches to musical sounds, most of them based on a mathematical description of sound sources, musical instruments, or enclosed spaces. Synthesis by physical models, begun as a method for studying the physical behavior of traditional musical instruments, is now one of the most useful methods of sound generation for music. Well-established models exist for several instrumental families and non-musical sound sources. For some other sources, effective and efficient models are still to be invented. Moreover, there are several structural and operational problems that need to be solved before physical modeling approaches the versatility and usability that are required by interactive and multimodal computer systems of the future. This is especially true when considering the problem of making the algorithms suitable for control purposes. The design of aids that facilitate an effective real-time control of the synthetic instruments should be considered as a task closely related to sound modeling and design. There should be an easy interpretation of the relationships among parameters, and physically-described sound models often lend themselves to this purpose. However, a higher level of abstraction is often desirable so that control can be exerted in terms of musical gestures.

Other interesting directions for future research are found in the abstraction of general computational structures, thus leveling up the idiosyncrasies of models of specific mechanical systems. These general structures will be useful to generate new classes of sounds and to relate them to existing classes that have thorough mathematical and acoustical descriptions. Another aspect that is emerging as crucial to achieve high sound quality is the perceptual evaluation of physical models, especially as far as the approximations introduced in model formulation and computational realization are concerned.

To conclude, we think that the joint research efforts of mathematicians, engineers, and physicists, together with active experimentation by musicians and psychologists, will enrich the expressive capabilities of virtual musical instruments and will give rise to new forms of human-computer communication via non-speech sounds. This scenario is a natural extension of the old tradition of cooperation and mutual intersection between science and music.

References

1. Avanzini, F.: Higher order numerical methods in physical models of musical instruments. A study on the clarinet. In: *Proc. Diderot Forum on Mathematics and Music*, pp. 11–19. Vienna, Austria 1999
2. Begault, D.R.: *3-D Sound for Virtual Reality and Multimedia*. Boston, MA: Academic Press 1994
3. Belevitch, V.: *Classical Network Theory*. San Francisco: Holden-Day 1968
4. Blauert, J.: *Spatial Hearing: the Psychophysics of Human Sound Localization*. Cambridge, MA: MIT Press 1983
5. Blinn, J.F., Newell, M.E.: Texture and reflection in computer generated images. Communications of the ACM, **19**(10), 542–547 (1976)
6. Borin, G., De Poli, G.: A Hysteretic Hammer-String Interaction Model for Physical Model Synthesis. In *Proc. Nordic Acoustical Meeting (NAM-96)*, pp. 399–406. Helsinki, Finland 1996

7. Borin, G., De Poli, G., Sarti, A.: Algorithms and Structures for Synthesis using Physical Models. Computer Music J. **16**(4), 30–42 (1992a)
8. Borin, G., De Poli, G., Sarti, A.: *Sound Synthesis by Dynamic Systems Interaction.* In: Baggi, D. (ed.): Readings in Computer-Generated Music, pp. 139–160. IEEE Computer Society Press 1992b
9. Borin, G., Poli, G.D., Rocchesso, D.: Elimination of delay-free loops in discrete-time models of nonlinear acoustic systems. IEEE Transactions on Speech and Audio Processing **8**(5). Accepted for publication (2000)
10. Borish, J.: Extension of the Image Model to Arbitrary Polyhedra. J. Acoustical Soc. of America **75**(6), 1827–1836 (1984)
11. Cadoz, C., Luciani, A., Florens, J.-L.: CORDIS-ANIMA: A Modeling and Simulation System for Sound Synthesis – The General Formalism. Computer Music J. **17**(1), 19–29 (1993)
12. Cadoz, C., Luciani, A., Florens, J.-L.: Physical Models for Music and Animated Image. The use of CORDIS-ANIMA in "ESQUISSES" a Music Film by ACROE. In *Proc. Int. Computer Music Conf.*, pp. 11–18. Aarhus, Denmark: ICMA 1994
13. Caussé, R., Dérogis, P., Warusfel, O.: Radiation of Musical Instruments and Improvement of the Sound Diffusion Techniques for Synthesized, Recorded or Amplified Sounds (revisited). In: *Proc. Int. Computer Music Conf.*, pp. 359–360. Banff, Canada: ICMA 1995
14. Chaigne, A.: On the Use of Finite Differences for Musical Synthesis. Application to Plucked Stringed Instruments. J. Acoustique **5**, 181–211 (1992)
15. Chaigne, A., Askenfelt, A.: Numerical Simulations of Piano Strings. I. A Physical Model for a Struck String using Finite Difference Methods. J. Acoustical Soc. of America **95**(2), 1112–1118 (1994)
16. Chowning, J.M.: The Simulation of Moving Sound Sources. J. Audio Eng. Soc. **19**(1), 2–6 (1971) Reprinted in the Computer Music Journal, June 1977
17. Chowning, J.M.: The Synthesis of Complex Audio Spectra by Means of Frequency Modulation. J. Audio Eng. Soc. **21**(7), 526–534 (1973)
18. Cohen, M.F., Wallace, J.R.: *Radiosity and Realistic Image Synthesis.* Cambridge, MA: Academic Press 1993
19. Cook, P.R.: Non-linear periodic prediction for on-line identification of oscillator characteristics in woodwind instruments. In *Proc. Int. Computer Music Conf.*, pp. 157–160. Montreal, Canada: ICMA 1991
20. Gabor, D.: Acoustical Quanta and the Theory of Hearing. Nature **159**(4044), 591–594 (1947)
21. De Poli, G.: A tutorial on digital sound synthesis techniques. Computer Music Journal **7**(4), 8–26 (1983). Reprinted in: Roads, C. (ed.): *The Music Machine* pp. 429–47. Cambridge, MA: MIT Press 1991
22. De Poli, G., Piccialli, A., Roads, C.: *Representations of Musical Signals.* Cambridge, MA: MIT Press 1991
23. Drioli, C., Rocchesso, D.: A Generalized Musical-Tone Generator with Application to Sound Compression and Synthesis. In: *Proc. Int. Conf. Acoustics, Speech, and Signal Processing*, pp. 431–434. Munich, Germany 1997
24. Fettweis, A.: Wave Digital Filters: Theory and Practice. Proc. IEEE **74**(2), 270–327 (1986)
25. Fletcher, N.H., Rossing, T.D.: *The Physics of Musical Instruments.* New York: Springer 1991

26. Florens, J.L., Cadoz, C.: The physical model: Modeling and simulating the instrumental universe. In: De Poli, G., Piccialli, A., Roads, C. (eds.): *Representations of Musical Signals*, pp. 227–268. Cambridge, MA: MIT Press 1991
27. Fontana, F., Rocchesso, D.: Physical modeling of membranes for percussion instruments. Acustica **83**(1), 529–542 (1998)
28. Fontana, F., Rocchesso, D.: Signal-theoretic characterization of waveguide mesh geometries for models of two-dimensional wave propagation in elastic media. IEEE Trans. Speech and Audio Processing **8**. Accepted for publication (2000)
29. Gazengel, B., Gilbert, J., Amir, N.: Time Domain Simulation of Single Reed Wind Instrument. From the Measured Input Impedance to the Synthesis Signal. Where are the Traps? Acta Acustica **3**, 445–472 (1995)
30. Gray, A.H., Markel, J.D.: A Normalized Digital Filter Structure. IEEE Trans. Acoustics, Speech, Signal Processing **23**(6), 268–277 (1975)
31. Haus, G.: *Music Processing.* Madison, WI: A-R Editions 1993
32. Incerti, E., Cadoz, C.: Topology, geometry, matter of vibrating structures simulated with cordis-anima. sound synthesis methods. In *Proc. Int. Computer Music Conf.*, pp. 96–103. Banff, Canada: ICMA 1995
33. Jaffe, D.A., Smith, J.O.: Extensions of the Karplus-Strong Plucked String Algorithm. Computer Music J. **7**(2), 56–69 (1983)
34. Karplus, K., Strong, A.: Digital Synthesis of Plucked String and Drum Timbres. Computer Music J. **7**(2), 43–55 (1983)
35. Kendall, G.S.: A 3-D Sound Primer: Directional Hearing and Stereo Reproduction. Computer Music J. **19**(4), 23–46 (1995)
36. Kuttruff, H.: *Room Acoustics.* 3rd edn. Essex, England: Elsevier Science (1st edn. 1973) 1991
37. Laakso, T.I., Välimäki, V., Karjalainen, M., Laine, U.K.: Splitting the Unit Delay—Tools for Fractional Delay Filter Design. IEEE Signal Processing Mag. **13**(1), 30–60 (1996)
38. McAdams, S.: Music: a science of mind? Contemporary Music Review **2**(1), 1–61 (1987)
39. McIntyre, M.E., Schumacher, R.T., Woodhouse, J.: On the oscillations of musical instruments. J. Acoustical Soc. of America **74**(5), 1325–1345 (1983)
40. Meerkötter, K., Scholtz, R.: Digital simulation of nonlinear circuits by wave digital filter principles. In *Proc. IEEE Intl. Symp. on Circuits and Systems*, pp. 720–723. Portland, OG 1989
41. Moore, F.R.: *Elements of Computer Music.* Englewood Cliffs, NJ: Prentice-Hall 1990
42. Moorer, J.A.: Signal processing aspects of computer music. Proceedings of the IEEE **65**(8), 1108–37 (1977)
43. Moorer, J.A.: About this Reverberation Business. Computer Music J. **3**(2), 13–18 (1979)
44. Moorer, J.A.: The Manifold Joys of Conformal Mapping: Applications to Digital Filtering in the Studio. J. Audio Eng. Soc. **31**(11), 826–840 (1983)
45. Morse, P.M.: *Vibration and Sound.* American Institute of Physics for the Acoustical Society of America, New York (1st edn. 1936, 2nd edn. 1948). 1991
46. Oppenheim, A.V., Schafer, R.W.: *Discrete-Time Signal Processing.* Englewood Cliffs, NJ: Prentice-Hall 1989
47. Pierce, J.R., Van Duyne, S.A.: A Passive Nonlinear Digital Filter Design which Facilitates Physics-based Sound Synthesis of Highly Nonlinear Musical Instruments. J. Acoustical Soc. of America **101**(2), 1120–1126 (1997)

48. Poggio, T., Girosi, F.: Networks for approximation and learning. Proceedings of the IEEE **79**(9), 1481–1497 (1990)
49. Press, W.H., Flannery, B.P., Teukolsky, S.A., Vetterling, W.T.: *Numerical Recipes in C.* Cambridge: Cambridge University Press 1988
50. Risset, J.-C.: An Introductory Catalog of Computer-Synthesized Sounds. Technical report, Bell Laboratories, Murray Hill, NJ, 1969
51. Roads, C.: Asynchronous granular synthesis. In: De Poli, G., Piccialli, A., Roads, C. (eds.): *Representations of Musical Signals*, pp. 143–186. Cambridge, MA: MIT Press 1991
52. Roads, C.: *The Computer Music Tutorial.* Cambridge, MA: MIT Press 1996
53. Roads, C., Pope, S.T., Piccialli, A., De Poli, G.: *Musical Signal Processing.* Lisse: Swets & Zeitlinger 1997
54. Rocchesso, D.: Modelli Generalizzati di Strumenti Musicali per la Sintesi del Suono. Rivista Italiana di Acustica **17**(4), 61–71 (1993). Paper invited by the panel of "Amedeo Giacomini" award, Associazione Italiana di Acustica
55. Rocchesso, D.: The Ball within the Box: a sound-processing metaphor. Computer Music J. **19**(4), 47–57 (1995)
56. Rocchesso, D.: Maximally-Diffusive yet Efficient Feedback Delay Networks for Artificial Reverberation. IEEE Signal Processing Letters **4**(9), 252–255 (1997)
57. Rocchesso, D.: Fractionally-addressed delay lines. *IEEE Transactions on Speech and Audio Processing.* Accepted for publication (2000)
58. Rocchesso, D., Scalcon, F.: Accurate dispersion simulation for piano strings. *Proc. Nordic Acoustical Meeting (NAM-96)*, pp. 407–414. 1996
59. Rocchesso, D., Scalcon, F.: Bandwidth of perceived inharmonicity for physical modeling of dispersive strings. IEEE Transactions on Speech and Audio Processing **7**(5), 597–601 (1999)
60. Rocchesso, D., Smith, J.O.: Circulant and Elliptic Feedback Delay Networks for Artificial Reverberation. IEEE Transactions on Speech and Audio Processing **5**(1), 51–63 (1997)
61. Rocchesso, D., Turra, F.: A Real-Time Clarinet Model on MARS Workstation. In: *Proc. X Colloquium Mus. Inform.*, pp. 210–213. Milano: AIMI 1993
62. Rodet, X.: Models of musical instruments from chua's circuit with time delay. IEEE Trans. Circuits and Systems-II **40**(10), 696–701 (1993)
63. Rodet, X.: Stability/instability of periodic solutions and chaos in physical models of musical instruments. In: *Proc. Int.Computer Music Conf.*, pp. 386–393. Aarhus, Denmark. ICMA.1994
64. Sarti, A., De Poli, G.: 1999 Toward nonlinear wave digital filters. IEEE Trans. Signal Processing **47**(6), 1654–1668
65. Savioja, L., Välimäki, V.: 1997 Improved discrete-time modeling of multidimensional wave propagation using the interpolated digital waveguide mesh. In *Proc. Int. Conf. Acoustics, Speech, and Signal Processing, Munich*, pp. 459–462
66. Scavone, G.P.: Combined linear and non-linear periodic prediction in calibrating models of musical instruments to recordings. In *Proc. Int.Computer Music Conf.*, pp. 433–434. Aarhus, Denmark: ICMA 1994
67. Schaeffer, P.: *Traité des Objets Musicaux.* Paris: Éditions du Seuil 1966
68. Schroeder, M.R.: Natural-Sounding Artificial Reverberation. J. Audio Eng. Soc.**10**(3), 219–233 (1962)
69. Schroeder, M.R.: Computer Models for Concert Hall Acoustics. American Journal of Physics **41**, 461–471 (1973)

70. Serra, X.: Musical sound modeling with sinusoids plus noise. In: Roads, C., Pope, S.T., Piccialli, A., De Poli, G. (eds.): *Musical Signal Processing*, pp. 91–122. Lisse: Swets & Zeitlinger 1997
71. Smith, J.O.: Viewpoints on the History of Digital Synthesis. In: *Proc.. Int. Computer Music Conf.*, pp. 1–10. Montreal, Canada: ICMA 1991
72. Smith, J.O.: Physical modeling using digital waveguides. Computer Music J. **16**(4), 74–91 (1992)
73. Smith, J.O.: Efficient synthesis of stringed musical instruments. In: *Proc. 1993 International Computer Music Conference*, pp. 64–71. Tokyo, Japan, 1993
74. Strikwerda, J.: *Finite Difference Schemes and Partial Differential Equations.* Pacific Grove, CA: Wadsworth & Brooks 1989
75. Truax, B.: *Handbook for Acoustic Ecology.* Vancouver: A.R.C. Publications 1978
76. Välimäki, V., Huopaniemi, J., Karjalainen, M., Jánosy, Z.: Physical Modeling of Plucked String Instruments with Application to Real-Time Sound Synthesis. J. Audio Eng. Soc. **44**(5), 331–353 (1996)
77. Välimäki, V., Tolonen, T., Karjalainen, M.: Plucked-string synthesis algorithms with tension modulation nonlinearity. In: *Proc. Int. Conf. Acoustics, Speech, and Signal Processing*, pp. 977–980. Phoenix, Arizona: IEEE 1999
78. Van Duyne, S.A., Pierce, J.R., Smith, J.O.: Traveling Wave Implementation of a Lossless Mode-Coupling Filter and the Wave Digital Hammer. In: *Proc.. Int. Computer Music Conf.*, pp. 411–418. Aarhus, Denmark: ICMA 1994
79. Van Duyne, S.A., Smith, J.O.: The 2-d digital waveguide mesh. In: *Proc. IEEE Workshop on Applications of Signal Processing to Audio and Acoustics.* Mohonk, NY: IEEE 1993
80. Van Duyne, S.A., Smith, J.O.: The tetrahedral digital waveguide mesh. In: *Proc. IEEE Workshop on Applications of Signal Processing to Audio and Acoustics*, p. 9a.6. Mohonk, NY: IEEE 1995
81. Verma, T.S., Levine, S.N., Meng, T. H.Y.: Transient modeling synthesis: a flexible analysis/synthesis tool for transient signals. In: *Proc.. Int. Computer Music Conf.*, pp. 164–167. Thessaloniki, Greece: ICMA 1997
82. von Neumann, J.: *Theory of Self-Reproducing Automata.* Urbana, IL: Univ. of Illinois Press 1966
83. Wan, C., Schneider, A.M.: Further improvement in digitizing continous-time filters. IEEE Trans. Signal Processing **45**(3), 533–542 (1997)
84. Warusfel, O., Kahle, E., Jullien, J.P.: Relationships between Objective Measurements and Perceptual Interpretation: The Need for Considering Spatial Emission of Sound Sources. J. Acoustical Soc. of America **93**, 2281–2282 (1993)
85. Wolfram, S.: Computational theory of cellular automata. Comm. in Math. Physics **96**, 15–57 (1984)
86. Zoelzer, U., Arfib, D., De Poli, G., Dutilleux, P., Rocchesso, D., Serra, X.: *Digital Audio Effects.* Chichester Sussex, UK: John Wiley and Sons, Ltd. 2001. To appear

Druck: Druckhaus Berlin-Mitte
Verarbeitung: Buchbinderei Stein & Lehmann, Berlin